Published by
The Hamlyn Publishing Group Limited
LONDON · NEW YORK · SYDNEY · TORONTO
Astronaut House, Feltham, Middlesex, England
© Copyright The Hamlyn Publishing Group Limited 1973
Seventh impression 1977

ISBN 0 600 30148 6

Printed and bound in Great Britain
by Hazell Watson & Viney Limited,
Aylesbury, Bucks

ACKNOWLEDGEMENTS

The author and publishers would like to thank the following for their co-operation in supplying pictures for this book:
Alcan Foil: page 129
Argentine Beef Bureau: page 166
Australian Recipe Service: page 184
Bejam Group Limited: page 9
British Bacon Council: page 107
Brown and Polson Limited (Mazola Corn Oil): page 90
'Carmel' Agrexco (Agricultural Export Company Limited of Israel): page 108
Dutch Dairy Bureau: page 72
Eden Vale Dairy Fresh Foods: page 71
Findus Frozen Foods: page 110
The Flour Advisory Bureau: page 89
Fruit Producers' Council: pages 36, 186
Grants of St. James's: page 223
Kellogg Company of Great Britain: page 183
KPS Freezers Limited: front and back cover
Lakeland Plastics (Windermere) Limited: page 35
Lawry's Foods International Incorporated: page 185
Mushroom Growers Association: page 147
The Nestlé Company: page 206
New Zealand Lamb Information Bureau: page 109
Plumrose Limited: page 130
Reveal Roasting Wrap: page 53
James Robertson and Sons: page 204
Sharwoods Foods Limited: page 205
Tabasco Pepper Sauce (Beecham Foods Limited): page 165
Taunton Cider: page 127
Trex Cookery Service (J. Bibby Food Products Limited): page 54
The Tupperware Company: pages 1, 83, 203
T. Wall and Sons (Ice Cream) Limited: page 224
John Wests Foods Limited: page 128
The White Fish Kitchen: page 148

Materials used for freezing were:
Polythene containers: The Tupperware Company
Polythene bags, sealing tape and blancher: Lakeland Plastics (Windermere) Limited
Foil containers and sheet foil: Alcan Foil

HOME ECONOMISTS
Christina Pitman
Audrey Smith

Colour photographs by John Lee
Roy Rich, Angel Studios
Illustrated by Hayward Art Group
Marilyn Day

Props kindly loaned by Heal's, David Mellor, Harvey Nichols, Casa Pupo and Wedgwood

Contents

Useful facts and figures

Notes on metrication

In this book quantities have been given in both metric and Imperial measures. Exact conversion from Imperial to metric measures does not usually give very convenient working quantities and so for greater convenience the metric measures have been rounded off into units of 25 grammes. The following tables shows the recommended equivalents.

Ounces/fluid ounces	**Approx. g. and ml. to nearest whole figure**	**Recommended conversion to nearest unit of 25**
1	28	25
2	57	50
3	85	75
4	113	100
5 (¼ pint)	142	150
6	170	175
7	198	200
8	226	225
9	255	250
10 (½ pint)	283	275
11	311	300
12	340	350
13	368	375
14	396	400
15 (¾ pint)	428	425
16 (1 lb.)	456	450
17	486	475
18	512	500
19	541	550
20 (1 pint)	569	575

NOTE: When converting quantities over 20 oz. first add the appropriate figures in the centre column, *then* adjust to the nearest unit of 25. As a general guide, 1 kg. (1000 g.) equals 2·2 lb. or about 2 lb. 3 oz.; 1 litre (1000 ml.) equals 1·76 pints or almost 1¾ pints. (In this book the conversion rate of 1 lb. to ½ kg. has been used.) This method of conversion gives good results in nearly all recipes; however, in certain cake recipes a more accurate conversion is necessary (or the liquid in the metric recipe must be reduced slightly) to produce a balanced recipe.

Liquid measures The millilitre is a very small unit of measurement and we felt that to use decilitres (units of 100 ml.) would be easier. In most cases it is perfectly satisfactory to round off the exact millilitre conversion to the nearest decilitre, except for ¼ pint; thus ¼ pint (142 ml.) is 1½ dl., ½ pint (283 ml.) is 3 dl., ¾ pint (428 ml.) is 4 dl. and 1 pint (569 ml.) is 6 dl. or a generous ½ litre. For quantities over 1 pint we have used litres and fractions of a litre, using the conversion rate of 1¾ pints to 1 litre.

Oven temperature chart

	°F	°C	Gas Mark
Very cool	225	110	¼
	250	130	½
Cool	275	140	1
	300	150	2
Moderate	325	160	3
	350	180	4
Moderately hot	375	190	5
	400	200	6
Hot	425	220	7
	450	230	8
Very hot	475	240	9

Metric capacity of freezers

The metric capacity of freezers is measured in litres. To convert cubic feet to litres multiply by 28·30.

Notes for American users

Each recipe in this book has an American column giving U.S. cup measures for the ingredients.
The following list gives American equivalents or substitutes for some terms used in the book.

British	American
Baked/unbaked pastry case	Baked/unbaked pie shell
Baking tin	Baking pan
Base	Bottom
Bicarbonate of soda	Baking soda
Cocktail stick	Wooden toothpick
Cake mixture	Batter
Deep cake tin	Spring form pan
Frying pan	Skillet
Greaseproof paper	Wax paper
Grill	Broil/broiler
Kitchen paper	Paper towels
Mixer/liquidiser	Mixer/blender
Muslin	Cheesecloth
Pastry cutters	Cookie cutters
Patty tins	Muffin pans/cups
Piping bag	Pastry bag
Piping tube	Nozzle/tip
Polythene bags	Plastic wrap
Pudding basin	Pudding mold/ovenproof bowl
Sandwich tin	Layer cake pan
Stoned	Pitted
Swiss roll tin	Jelly roll pan
Whisk	Whip/beat

NOTE: Where baking powder is listed, use the double-acting type.

The British pint measures 20 fluid ounces whereas the American pint equals 16 fluid ounces.

Foreword

Many helpful organisations have assisted in compiling the information presented in this book, and ensuring that the latest equipment and techniques have been included.

The author gratefully acknowledges the assistance of manufacturers of freezers and freezing accessories who have given their latest products for experimental use, and the co-operation of her colleagues, the Editor of *Freeze* and the Managing Editor of *The Freezer Family*. These authoritative magazines are immensely helpful in keeping freezer owners up to date with new developments in the field. The author also wishes to thank Myra Street, Cookery Editor of the Hamlyn Publishing Group and her staff of trained home economists in the Hamlyn Group Kitchen who have put many recipes and freezing techniques through exhaustive tests; her editorial assistant Christine Curphey who has carried out comparable tests from the housewife/freezer owner's point of view, and the Electricity Council and Food Freezer Committee whose advice on nation-wide freezing trends has been invaluable.

Introduction

It's come at last. The new Ice Age has arrived, and those tough crusading days to get the home freezer accepted as a vital piece of domestic equipment are over. I have enjoyed being a leader in the fray. But now, almost every housewife has at least one good reason why she needs her freezer.

She's a woman who hates cooking in penny numbers, but will gladly spend hours in the kitchen producing a month's supply of cakes or casseroles when the mood takes her. She's a woman whose children see something they covet on every counter and turn a peaceful shopping trip into an assault course. She's a woman with a worthwhile job who no longer needs to worry about the shops being shut if she's late leaving the office. She's a woman with an executive husband who has unfailing faith in her ability to produce an impressive meal for guests, either with or without warning. She's a woman who carefully tots up the household bills and gets a real thrill out of making substantial savings. She's a woman who is frail or elderly or isolated and simply cannot get out to the shops very often. She's a woman who has a special diet problem in the family to be catered for, or just likes to feel that if her husband fancies a pizza, or a sorbet, or wheatmeal rolls for dinner – it's right there in the freezer.

I have been quoted many other reasons why life without a freezer would now be unthinkable. If washday blues virtually vanished with

the advent of the automatic washing machine, the day of the housewife's true liberation certainly dawned with the coming of the freezer. I do not promise you as glibly as some people may do, that you will now live more economically. Readers tell me (and when I lecture on freezing so do members of the audience) that they find themselves living better rather than saving money. Since the quality of our living is recognised as being so important, this is, in itself, a valid argument for having a freezer. But if your purpose in buying a freezer is to save money on food – and you shop and plan meals with economy in mind – dramatic savings can be made. Bulk buying and seasonal buying are boons to the thrifty housewife. Even when your freezer is filled on a strict budget, you will probably find that family meals become more appetising and varied. And you will still enjoy the saving which has nothing to do with pounds and pence, but which concerns your output of time and energy.

Living better, saving money, having more leisure and not being too tired to enjoy it! If you think this too extravagant a claim for freezer ownership, you have obviously not bought one yet. Or maybe you are not using it to the very best advantage. That is why I have written this book. It has been carefully planned so that you may find the technical information, general advice, and recipes you need almost at a glance. The subjects you'll want to refer to most often appear throughout on the Golden Guide pages, which give you precise instructions on dealing with specific problems, like storage and defrosting. Every other aspect of freezing has been explored and explained in the logical sequence in which the new freezer owner will usually require the knowledge.

Start freezing now! There are exciting technical developments just over the horizon to make a housewife's life a happier one – automatic freezer defrosting for instance, and microwave cookery (which will enable us to take suitable containers straight from the freezer, pop them into the microwave oven to defrost and reheat in a matter of minutes). I have seen this demonstrated convincingly in Tupperware containers. Yet more developments are on their way.

The freezer is the most exciting and revolutionary influence to hit the home front since I first started writing cookery books, and I sincerely believe that learning to use your freezer well is an entirely new domestic art which housewives of the future will all want to acquire. I shall be happy if my book helps to open up for you this whole new world of easy catering for your family and friends.

Audrey Ellis

Part 1 *Learning to make a friend and ally of your freezer*

Facts about freezing

Frozen food is like the fairytale princess who fell asleep at the peak of her youthful beauty for a hundred years, and reawakened without having aged one day. Freezing halts all the natural processes of decay in food, and they remain in a state of suspended animation until the kiss of returning warmth allows these processes to continue.

All food is subject to spoilage and will eventually become unfit to eat, if it remains at a temperature where enzymes are active and micro-organisms can survive and multiply. Age-old methods of preservation such as drying, salting and smoking create conditions which impede these processes but drastically alter the texture and taste of the food concerned.

How freezing works to preserve food

Freezing is the best method of preserving food in its natural state. In many cases it is impossible to tell, after it has been defrosted, that the food has ever been frozen.

How is this miracle achieved? Freezing deprives micro-organisms (bacteria, moulds, etc.) of the conditions they need to thrive, and inactivates enzymes. These destructive elements remain dormant until a rise in the temperature restores more congenial conditions. Then, enzymes are reactivated. Harmful bacteria once more begin to multiply, unfortunately with renewed vigour. The natural course of decay, halted as long as the food has been in the frozen state, then continues.

The temperature range involved is a wide one. To start with, your refrigerator cabinet is regulated to a temperature around 47°F (8°C) at which bacterial activity slows down considerably. This allows you to store food, for at least a week, which would become uneatable if left, for even one day, at average room temperature. In the kitchen this may be as high as 75°F (24°C) for much of the time. Refrigerators also have a frozen food storage compartment maintained at a temperature level below 32°F (0°C), the freezing point of water. Frozen food can be safely stored here for a limited period. The length of time is governed by how cold it is inside this compartment, which you can judge by the star rating on your refrigerator.

Since water freezes at a comparatively high temperature, ice cubes can be made even in a one-star storage compartment. All food contains a certain amount of water but its other constituents are not so accommodating. To reach the fully frozen state, they require reduction to a much lower temperature. For example, a pastry pie with a sweet filling might have to be reduced to about 15°F (−10°C). By the time they reach 0°F (−18°C) all foods are fully frozen.

Although bacterial growth *virtually* ceases at this point, some minute changes continue, due to enzyme and other chemical activity which happily cause no noticeable change in the food. However, even the slightest rise in temperature may cause damage, as bacteria in particular tend to go into action, refreshed by their slumber, as soon as food 'warms up' above the critical freezing point. Each time the food is again fully refrozen there is a slight build-up of bacterial contamination. That is why temperature fluctuations always ought to be avoided.

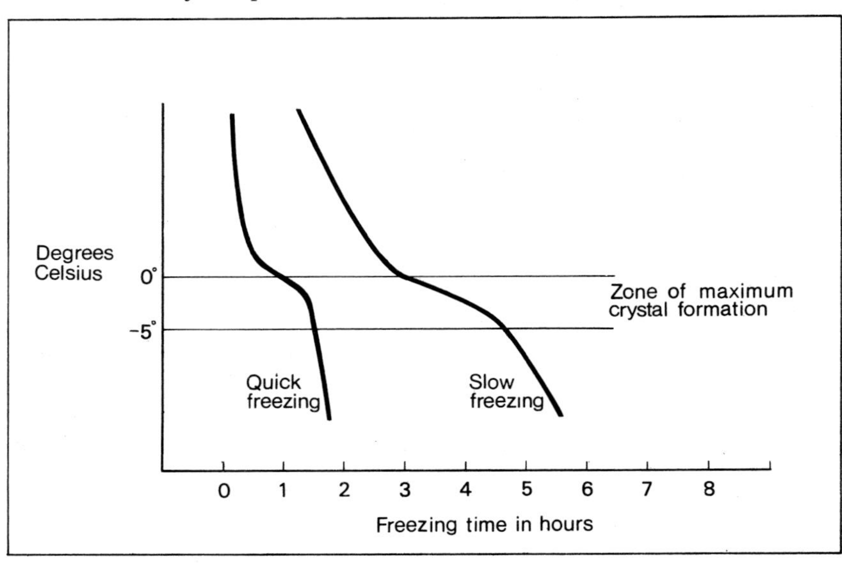

Hygienic care

Food for freezing must be perfectly clean, and handled hygienically. Materials used for wrapping and packing must also be scrupulously clean.

All food, from the chemist's point of view, contains some harmful bacteria, although it might be quite safe to eat. A lifetime of exposure to this hazard makes us partially immune. Babies have not yet built up this immunity and tainted food can seriously upset them. Cooking, by applying heat, destroys most of these bacteria, and we eat a great deal of freshly cooked food. Since freezing does not destroy micro-organisms as heat does, avoid leaving food intended for freezing in warm, dark, moist conditions, which encourage existing bacteria to thrive. They are invisible to the naked eye, often impossible to detect by taste or smell. A warm oven, for instance, is not a safe storage place to keep food – it is an incubator for germs. Uncovered food elsewhere in the kitchen is a danger – I've seen a housewife leave the kitchen, and return a few minutes later to continue packing operations, blissfully unaware that the family cat had walked all over the preparation table and sniffed at the food in her absence.

Perfect results

Food, if not harmed, is definitely not improved by freezing so it is not worthwhile to give your time, effort and freezer space to anything but the best. In fact, even good food can be spoiled by bad handling, incorrect packing and freezing methods, or lack of care in defrosting. Our aim as careful cooks is to choose the best food available and ensure that it suffers the least possible loss of appearance, texture, flavour and food value while frozen. Some vitamins, especially B and C, are extremely fugitive and could vanish almost entirely if the right rules are not observed. Our motto ought to be, freeze fresh food fast. That's real four-star freezing!

What damages food?

Undoubtedly, careless packing and slow freezing are the twin culprits when disappointing results emerge from your freezer.

1. **Careless packing** How to pack food correctly is described in detail in chapter 4, but the basic principle, which cannot be repeated too often, is that food has to be protected from the extremely cold, dry air of the freezer cabinet. Remember you are preserving food much longer, and therefore at a much lower temperature than by any method you have used previously. Cold air is dry, and the colder it becomes the drier it gets. It is therefore capable of absorbing moisture from food, causing dehydration and much damage. Also, cross flavours and aromas from badly packaged strong-tasting food can invade other packs. Only wrapping and containers which are fully moisture-vapour-proof and airtight can prevent this.

2. **Slow freezing** A reader once disagreed with me that quick freezing was essential. She felt it must be such a shock to the food! But I explained to her carefully that most food has a high water content (even meat contains 70% water and strawberries as much as 90%). When this water content freezes, it expands by 10% of its total volume, ruptures the food cells, in the form of jagged ice crystals, conglomerates outside the cells to form larger crystal masses and ruptures the cells

again from outside. The more slowly the food passes through the temperature range around 32°F (0°C) at which water freezes, the greater the damage. This is not apparent until the food is defrosted and the cell structure collapses. No harm of course occurs when ice cubes are frozen slowly in your refrigerator, but if you try the experiment with strawberries the result, on defrosting, will be a mushy disaster. An efficient freezer will have at least one compartment which can be reduced to −5°F (−21°C) or lower for quick freezing.

To show you how long your food is at risk compare the length of the period in which crystals form during slow and quick freezing.

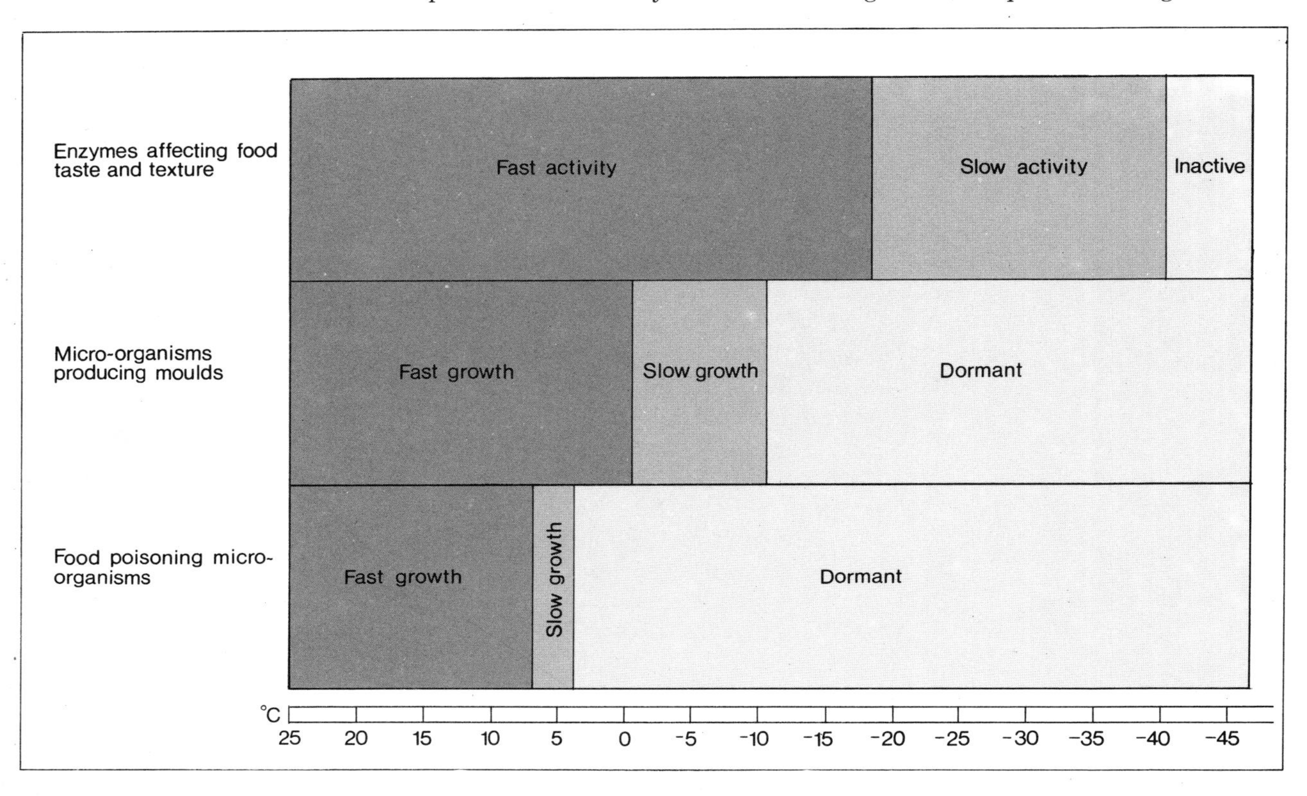

A chart showing how long the enzymes and micro-organisms in food remain in the various stages during slow and quick freezing.

Becoming a freezer owner

You now know how freezing works to preserve food and if you have not got one already, you've decided to become a freezer owner. That is the right way to start, so that you have absolute confidence in your investment. Confidence that a big, valuable stock of food will be safely cared for by the gentle white giant, some of it for as long as a year, before it emerges to grace your table. But whether or not you are satisfied with your freezer depends very much on making a wise choice of the type and model most suited to your needs. It really pays to investigate the freezer market and to understand thoroughly what you are shopping for, before signing on any dotted line. First of all – is it really a home freezer you are buying?

When is a similar piece of equipment NOT a freezer?

1. A conservator or 'cold store' is *not* a freezer. You may be offered a useful storage chest at a very reasonable price which the salesman will tell you is a home freezer, and is in fact a conservator. This is intended to store commercially-frozen foods for whatever their storage life may be. But it is not intended for, or capable of, freezing down fresh food quickly. Unless there is a switch or dial on the cabinet which can be adjusted to reduce the temperature, temporarily, to *below −5°F (−21°C)*, you cannot expect to get the best from your home freezing efforts.

2. A refrigerator with a frozen food storage compartment is *not* a freezer. The criterion is really whether any part of a combination refrigerator with a separate door bears the four-star rating. Star marking up to three on such a compartment indicates the length of time for which ready-frozen foods can be stored. Even a three-star rating only protects food, in the frozen state, adequately for 3 months.

Do not be deterred, by the way, from using the frozen food storage compartment of your fridge for making ice cubes (water as you have seen, freezes at a relatively high temperature and crystal formation is unimportant), or ice cream. Most recipes prescribe beating the cream as it freezes to break down crystal formation. But do not be surprised if you fill the compartment with fresh food and find it still only partially frozen days later. The spirit may be willing, but the capacity to freeze down is far too weak!

Here is a useful chart of comparative storage temperatures in fridges and freezers, showing how the star rating system works.

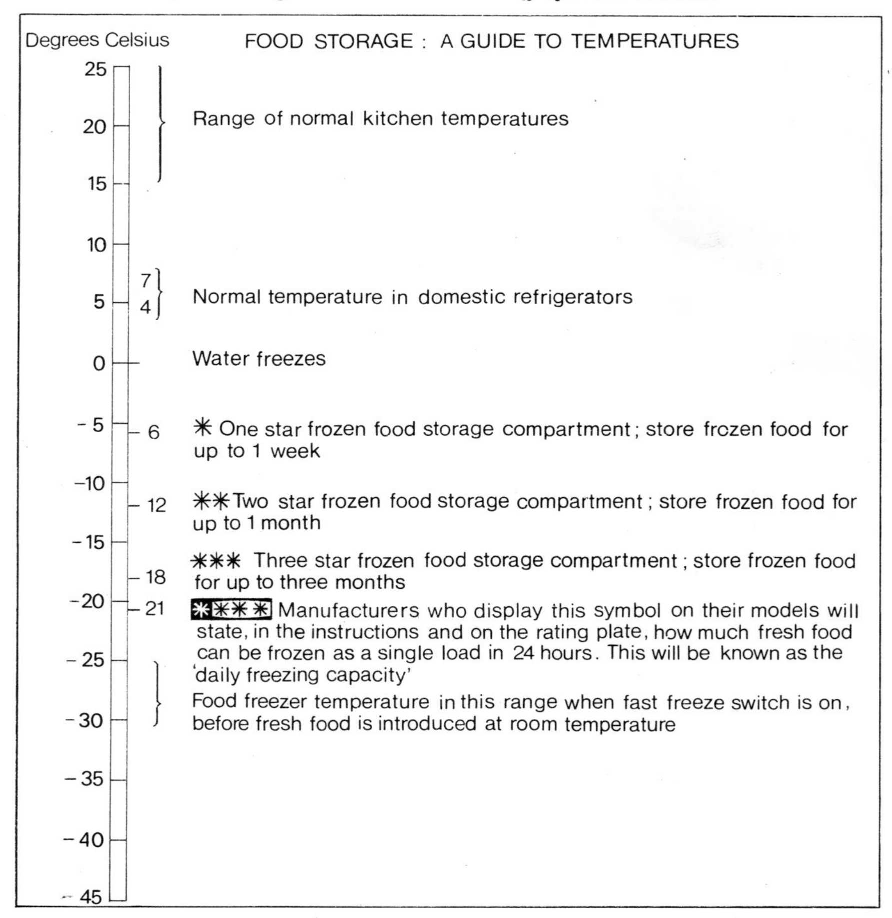

How efficient is your freezer?

Before we discuss the relative advantages of a chest or upright freezer for your purpose, it is important to understand that some chest freezers will be more efficient than others, and that this also applies to upright freezers. The cooling tubes which are frequently invisible when they run within the shell of the chest freezer, may not be in the bottom as well as round the sides of the cabinet. If there is a fast-freeze compartment, it does not mean that it has a special separate cooling system capable of reducing this compartment to a lower temperature, merely that the cooling tubes do run closely together around the sides of the compartment. In an upright freezer, the cooling tubes will in fact form some or all of the shelves. Freezing down will be faster on these shelves, just as it will be if packs are placed against the walls of the cabinet which house the cooling tubes in the chest freezer. The nearer the food comes into contact with the cooling tubes, rather than other packs of food, the lower the temperature and the faster the freezing. Inner flaps which protect the contents of the shelves against the loss of cold air when the door is open, increase efficiency.

Is the salesman your best adviser?

Unfortunately, a salesman's first aim is to make a sale, and it is up to you to shop around and become knowledgeable about freezers generally so that you do not have to accept his advice as the gospel. Although genuinely anxious to be helpful, many salesmen are not really qualified to advise on the technical points which are so important, and do not even have a sound knowledge of their legal responsibility towards the customer. Be sure, if you are offered a bargain, that you are not being offered a used freezer, or one of 'second' quality, with faults. It is the company's responsibility to ensure that the freezer works and does its job to any purchaser's reasonable satisfaction, and you are quite entitled to expect this, even if it goes as far as replacing the freezer or refunding the cost. However, be careful before accepting delivery of the freezer to inspect it for faults which are immediately apparent, as it might be argued that you must have been aware of the faults and accepted these drawbacks when deciding to purchase the freezer.

Another point where you need to make a careful check on the salesman's information is on the capacity of the freezer. Is he quoting you the gross or net capacity? This could vary considerably, e.g. a large upright freezer of 11·1 cubic feet gross may have only 10·7 cubic feet net capacity.

Other useful extras to look for

Internal light Important in a badly lit site, like a cupboard.
Lock Essential where children may open freezer without permission.
Condensation guard An anti-condensation device to be used when the freezer is sited in a moist ambient atmosphere.
Thermostatic switch To be raised when the ambient temperature drops seasonally, i.e. the freezer is not sited in a room which has a comparatively steady temperature.
Front shelf flaps To retain cold in an upright freezer when door is open.
Drop-front grids To retain packs in an upright freezer when it is fully packed.
Pull-out baskets Instead of all fixed shelves, in upright models.
Baskets with reversible handles To hang across top of chest freezer.

Roast potatoes, French fried potatoes, duchesse potatoes (see page 179).

Warning lights A colour-coded light panel showing that freezer is connected, fast-freeze switch is on, or warning light when temperature rises above safety level. At present there is no conformity in colour coding – make sure you understand your freezer's colour code.
Warning device This remote control panel can be fitted in the kitchen to indicate whether all is well with a freezer sited elsewhere.
Sliding transparent cover on control panel Keeps inquisitive hands at bay. Useful where toddlers are able to reach the panel.
Drainage outlet Makes defrosting easy as the water drains into a tray for disposal.

Super star freezing

Look for this four-star symbol when you buy a new freezer. It should be prominently and permanently marked on all home freezers and the weight of food which can be frozen daily should be stamped on the rating plate. This is an international symbol, not restricted to British manufacturers, and is not merely an extension of the one, two and three star refrigerator symbols. From the drawing you will see that the new star is larger and of a different colour. This is the mark which distinguishes genuine freezers from appliances which look similar but are only capable of storing ready-frozen foods.

The existing British Standard, BS 3999 part 10, defined the freezer as being suitable for freezing a stated quantity of food from 77°F (25°C), to 0°F (−18°C) or below within 24 hours, and for the storage of pre-frozen food at this temperature. The new British Standard, which is a revision of that part of BS 3456 dealing with domestic freezers, will more fully define the requirements to earn that precious fourth star. The freezer will be required to reduce a specified weight of food daily to 0°F (−18°C), without lowering the quality of the food already frozen and stored. This pre-supposes that it is a piece of equipment which can be reduced considerably below 0°F (−18°C).

There should be no confusion in your mind when you purchase a refrigerator/freezer as the separate doors will each be marked with their respective star rating.

How a freezer works

There is no magic wand which waves over the food in your freezer to preserve it for a much longer period than in a refrigerator. Some housewives look on the freezer as simply a super-fridge, and entirely fail to appreciate the difference between for example, a domestic freezer and a conservator or their three-star frozen food storage compartment. Make sure before buying a freezer that it has the four-star rating described above. But before choosing the model, it is worthwhile reading this section to make certain that you know how it works, and just what you will be getting for your money.

The freezing cycle

Imagine a central heating system. A boiler heats the water which a pump circulates through a system of pipes and radiators, where some of the heat is given out to warm your rooms. This cools the water, which is then pumped back to the boiler for reheating, and the whole cycle is repeated. A thermostat ensures that the temperature in the rooms is kept up to the desired level by switching the boiler on and off, as required. Your freezer works roughly on the same principle, but the cycle is intended to produce the opposite effect. Instead of water, a refrigerant liquid is used, which boils and vaporises at a very low temperature. The cooling system is also governed by a thermostat, to keep the temperature inside the cabinet reduced to a safe level for long term food storage. This is how the system works. It consists of three components.

1. The condenser.
2. The evaporator.
3. The compressor.

At the beginning of the cycle, the refrigerant has been returned to its liquid form in the *condenser*. From here it is forced through a very narrow capillary tube into the system of large *evaporating* coils, where the resultant fall in pressure makes it boil. In the case of a chest freezer these coils are enclosed within the 'shell' against the *inner* walls and sometimes the bottom of the freezer cabinet. In upright models the coils usually form some or all of the shelves. Here the refrigerant expands into a vapour, a process that causes it to absorb heat from the food in the cabinet. The warmed vapour is then drawn into the *compressor*, which in principle is nothing more elaborate than a simple pump operated by a motor. Here the vapour is compressed, which makes it even warmer. It then passes back into the *condenser* and becomes a liquid again. The condenser consists of a system of long pipes. In the chest freezer they are concealed in the 'shell' under the *outer* walls of the cabinet. Upright models are usually called static plate machines and the pipes are painted black and can be seen *at the back* of the cabinet. This is the stage at which the heat absorbed by the vapour has to be given out. To make it more efficient, the process of giving out the stored heat may be assisted by a fan. If yours is a fan-cooled freezer, the heat will be expelled from a grill, usually placed at the side or back of the cabinet, and this is why you must allow air to circulate freely at that end of the freezer. Freezers with skin-type condensers dissipate the heat through the walls of the cabinet, which explains why the outside of the cabinet may feel warm to the touch when the compressor is running. In this case, air must be able to circulate freely all around the freezer, or the ambient atmosphere will become over-heated.

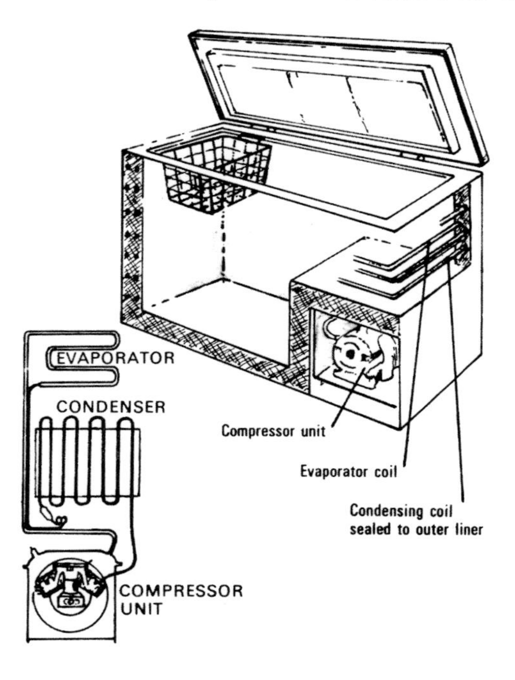

Why not call it a deep-freeze?

And why not call the process deep freezing? These vexed questions trouble most housewives who feel they may be missing out in some way if they are not deep-freezing food. Once again, it's not a question of technical mystique. Only when women began to pioneer techniques of freezing down fresh food, instead of filling their freezers to the brim with bought frozen foods, did the problem of the correct description arise.

The commercial processes which produce, for example, very good commercially-frozen potatoes, require more powerful machines and complicated processes than we can achieve in our domestic freezers. Freezing food by immersing it in liquid nitrogen, for example. Or blast freezing, in which the food, such as sliced green beans, travels through a freezing chamber on a conveyor belt and a blast of super-cold air from beneath keeps the beans dancing on a cushion of air during their entire journey. It is understandable that firms which produce results we cannot match, and can store the food at much lower temperatures for much longer than we housewives can achieve, prefer to reserve the title of deep-frozen food for their products. But the basic method of food preservation is the same. Food which is frozen almost fantastically quickly, and stored at temperatures as low as −40°F (−27°C), has a high quality, as there is so little damage by crystal formation, and a long storage life. Even the minute changes referred to on page 12 are much retarded by the lower temperature. But all frozen food, so to speak, lives in a state of suspended animation. And every housewife who carefully arranges strawberries on a tray, so they do not touch, and open freezes them in 2 hours, instead of in 4 hours crowded together in a bag or container, is employing, in essence, the methods and efficiency of the commercial quick-freezing experts.

Comparing prices of commercial and home-frozen food

In a mood of enthusiasm you may tell your husband that his home-grown vegetables have cost *nothing*, compared to so many pence per pound for a bought pack of frozen ones. Just a word of warning, that is not strictly true. The cost of fuel for blanching and keeping that area of the freezer reduced to a sub-zero temperature which the pack occupies for 6 months or more, must be *something*. So must be the cost of the throw-away bag, or a percentage of the cost of a durable polythene container. Labels and twist ties cost a little too. When you have your freezer serviced this adds, minutely, to the cost of storing each pack. If you pack far too many runner beans, just because they are in the garden for the picking, and they are still cluttering up the freezer a year later, they are not such a bargain compared with a commercial pack you know you will need within a couple of months. Cheaper, yes, but not an entirely free bonus for freezer owners.

Choosing and using your freezer well

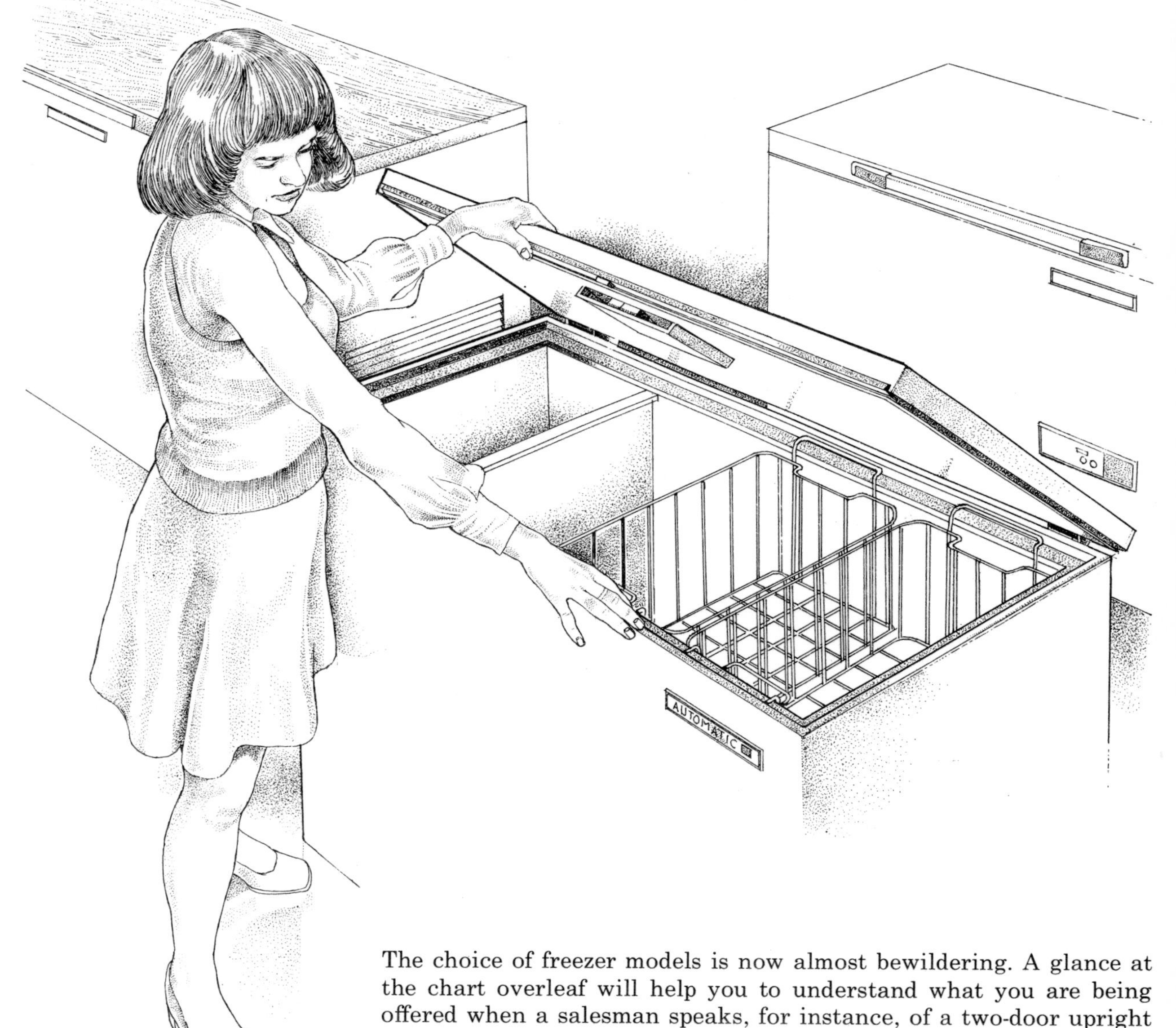

The choice of freezer models is now almost bewildering. A glance at the chart overleaf will help you to understand what you are being offered when a salesman speaks, for instance, of a two-door upright freezer or a refrigerator/freezer.

Upright freezers

Climbing the wall is an exercise we shall all have to get used to with the trend to keep modern kitchens small, while more and more useful and necessary pieces of domestic equipment are added to the existing range. This applies particularly to freezers where so often the only possible wall space in the kitchen is already occupied by a refrigerator. The

1. Refrigerator/freezer

2. Small one-door upright freezer

3. Large one-door upright freezer

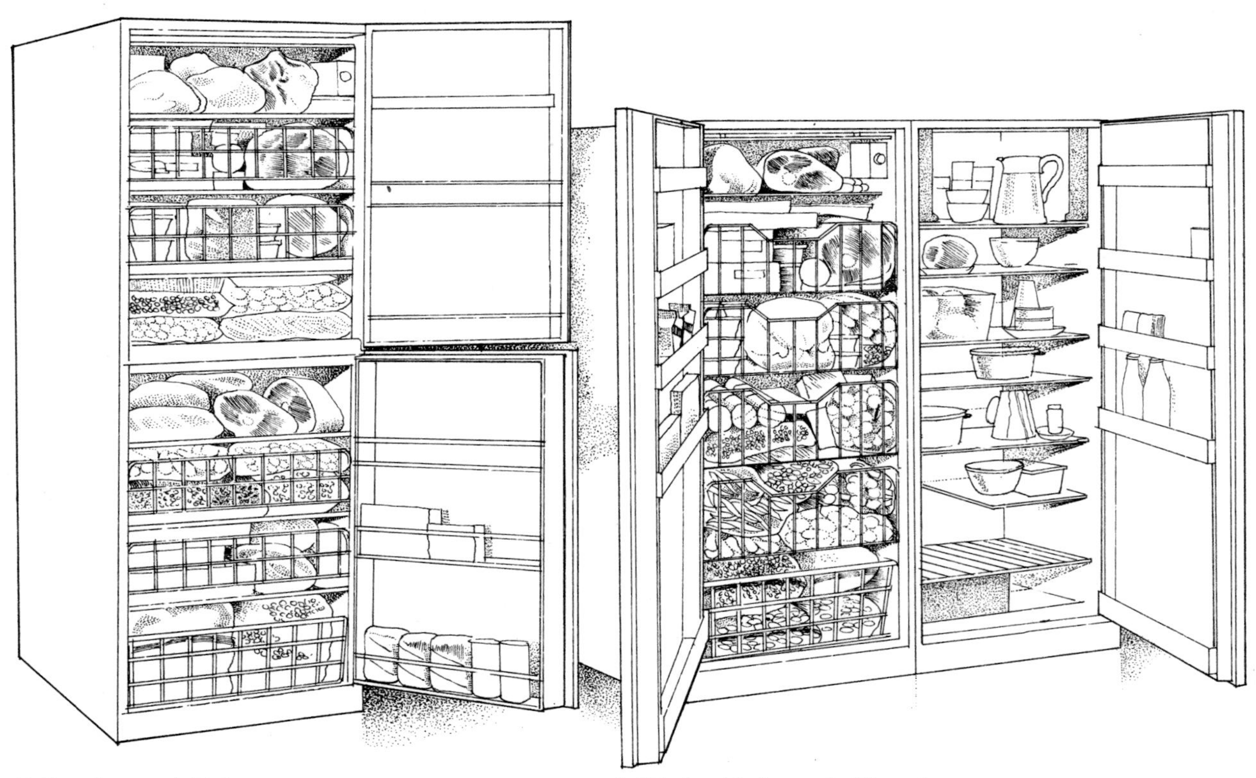

5. Two-door upright freezer

6. Side-by-side freezer/refrigerator

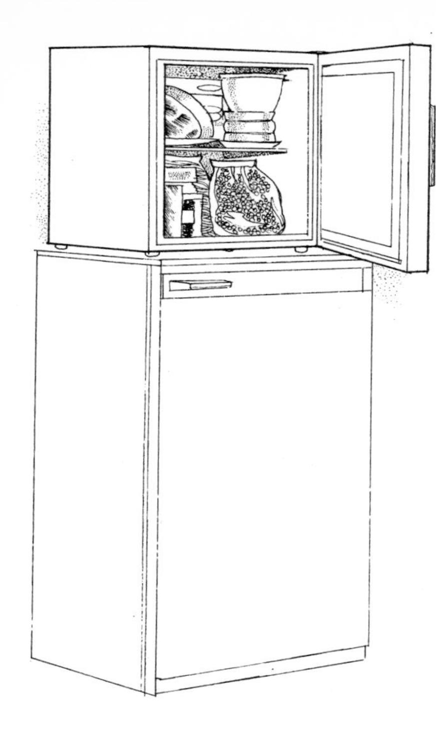

4. Small top-of-the-cabinet freezer

upright freezer is essentially a box with a front-opening door. It can be less than 2 feet (60 cm.) high, or soar to the ceiling.

1. **Refrigerator/freezer** The advent of four-star freezing has done much to prevent the common mistake that if your refrigerator has a separate frozen food storage compartment, it is a small freezer. But where floor space will only permit one piece of equipment the stacked refrigerator/freezer is the answer. The relative capacities of the freezer and refrigerator compartments are all-important. An early tendency in design was to allot equal space to both, but experience has proved that possession of a freezer reduces the requirement for ordinary cold storage space. The ideal ratio would seem to be a 4 cubic foot refrigerator above a 10 cubic foot freezer (Biggs Waterfall 'Niagara'). If the area available is only 20 inches square, there is a slimline model to fit. If you have a fairly new refrigerator, there may be just the right model of freezer available to turn it into a fridge/freezer. New ranges include freezers which stack on top of existing refrigerators, or can be used as free-standing units. A stacking kit can be supplied which facilitates this arrangement and both appliances run off one lead.

2. **Small one-door upright freezer** Small upright freezers can fit under a work surface (a tuck-away 3·8 cubic foot freezer is only $33\frac{1}{2}$ inches (110·5 cm. high) or provide their own built-in working surface on top. This means you need only sacrifice one cupboard unit to accommodate it. In this case, door storage shelves would be a handy asset, but are not always available. Before you buy, consider seriously whether anything less than 5 cubic feet will really be enough. Make sure at least one shelf is removable and not part of the cooling system, or you will have difficulty storing a turkey or other bulky packs.

3. **The large one-door upright freezer** Separate hinged shelf flaps to conserve cold air would be a real advantage in this type, as such a large area is exposed to invasion by warm air each time the door is opened. For the short woman who can only reach in comfortably if the top of the cabinet is no higher than eye level, this is a choice which limits storage space to about 10 cubic feet – an average family requirement.

4. **Small top-of-cabinet freezer** The babies of the market (these vary from 1–2 cubic foot capacity) continue to sell briskly to flatlet dwellers and those with only a kitchenette. For people who live with space at such a premium, even a little freezer storage space is better than none. Its use can be better exploited if it stands on top of a refrigerator with a large three-star frozen food storage compartment, so that packs frozen down in the small freezer can be transferred for storage for up to three months.

Chest freezer

5. **The two-door upright freezer** This type seems to meet universal approval. It gives very capacious storage on a small floor area, and since only one door need be opened, the intrusion of warm air is not great. A low stool, with a very steady base, ought to be handy for reaching the top shelves. Remember that household accidents through falls account for a large proportion of all accidents, and don't climb on a rickety chair. Also, verify that the floor in your kitchen will stand the weight of the loaded freezer, which is considerable – could be more than 500 lb. (250 kg.)! Plan how to stock it so that bulk-bought meat

goes on the higher shelves, with a selection of cuts for day-to-day use lower down where they are easily available. Some models are completely divided, others merely have double doors, and cold air circulates through a single cabinet. Two-door freezers of the former type would certainly conform with Jewish dietary laws for completely separate storage of milk and meat foods.

6. **The side-by-side freezer/refrigerator** This is a lovely, luxurious piece of equipment for the woman with space along the kitchen wall to fill. Not cheap, but the convenience is worth it. The Electrolux model has a refrigerator with a chiller zone as well as storage space at the normal temperature, and there is more storage inside both doors. This gives you access to an area of frozen, chilled or refrigerated food, visible right across an area four times as wide as the average fridge door, and if that isn't luxury I don't know what is. Electrolux also make matching freezer and refrigerator units – the model 60 freezer with a capacity of 6·2 cubic feet and the model 78 refrigerator with a capacity of 7·5 cubic feet – which can be planned in, at a convenient height off the floor, using base units from their Scandinavian range.

The chest freezer

This is essentially a box with a top-opening lid. As it is slightly less complicated to manufacture, it is a better bargain in storage capacity than the upright. It is also more economical to run because there is very little loss of cold air when the lid is lifted and this means that the lid can be left open for some time without partial defrosting of the exposed surfaces of frozen packs. Extra refinements may add to the cost but may make the chest more convenient to use; for example, a specially efficient fast-freeze compartment, such as the Bosch model has, with its three light warning system and the Osby which has a free-standing evaporator. You simply lift this out to defrost in 15 minutes, leaving the food where it is – protected in the cabinet.

warm air

warm air

cold air

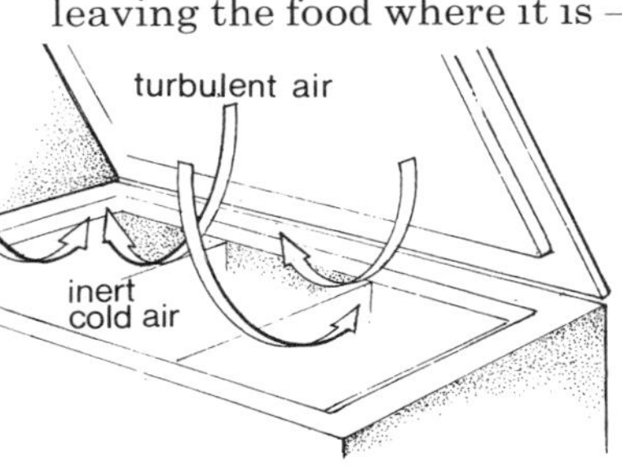

ARGUMENTS FOR AND AGAINST

Upright freezers

Door storage space enables small items which might tend to be forgotten if put at the back of a shelf, to be easily accessible. However, some authorities feel that these packs are stored at a higher temperature than the contents of the freezer cabinet, often exposed to fluctuations of temperature when the door is opened, and ought not to be food easily damaged in quality by semi-defrosting and refreezing.

Pack retainers take several forms, including a front control guard which makes a fold-out retaining rack. Sometimes this acts as a platform on which packs can rest while you reach inside. Some models have separate spring-hinged doors to each shelf, which conserve cold air even when the outer door is wide open. Bosch produces a two-door

model with a pull-out shelf between the doors on which to rest packs while you rearrange the interior. The lack of somewhere to put packs, other than on top of the freezer, is a disadvantage unless other pieces of equipment are arranged to give you an adjacent working surface.
Cold air loss occurs far more seriously in the upright than the chest. Each time the door is opened there is an inrush of warm, moist air from the surrounding atmosphere. The cold air inside which is relatively heavy and inert tends to slip out and be replaced by this much lighter, warm air. The moisture condenses and forms a thick layer of frost inside the cabinet. All the packs exposed by opening and closing the door will be continually defrosting and refreezing to some degree.
Defrosting cycle must definitely take place at least twice a year in an upright as temperature fluctuation causes far more frost formation in the cabinet than in the chest freezer, and the cooling tubes which are exposed and form part or all of the shelves tend to frost up considerably.
Automatic defrost is an expensive extra but some models without this facility do have a drainage outlet to make defrosting easier. Automatic defrosting (and there is an Osby model which automatically defrosts every 24 hours) prevents any build-up of ice on the freezer walls or food. However, I find it essential to remove all the packs from the freezer for a good sort-out now and again and defrosting gives me the ideal opportunity.
Accessibility is clearly greater in the upright freezer than in the chest but there are disadvantages with both types for the very short woman. You may find that if you choose a chest (such as the Bosch which is only 33½ inches (110·5 cm.) high) instead of the usual 36 inches (107 cm.) it will be less trouble to reach down to the bottom, than to be forced to climb on a stool to reach the top shelf of a tall upright freezer.

Chest freezers

Layered storage, which is essential in a chest cabinet, makes it impossible to see or reach more than one-third of the contents without removing the top layer of packs. For this reason it is better to use sliding baskets which hang suspended on the rim, and can be moved to and fro for easy inspection of packs beneath. These should also be accommodated in baskets or plastic carriers or you will spend endless time removing single packs to reach an item stored on the floor of the cabinet. The fast-freeze compartment need not be kept empty at any time, but a basket which fits it conveniently should be used so that it can be transferred to the main cabinet and replaced by another basket of fresh food to be frozen down.
Large packs or oddly-shaped packs are much easier to accommodate in a chest than in an upright model, where the space between shelves is so limited. Items stored at the bottom should be those which are not frequently required, e.g. a year's supply of French beans in 5-lb. (2½-kg.) packs. I usually take out one of these large packs and divide it between small family-size quantities every few weeks.
Sorting shelves are not frequently required as baskets can be temporarily balanced across the top of the open cabinet. But some baskets have a hinged flap which can be opened outwards to act as a sorting shelf.
Defrosting cycle need only take place once a year, if all food is securely packed and the lid is not frequently left open for long periods. Unfortunately a bad closure of a door or lid quickly causes a frost build-up which actually prevents full closure until it has been removed.

The main problem is to maintain the stock at a low temperature while defrosting the cabinet. A drainage outlet and water collection tray simplify the removal of defrost water for those who really cannot reach the bottom of the cabinet conveniently. The chest freezer actually defrosts more quickly when empty than an upright of comparable capacity.

Installation and running costs

Installation is a matter of plugging into a three-pin 13 amp earthed socket, so unless you have to provide a new point there should be no installation charge; but it is wise to have the supplier or an electrician approve the site. A lighting point would, of course, be quite unsuitable. The freezer should be close to the socket because a trailing length of flex is an invitation to an accident. Someone could trip over it, or could inadvertently jerk the plug out of the wall, just far enough to break the connection, without it being apparent. As it is so important not to switch off the freezer by mistake, it is better not to plug into an adaptor (although another piece of equipment could be run off the same plug). As an extra precaution, tape the switch down.

Running costs will vary a little according to the site but you can expect your freezer to cost only a little more to run than a refrigerator of comparable capacity. Throughout the year your freezer will use something between 1½ and 2 units of electricity per cubic foot per week. While a small freezer will cost around 6p a week to run, the popular 12–14 cubic foot models only cost from 15–20p, so a large freezer is relatively economical to run.

Siting

The most convenient and accessible site is undoubtedly the kitchen. If the temperature is higher than average (as it often is in a small kitchen) the compressor will have to come into action more often to keep the temperature inside the cabinet down, and this increases the cost of running the freezer. Also the inrush of very warm air (moist with steam from cooking) each time it is opened will cause heavy frost formation on the inside walls and entail more frequent defrosting.

Experience shows that it really is best to have the freezer somewhere inside the house. Who wants to run out to the garage in the rain to fetch in a pack of pork chops? An airy pantry, lobby, end of a corridor, inconspicuous corner of the spare bedroom – all are suitable. Less good would be a small cupboard under the stairs or converted larder. In a confined space the heat generated by a 12 cubic foot freezer would be equivalent to having four 100-watt light bulbs burning all the time, and if the heat could not be easily dissipated the ambient atmosphere would soon become overheated and overwork the motor, as much as if the freezer was in a hot steamy kitchen. Siting in the garage or any unheated building is not ideal. It is almost bound to be damp, and this will cause rusting long before the normal life of the freezer expires (less than 10 years instead of a possible 15) and extreme cold surrounding air is as unfavourable as very hot air for the operation of the motor. If you have a condensation problem in a brick-built garage or outhouse, the walls could be covered with ¼-inch (½-cm.) asbestos sheeting, providing you batten the walls to allow an air space between the bricks and the asbestos. A ventilated room is preferable to an unventilated one, and when damp is present it is particularly important to ensure that the electrical installation is carried out to a professional standard. It is

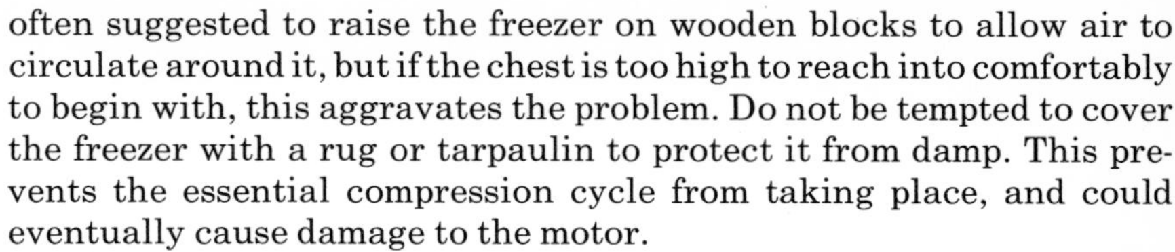

often suggested to raise the freezer on wooden blocks to allow air to circulate around it, but if the chest is too high to reach into comfortably to begin with, this aggravates the problem. Do not be tempted to cover the freezer with a rug or tarpaulin to protect it from damp. This prevents the essential compression cycle from taking place, and could eventually cause damage to the motor.

If the freezer is out of sight it may be out of mind. This could be dangerous if you leave the lid open, or inquisitive children take a peep. A warning system linked to the kitchen (visual by a winking light or aural by a buzzer) is expensive but worth considering.

Be sure the site is well lit. You must be able to identify packs at a glance. If good lighting is impossible and the freezer has no interior light, keep a powerful torch handy, as well as a pair of oven mitts or gloves to slip on (frosted fingers sting badly after a few minutes' rummage) and your freezer log and pencil. I keep freezer labels and a writing implement at the ready so as to have no excuse for *ever* putting an unlabelled pack in the freezer.

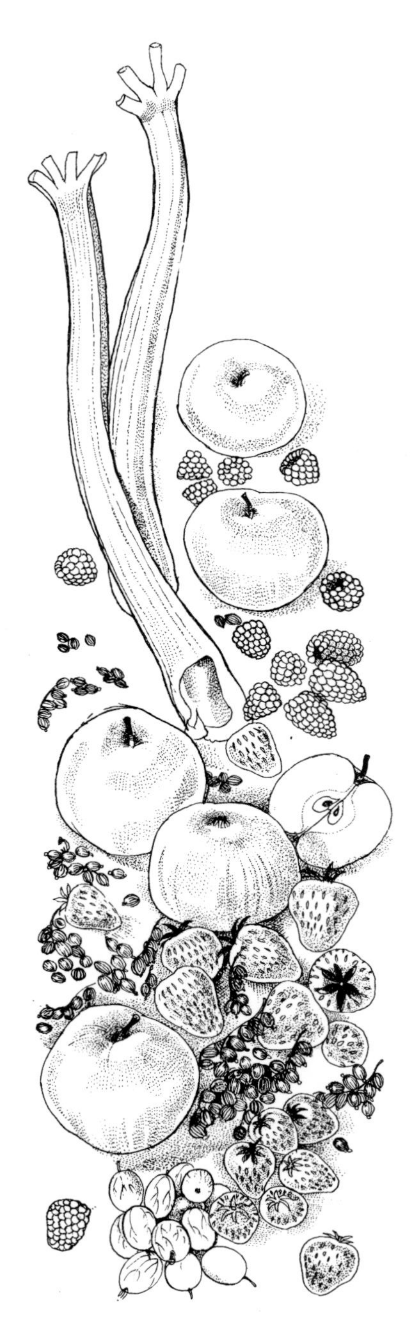

The problem of noise

Are you very sensitive to background noise in the home? For instance, a conservatory or extension room to your home often has walls which contain large areas of glass. This is bound to be very hot in summer, although otherwise a good site for a freezer. A fan-assisted cooling system on the freezer would be essential to cope with these extremes of heat, but it is relatively noisy. This disadvantage may not worry you particularly and you may even feel it is worthwhile accepting if your kitchen is very hot or if the freezer has to be sited near the cooker or boiler. It is a fact that the fan-assisted type is noisier than the skin-condenser type so bear this in mind if you are unable to keep a door shut between you and the freezer, and you are very sensitive to noise. There is another type, the static machine, which makes little noise and has a degree of power somewhere between that of the skin-condenser and fan-assisted models. It may help you to make your choice if you remember that the skin-condenser type is less prone to condensation and therefore more suited to the humid conditions likely to be found in a garage (these conditions are aggravated if you drive the car in and out several times a day).

Choice

It is *cheapest* to buy for cash as you nearly always pick up a bargain offer, or get a good discount. It is *easier* to buy on hire purchase because you can spread out the investment and may be *wiser* because you can insist that the deal covers insurance and maintenance. Freezers are remarkably sturdy, there are so few moving parts to go wrong. But if you need service, you need it fast – within 24 hours or your stocks will be damaged.

Only you can weigh up the relative advantages of chest and upright models; how important it is to you, personally, to have easy access to a shelved upright in the kitchen, or a bigger and probably cheaper chest, far less accessible in the garage (which may or may not be attached to the house). Second hand freezers are available almost always because people want to buy a bigger one. But since they have a considerable 'mileage' clocked up already, try and take over service facilities, or get new ones organised ready for the day of need. If the freezer is pre four-star rating, make sure it *is* a freezer and not a conservator.

Taking care of your freezer

Is it brand new? A new freezer should be washed out with warm (not excessively hot) water and thoroughly dried before switching on. Check the instruction leaflet and make sure whether the temperature control has been set before the freezer left the factory or whether you need to set it yourself. Be particularly careful about fitting the plug to the flex, and switch on. Leave the freezer empty, closed and operating for at least 12 hours before putting any food in it. Do not be surprised if frost begins building up on all the inner walls almost at once, but you need not remove it until it is at least ¼ inch (½ cm.) thick. This should not happen more frequently than three times a year at the most in an upright freezer or twice a year in a chest. Most uprights only need defrosting twice a year, and the chest type once a year. If frost builds up more quickly, make sure that the magnetic seal on the door is completely effective and if it is, make a good resolution to open the freezer less often. Fortunately defrosting is usually necessary just about the time you need to sort out and restack the contents. If you leave it for too long, food packs begin sticking together, the shelves may become immovable and baskets adhere to the walls as though stuck with glue. Of course, defrosting ought to come at a time when your freezer stocks are likely to be at low ebb – for example just after Christmas.

The defrosting process There is no need to make a tiresome chore of defrosting, in fact it always takes less time than one expects. Switch off, remove the contents, pack them closely into insulated bags or cardboard cartons and cover the whole pile with a blanket, previously chilled in the freezer, if you are particularly conscientious. The fact that the door or lid is open all this time will mean that the frost has already softened. Knock it off the shelves or scrape down the walls of a chest model with a plastic paddle scraper. The more frost you can remove in the solid form, the less water you have to mop up. Manufacturers object to applied heat as they feel this damages the interior surface, but a plastic bowl of hot water placed inside the cabinet or a discreet 'blow' with your hair drier will speed up the process. (As soon as it is empty, I place a rectangle of foam plastic in the bottom of my chest freezer and can often roll up and remove this completely with all the frosty fragments I have scraped down, and dispose of them before any water forms.) Wipe out the interior carefully with a solution of bicarbonate of soda – 1 tablespoon to a quart (generous litre) of hot water. Then wipe dry with a clean absorbent cloth and switch on again. Either replace the food immediately or leave the freezer running for an hour before doing so. In any event, even for a large freezer, the whole process should not take more than two hours altogether.

The exterior of the freezer keeps its pristine freshness and will be protected from rusting if it is occasionally polished with white furniture cream or car polish.

If at any time strong-smelling food has been open frozen or packed in a material which is not moisture-vapour-proof there may be traces of a lingering unpleasant smell in the cabinet. If this is so, clean the inside with a solution of 1 tablespoon bicarbonate of soda and ½ pint (3 dl.) vinegar to 1 gallon (4½ litres) of water. Remove and use up the offending packs immediately. Many readers have complained that Brussels sprouts and sprouting broccoli are particular offenders in this respect.

Regular servicing once a year, while desirable, is not usual. It is better to have a good maintenance engineer on 24-hour call.

An ideal freeze plan for every housewife

Giving up your old day-to-day shopping habits may be a bit of a wrench. But large scale shopping soon becomes an agreeable habit.

When you buy more at a time you realise that you are playing a fascinating game. Look through the brochures, check prices of different suppliers, and then against the same volume of food bought in a series of small packs. One firm may supply boiling fowl (which you enjoy) very cheaply, another may not stock it at all. You may be ordering food delivered to you without realising that you could save by visiting a cash-and-carry freezer food centre, or be making a tiresome journey to one when it could suit you better to have food delivered.

In the south-eastern area of Great Britain for example there are four main ways in which you can buy to stock your freezer. Similar services exist in other areas.

Cash and carry freezer food centres

Bejam Bulk Buying, a big multiple firm with many branches, will offer you outstanding service. First, in buying the freezer. You can buy a top quality freezer at a price well below that listed – not an out-of-date model – for cash on delivery, by post-dated cheques, bank card facilities or on hire purchase over one year. Choose from a wide range, currently 40 models. At the same time you can buy insurance, a nominal sum for the first year, and a reasonable amount for successive years to cover any necessary service, replacement, or loss of food other than by your own negligence – turning off the switch by mistake and not noticing, for instance. Such a firm wants you as a frozen food customer, so makes sure you get a good freezer at a rock-bottom price, and that it keeps on freezing. They also run a budget account for your purchases of frozen food. This is how it works. The credit limit is 15 times the monthly payment. Say you decide to spend £10·00 per month on freezer foods, you would receive £100·00 worth of food vouchers right away and more at regular intervals. A service charge of 1½% is made only on the balance outstanding. The stock list is fantastic, around 500 items to choose from and this firm is always pioneering new ones – frozen tomatoes, for instance, when no-one else has dreamed of offering customers frozen tomatoes. They are also wise enough to realise that women make expensive and disheartening mistakes in stocking their freezers through lack of experience, and each branch employs a home economist who can advise on problems of choice, and help to explain and avoid freezing failures. She is there to be asked – no need to feel awkward about wanting help! The centres also sell a small but well-chosen range of accessories for home freezing, because everyone knows the experienced freezer owner freezes down her own produce as well as buying commercially-frozen food. And when you have filled your trolley from the rows and rows of freezers, it is packed into a big insulated bag which gives it protection for 3 hours until it can be installed in your own chest or upright cabinet. This service works well if you have car transport and live within driving distance of a Bejam branch, of which there are 60, and the firm is rapidly expanding.

Home delivery frozen food service

Some department stores like Harrods, operate a van service which delivers frozen food (packed with dry ice) within a wide radius of the town in which they are situated. No delivery charge if the order is £5·00 or more. The emphasis is more on quality than economy and personal service (meat cut up to your personal requirements and a telephone advice service on how to cook it, for instance). If you are an account customer at a big store this is a convenient way to buy, especially if you want medium-size rather than monster packs. The Birds Eye Home Freezer Service is well known throughout the country. It supplies, through 40 depots, all over the country. A telephoned order brings you refrigerated delivery of their range (which includes Smethursts, Tempo and Country Kitchen). If the housewife is out and has given instructions to do so, the order will be left, packed with dry ice, at the customer's risk. Again, not so much enormous packs at huge savings, but quality foods for which these household names are famous.

Dual delivery/self collection service

Sometimes a big firm like Brake Brothers of Tonbridge, in an area with small urban concentration and large semi-rural cum suburban surroundings, finds it essential to give a delivery service. Again a big stocklist to choose from, seasonal offers, regular postal contact, with the latest price list. You can buy freezers, accessories and books from them too. They will also give you a cash and carry service.

A similar service offered by smaller firms needs a careful check on the quality. Unlike the giants who have their food grown and packed for them in the country of origin, these firms often buy where supplies are cheapest. There is a world of difference between top grade and low grade peas. Remember frozen food varies in quality. If it costs *more* possibly the reason is that it is *better*.

Wholesale butchers

A bulk purchase of meat is one of the most important and expensive investments a freezer owner makes. You may find it more satisfactory to patronise a wholesale butcher who will cut up the meat just as you want it rather than buying packs from a freezer food centre where you have no choice. Although many retail butchers supply meat for the home freezer they do not specialise in it. You will get a better bargain from a butcher who has a retail outlet but also operates a wholesale business, like Price Brothers of Edgware. This allows you to make personal contact and discuss your requirements and preferences. Or, go straight to the market.

Decide to buy a freezer and you will be one of the most sought-after women in town. The manufacturers of freezer, and all the kit needed for home freezing, the vendors of frozen food, all will court you, for freezing is big business.

One housewife's own freezer story

"I know I'm rather calamity prone, but when I bought my first freezer I must have made every mistake you can imagine. My husband only agreed on the understanding that, including the initial stock of food, the whole investment would not exceed £100. I bought a huge and unwieldy chest freezer at a so-called giveaway price, which was too long for the kitchen wall space available as I had not allowed for the necessary ventilation space at one end; so it had to go in the garage. This would not have mattered if the garage had been reasonably dry and accessible to the house, but a visit to the freezer entailed a two minute trot round the back, fully exposed to the elements, and as the garage is damp, rust unavoidably damaged the freezer. Also, it was really a commercial model and had no fast freeze compartment, the lid was not counter-balanced and there was no interior light. This last was a serious inconvenience as the interior of the garage was always dark in the freezer corner and the one light bulb badly placed. I had to keep a torch handy as well as oven mitts to prevent frost-frozen fingers. I am short, and the freezer was 35½ inches (92 cm.) deep – the precious 2 inches (5 cm.) lower on other makes of chest would have saved me from taking a header into it on more than one occasion. My husband raised the freezer on blocks to prevent damp rusting the works, which meant I could hardly reach even the top baskets. He had to build me a duck board to stand on each time I opened the lid. There was the black day when I started the car, put it into reverse by mistake and dealt the end of the freezer a mighty blow. A bit harder and it would have been mortal!

I entirely underestimated the importance of keeping a freezer log from the start so that I never knew where anything was and spent hours unpacking and repacking by the light of my torch. I constantly found myself thawing a whole 10-lb. (5-kg.) bag of stewing meat or pack of chops to extract just the amount needed for one meal. If only I had packed it in convenient meal-size portions! I failed to appreciate that wet packs or semi-defrosted packs would refreeze and stick together. I did not have enough baskets and none of the kind with reversible handles to hang and slide to and fro as the top 'shelf' so to speak.

For months I was depressed by the vast empty space in the freezer as I could not afford to stock more than a quarter of its capacity immediately. The instruction book the salesman gave me was totally inadequate and for ages I packed food in thin polythene bags which the children had always used for sandwiches, imagining that those gave adequate freezer protection. After hearing from a friend how much open freezing improved quality of fruit, I blanched a huge quantity of Brussels sprouts and open froze them before bagging. The smell lingered in my freezer and hit me full blast each time I opened the lid for months afterwards. I made lots more classic mistakes, locked the freezer and lost the key, switched off the electricity at the main before going on holiday (thus switching off the freezer in the garage) and . . . I could go on and on.

Even so, I loved being able to produce a choice of vegetables or ice cream or my own chicken pies, whenever I wanted.

It wasn't until I visited a friend who had bought one of your books, Audrey, along with her freezer, that I realised I was a bit lacking in expertise. Her freezer/fridge had taken the place in her kitchen of the old refrigerator and occupied exactly the same floor space. Of course it cost much more than my chest had done but as the purchase was tied up with insurance against mechanical failure and accidental loss of food, I reckoned that the score was about even. When we moved I told the new owner frankly that the freezer probably had only a few years of useful life and he took it over on that understanding.

Although I pride myself on the well-packed and stacked contents of my new upright freezer, and it is marvellously convenient to have it in the kitchen, I miss one thing – the capacious storage space for awkwardly shaped large packs – turkey, big packs of meat for instance. All the shelves seem far too close together and cannot be moved. Even by taking out the one removable drawer, my turkey bulged out of its allotted space and I am sure put a strain on the magnetic seal of the door. Also, handy as the shelves in the door are, my kitchen is hot because I have an Aga cooker and the packs never seem completely frozen and solid. Perhaps I keep the door open too long but I must open it sometimes. On balance though, the convenience of having the freezer under our own roof and being able to take any item from the shelf mentioned in my freezer log without disturbing more than one or two other packs, greatly outweighs those minor inconveniences. Mine is a 2-door freezer by the way, I have the old refrigerator under the lowest shelf in my walk-in larder which gives me 12 cubic feet frozen food storage space and 5 cubic feet refrigerator space; just about perfect I reckon for the needs of a family of four. I must admit I cook only savoury dishes for my freezer – stews, pies, casseroles and flans. My cakes are not worth eating so they are not worth freezer space."

1 Plunging blanched French beans into a bowl of iced water.

2 Packing well-drained beans into coloured polythene bags for easy identification.

3 Putting a bagged meal-size portion into the freezer basket.

1

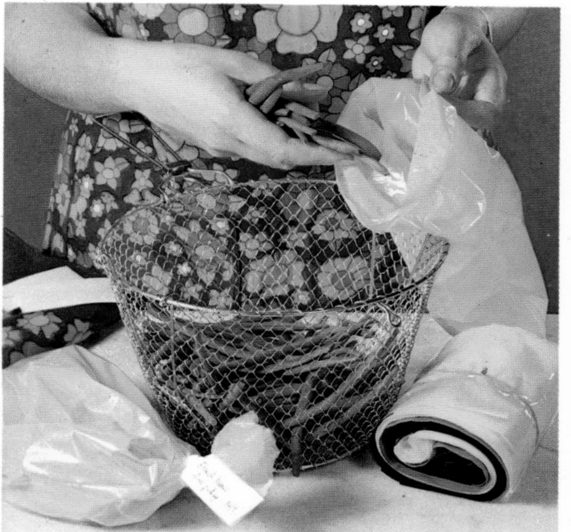
2

Frozen vegetables cooked by the conservative method in the minimum amount of water.

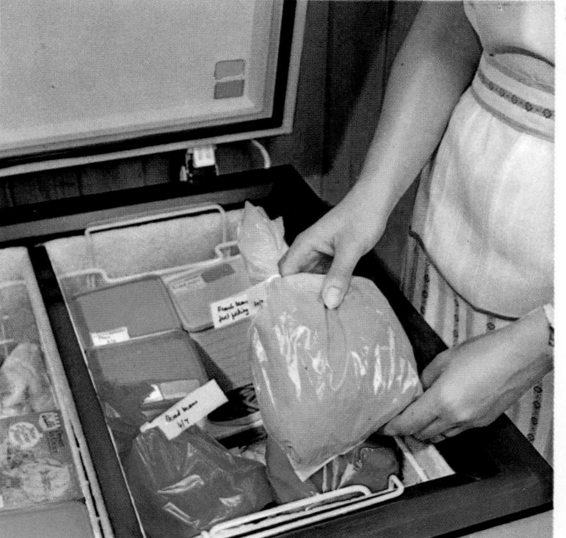
3

Packing materials and methods for various foods

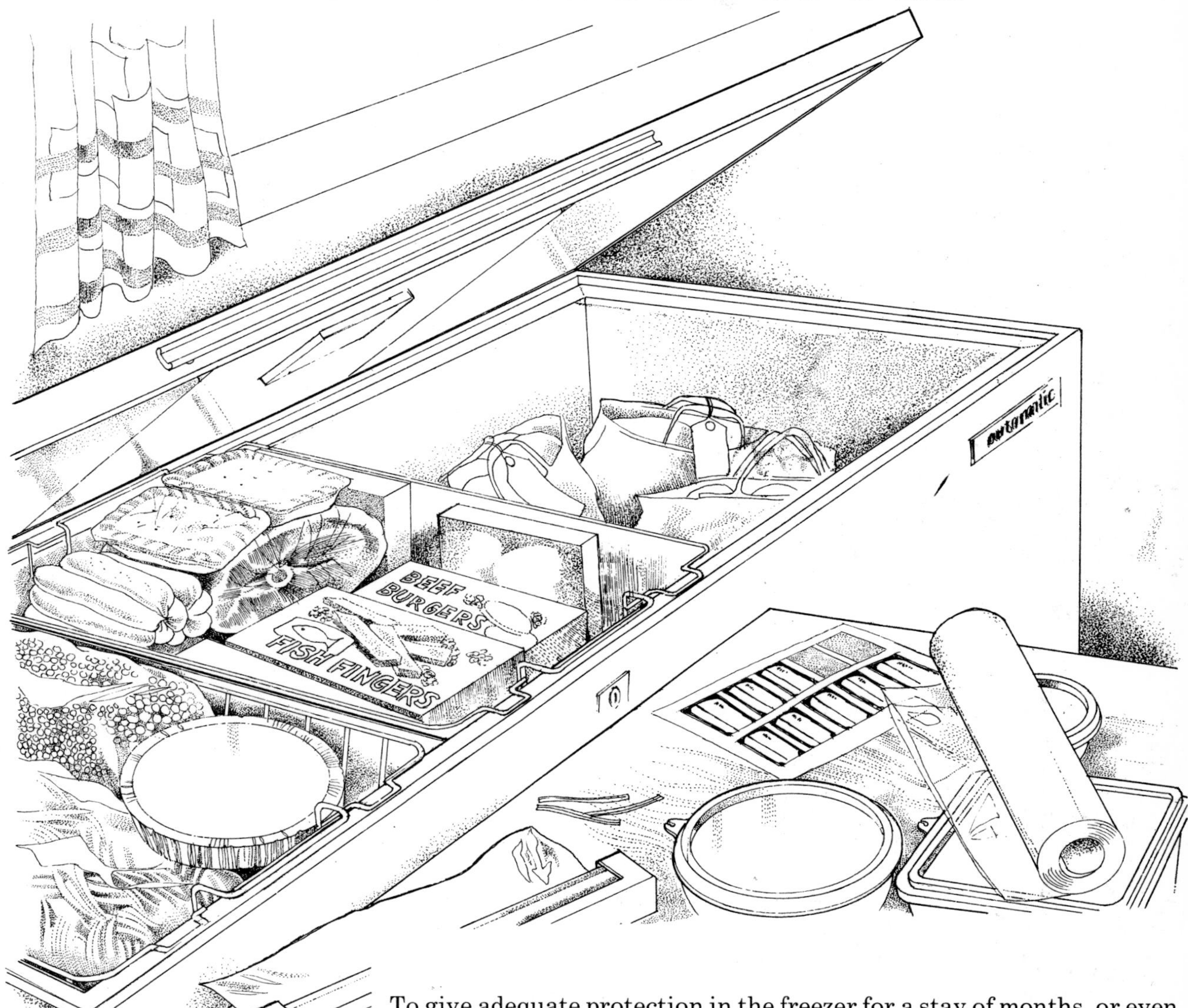

To give adequate protection in the freezer for a stay of months, or even a few weeks, all food must be sealed into an airtight wrapping or container made of material that is moisture-vapour-proof. Cheating on one or other of these vital requirements may lead to disastrous results – dehydrated food lacking in flavour, or worse still, tasting and smelling of something completely different. While the use of freezer tape *can* transform a parcel or waxed carton into an airtight pack, this takes time and trouble, and is not completely certain as a tiny air inlet may not be apparent. That is why foil, which moulds or crimps into a closure, polythene bags with twist ties and polythene containers with a seal are my favourite freezer packs.

Pear condé, gingered pear coupes (see pages 210, 158).

Moisture-vapour-proof materials

Here you may be in for something of a shock; not all polythene containers, bags and aluminium foil are moisture-vapour-proof.

Polythene containers Must be able to withstand sub-zero temperatures without splitting or warping. Tupperware containers are not only guaranteed for 10 years' use, but proof against reduction to −70°F (−60°C). Cheaper containers may split or warp after one use, so that the lid which gives an airtight seal is never again a perfect fit. The seal should make the container airtight.

Coloured polythene containers An aid to easy identification without getting at close range to the label. Coloured lids are helpful but it is better still for the container itself to be coloured. Use warm colours for fruit and sweet sauces, cool colours for vegetables and savoury sauces for example.

Polythene bags Must be of thick gauge (200–250) to be moisture-vapour-proof. To give them an airtight seal it is necessary to close them with a tightly tied plastic or paper coated wire tie (twist ties) or with freezer tape. Ordinary clear adhesive tape is not suitable because it peels off at low temperatures. Gussetted bags can be filled to make a square-standing pack, sealed with a twist tie, freezer tape or simply tied in a knot. Plain bags can be sealed in any of these ways, or across the open end with a heat-sealer. These packs can be laid flat and stack very conveniently, rather like commercial packs.

Coloured polythene bags Come in six different colours and are useful in that you can create your own colour code when using them. Once you have decided that you use green for vegetables, white for fruit and black for the dog's dinners, it is no particular disadvantage that the packs are opaque. But for those who, like me, have a phobia about it, there are 'stripey bags' in various sizes, which have alternate stripes of clear and coloured polythene. Batching bags permit a number of similar items to be packed together. In this case it is really better to use clear polythene bags inside the batching bag, or for example use a large red bag for meat, and inside, package beef in small red bags, pork in white, lamb in blue and so on.

Boilable bags These come in various types. A heavy gauge bag (which you can make yourself from sleeve polythene with a heat sealer, or you can simply seal at one end by heat) has been available for some time. New arrivals on the freezer scene are high density polythene bags, which can be closed with twine or twist ties, used for freezing and can be placed straight from the freezer into boiling water to defrost and reheat the contents. Vegetables can be blanched in the bags but it is a lengthy process and generally unsatisfactory. Cooked foods, blanched vegetables, and raw smoked fish have all given good results in these bags, although it is a wise precaution to pack a number of them all together in some container for protection against damage in the freezer. A special advantage, for example with vegetables, is that if they are packed with a knob of butter and a little seasoning, there is no loss of the moisture and nutrient elements at the reheating stage, because they do not come into direct contact with boiling water. If carefully handled and washed, the bags can often be used more than once.

Intrepid freezer addicts may like to experiment with the blanching-freezing-reheating process; such experiments are all part of the fun of owning a freezer.

Sheet wrapping Can take the form of rolls of polythene sheeting (which needs sealing with freezer tape) or self-cling polythene film. The latter is very good for items which need close wrappings, such as wedges of cheese, but the packs really then need overwrapping or placing together in a container. Heavy gauge aluminium freezer foil makes a self-sealing close wrap and ordinary kitchen foil (used double) is also suitable for the same purpose. Foil has the advantage that it needs no freezer tape if the edges are crimped together or smoothed close to the surface; in fact an entirely airtight and watertight pack can be achieved with sheet foil. Another use is to line a favourite container, such as a pretty ovenproof casserole, or a family-size saucepan with a good heavy base. Foil can be moulded to fit the container, filled with a cooked stew, and the surplus foil brought together in the centre when the contents are semi-frozen, to make a compactly-shaped pack. The advantage of this is that if the pack is held under warm running water for a short time, the foil can be peeled off and the frozen food fits neatly into the casserole for oven heating, or the saucepan for top-of-the-cooker heating. I also use foil freezer bags successfully for curries and stews. They are not cheap, and in my experience can only be used once. Then you must handle them adroitly as they tend to disintegrate if handled carelessly when lifted out of the saucepan after reheating.

Shaped foil containers These now come in many shapes and sizes. They are useful for freezing cooked dishes to be reheated in the oven. Some have covers supplied which must be removed before the containers are put in the oven. Foil pie plates, in which pastry can be frozen and then cooked, are also extremely handy. Sturdy shaped foil dishes can be sterilised in boiling water or Milton solution (not such a chore when Mums remember it from baby's early days). Look out, too, for the small foil containers with see-through lids. Sheet foil has another bonus; because it is so strong a narrow strip folded and placed under a cake helps you to lift it out by the protruding ends. This prevents any piping or decoration on the cake from being damaged.

Waxed cartons, glass jars and aluminium boxes Are, after long experience, my least favourite containers although I do use them. Jars with straight necks and screw tops are the best but since glass is always suspect in case it may shatter when reduced to a low temperature, test jars by filling them with water, leaving a good headspace. Put them inside a strong polythene bag. Leave in the freezer on test for several days and if the jar is still intact, that particular type can be considered trustworthy. Waxed cartons have the disadvantage that even with care they can only be used a few times, and the lid does not give an airtight seal, so has to be taped round completely. So do aluminium boxes, but they do conduct heat rapidly and freeze and thaw contents quickly.

Pyrosil A toughened ceramic which withstands freezing, oven cooking or cooking over direct heat, straight from the freezer. Too expensive for a long-term plunge perhaps, but great for a dinner party dish prepared a few days ahead to come smartly to the table.

Other freezing accessories

Freezer tape Any adhesive tape which sticks under freezing conditions can be used. One which resembles masking tape has the advantage that you can write on it with a lead pencil or biro and it can be ripped off without taking half the bag with it.

Twist ties The cheaper ones are paper-covered wire; the better ones, which I use, are covered with plastic and come in many colours. They can be used more than once and in themselves are a handy form of colour identification.

Labels Must be self-adhesive or they will drift away from the pack. They come in various sizes, very useful if you want to write full defrosting and serving instructions on a large one, and again come in various colours. These can colour-identify in two ways. Either label all vegetables with green labels, or pastry products with yellow, etc. Or use a seasonal rotation system. Green for the first three months of the year, yellow for the next three months and so on, so that you can tell at a glance roughly how long any pack has been in the freezer. Tupperware labels come in two shades of blue and are marked simply 'Food' and 'Use by'. The backing sheet indicates safe storage times. All self-adhesive labels should be peeled off while the container is dry. One type of twist tie comes complete with a flag-like label at the end and I find this a boon if I forget to label a polythene bag before filling it. It is very hard to stick a label on a bag filled with knobbly food, and all but impossible to stick it on a frozen pack.

Punch-printed labels A gadget, called Dymo, punches out a label which is so legible that short-sighted freezer owners can pick out the lettering at a fair distance. No more glasses falling off your nose into the freezer!

Heat sealers If you own a big freezer and fill it with garden produce, you will find this a handy gadget. You won't get a better one by paying a fancy price. All it does is generate heat through teflon-coated surfaces which allow the polythene sheets to melt and combine, without sticking to the source of heat.

Blanching equipment A blanching basket is quite heavy, takes a lot of space to store and is slow to heat through. So, each batch of vegetables keeps you waiting before it comes to the boil. A light collapsible blanching basket packs flat to store, and makes it easier to gauge blanching time accurately. Nylon muslin bags or metal strainers will substitute. If you have no blanching equipment, boil vegetables in the open pan for the specified time and strain off through a colander. NOTE Re-using bought yogurt and cream cartons after the contents have been eaten can only be recommended if they are very thoroughly washed, rinsed and dried. Some come with their own snap-on lids which may not be completely airtight. In most cases a cap of foil smoothed very closely down the sides of the container gives a better seal than re-using the snap-on lid.

Non-essential impedimenta

A pair of washable oven mitts really are essential to handle frozen food comfortably. A rather fearsome serrated knife, such as the Pyrex Slicicle makes it possible, if not easy, to cut up frozen food, even meat. To keep your fingers intact, put the board on a tea towel so that it does not slip about on a shiny table surface. Then put a layer of folded kitchen paper between the food and the board so it will not slide away from you. There are two thermometers which might be useful, one to keep in the freezer to check the exact temperature at which it is maintained (something impossible to judge from the numbered dial) and another to use when roasting meat from the frozen state. A meat thermometer has always been considered indispensable in American households.

Keeping a freezer log

Nothing is more certain than this; if you decide you will rely on your memory for the exact contents and disposition of same in your freezer, you will regret it. After weeks and months of diving unsuccessfully into the dark recesses it is utter bliss to open a little book and see what you have in store, and where, at a glance.

There are various columns into which each page can be divided. In my opinion the smallest really adequate number is four. These should be headed – Date in, Description of pack, Number of packs, Shelf or basket. In this case the number should be indicated by separate strokes rather than the figure itself, each of which can be crossed through as a pack is removed. If a fuller record is required, an extra column can be headed 'Date out' and three extra spaces allowed so that the date of withdrawing each pack can be recorded. Keep the freezer log with a biro or pencil near the freezer, *not* for example in a kitchen drawer if the freezer lives in the garage.

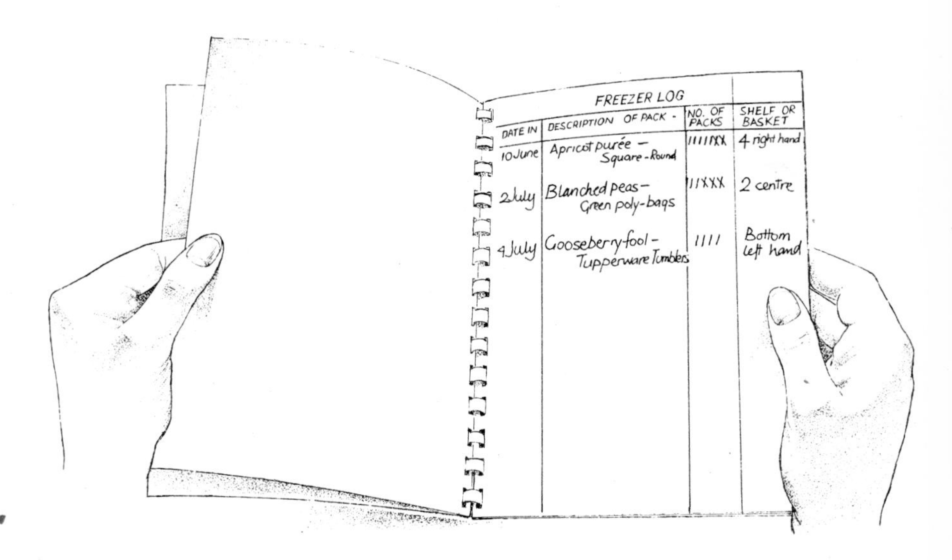

The problem of packing delicate foods

Some foods (such as decorated cakes and soufflés) are so delicate that even when frozen they can easily be damaged by pressure from outside the pack. Others, such as soft berry fruits and blanched vegetables, are liable to be squashed by their own weight inside the pack itself. For foods in the first category, the answer is to follow the open-and-shut freeze method:

1. Open freeze until completely firm. Do not leave longer than necessary or the food may dehydrate or absorb cross flavour.

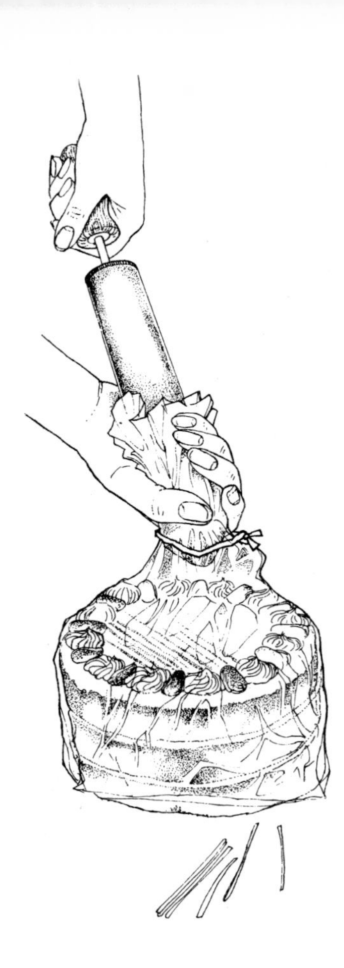

2. Bag wrap and withdraw air using a special pump or straw, so that no air spaces are left. This gives protection against dehydration but not against pressure, so put the bagged cake or soufflé inside a round container, or whatever shape best fits it. It should stand clear of the sides as it is already protected from direct contact with air and is easier to remove if adequate finger spaces are left. One trick is to place the party special on a Tupperware seal and invert the container over it. This I recommend especially for short-term storage, but be careful when removing the container from the freezer to slip your hands under the seal to take the weight. Cardboard cake boxes, which fold flat when not required, can also be saved for this purpose. I sometimes open freeze the cake on the serving dish and simply invert a plastic basin or Pyrex Rainbow over it.

Secondly, those delicate berries and newly blanched vegetables! Packing in bags forces the food down and squashes it when the bag is drawn together to eliminate air spaces and secure a good seal. This damage could be considerable and would not be noticed until defrosting took place. For this reason I pack raspberries and blanched peas in rigid-based containers with foil dividers, cut out round the base of the container, placed half-way up. This relieves the pressure as well as distributing the weight and prevents the bottom layer from getting mushy as the contents defrost. Fruit eaten while still chilled has a better texture than the fully defrosted equivalent.

Leaving a headspace

Vegetables and fruit require hardly any headspace because there are tiny spaces all round them to allow for expansion. Liquid or semi-liquid foods (soups, sauces and stews) contain a lot of water so some headspace must be allowed, remembering that water expands by 10% of its total volume as it freezes. If packed in a shallow container with a comparatively large surface, $\frac{1}{4}$–$\frac{1}{2}$-inch ($\frac{1}{2}$–1-cm.) headspace usually suffices. In a deep container, such as a polythene tumbler, the surface is small and $\frac{1}{2}$–$\frac{3}{4}$ inch (1–1$\frac{1}{2}$ cm.) would be a safer allowance, according to the capacity of the container. It will be observed that the surface does not expand upwards evenly. Frozen liquid looks more like a moonscape and that awkward excrescence in the middle could force off a seal if not allowed for. However, since unnecessary air spaces only encourage dehydration, don't leave more than is needed or you will have some of the food's essential moisture coyly nestling in the form of frost crystals under the seal.

Getting rid of those air spaces

It is not easy to find any sort of container which exactly fits awkwardly-shaped foods. When you try to fit a chicken-shaped chicken into a bag-shaped bag you may end up with unwanted pockets of air which cannot be smoothed out and eliminated before fastening the twist tie. There are two ways to solve this problem:

Drawing out air with a straw A pump is available (The Icicle) to do this for those who are lazy or disapprove of the straw method. Either way, fix a twist tie loosely around the neck of the bag, insert pump or straw, remove air until bag snuggles cosily round the food, withdraw the pump or straw and tighten up the tie. Never breathe out through a straw into the bag as this will treat the food inside to a full blast of any

germs you may be harbouring. Remember you might go down with a cold tomorrow. Empty your lungs, then suck in the air from the bag.

Displacing air by immersing in water Another way is to plunge the bag up to the neck into a bowl of cold water, forcing out surplus air. Hold the neck of the bag above the water and fix a twist tie. Tighten, then lift the bag out of the water. One snag is that the outside of the bag must be wiped dry or you will have a mass of oddly-shaped objects shrouded in ice and sticking together in the freezer. The drying process is not easy with odd-shaped packs.

Speed and hygiene

Why is a quick journey from fresh to frozen just as important as quick freezing itself? Food is at risk by contamination from the time it is picked, killed or cooked, until it is safe in the freezer cabinet. The twin dangers are the continuing processes of natural deterioration and the inevitable exposure to more contamination by handling in every-day situations. We do not live in sterile conditions – if we did the kitchen would resemble an operating theatre!

Cooking or blanching involves raising food to a high temperature and allowing it to pass through a zone of intense bacterial activity as it cools. Slow cooling is as bad as slow freezing! Food must pass quickly down through the crystal formation zone to avoid damage to the cell structure. It must also pass quickly through the cooling zone to allow bacteria the least possible chance to multiply.

It is only common sense to pick out any bruised or damaged specimens from fruit or vegetables for freezing and use these up at once and to cut away imperfect parts and turn the remainder into purées or use in soups.

Vitamin loss

When vegetables are placed in boiling water to blanch them, there is some inevitable loss of vitamin content, particularly of vitamins B and C. You may lose up to 30% of the vitamin C content, the most elusive of all, but providing you don't repeat the boiling process when defrosting and serving, this loss is no more than you would experience in cooking fresh vegetables. Commercially-frozen vegetables are often frozen, literally, within a few hours of picking. Those you buy in the greengrocer's may have lost 40% of their vitamin C already, having been so much longer on their journey from plant to cooking pot, and if you freeze your own from the garden, you can imitate the commercial process in speed at least. The younger, fresher, more vitamin-rich you catch and freeze them, the better all fruit and vegetables emerge from the freezer. Vegetables ought to be just at maturity (they may taste a bit mealy and oversweet if over-mature) and fruit at the same stage. Slightly over-ripe fruit, and vegetables to some extent, are better frozen as purées, mousses and soups.

Vitamin C is also a useful ally in preventing vegetables from becoming discoloured. A few vegetables, notably potatoes and Jerusalem artichokes, become discoloured when peeled and exposed to the air. Fruits are more vulnerable. Apples, pears, peaches and apricots are the worst offenders. Soaking fruit in lemon juice as you prepare it prevents some discolouration but does make the fruit sour. A better way is to add a mild solution of ascorbic acid (see page 59) which you can buy from the chemist.

Golden guide number 

to preparing vegetables for freezing

Most vegetables keep well for a full year. For long-term freezing, blanching is the all-important step, as it virtually halts the enzyme action which causes spoilage. Vegetables, unlike fruit, are a non-acidic food and once picked start to deteriorate, even in the frozen state. Brussels sprouts for instance show an appreciable deterioration after a week of frozen storage, if they are not blanched. Certain vegetables, marked thus *, are suitable for short-term storage without blanching.

Blanching instructions

Use a rigid or collapsible blanching basket. Immerse the prepared vegetables, 1 lb. ($\frac{1}{2}$ kg.) at a time, in a large saucepan of rapidly boiling water. Time the blanching from the moment the water returns to the boil. Remove and cool under running cold water or dunk up and down in a large bowl of iced water. Chill quickly, drain thoroughly, spreading out to dry on absorbent kitchen paper, if necessary and pack.

Vegetable	Preparation	Blanching time	Packing	Extra points
Asparagus	Wash, trim, blanch, grade and tie in bundles.	Thin stems 2 minutes Thick stems 4 minutes.	Rectangular polythene containers	Buy at their cheapest – use stalk trimmings for soup.
Artichokes Globe	Remove outer leaves, trim tops wash well, add lemon juice to blanching water, drain and cool.	7–10 minutes	Polythene bags	
Jerusalem	Cook and freeze as purée for soups.		Polythene containers	
Aubergines	Peel, cut into 1-inch (2·5-cm.) slices, blanch, cool and pack in layers using non-stick paper.	4 minutes	Foil trays	
Avocados	Most successful as purée. with seasoning and lemon juice added.		Polythene containers, freezer film over surface to prevent discolouration	
Beans				
Broad	Pod and blanch.	3 minutes		
French*	Top and tail, slice or leave whole.	3 minutes	Polythene bags	Freeze whole unblanched; store 3 months only.
Runner	Trim, slice thickly.	2 minutes		
Beetroot	Use small ones, cook until tender, cool, peel, slice or dice.	Boil 40–50 minutes	Polythene containers	
Broccoli	Trim, wash in salted water, cut into sprigs, blanch, drain well.	3–5 minutes depending on size	Polythene bags	
Brussels sprouts	Peel, trim and wash. Grade, blanch, cool and drain.	3–5 minutes	Polythene bags	
Cabbage	Wash, shred finely or cut into wedges, blanch, drain well.	$1\frac{1}{2}$ or 4 minutes	Polythene bags	Chinese cabbage – blanch shredded for $1\frac{1}{2}$ minutes – serve with oriental dishes.

Vegetable	Preparation	Blanching time	Packing	Extra points
Carrots*	Trim, peel or scrape and wash, slice or dice.	4 minutes	Polythene bags	Freeze young carrots whole unblanched; store 3 months only.
Cauliflower	Trim into small sprigs, wash and blanch in salted water, drain well.	3 minutes	Polythene bags	Choose tight, white heads with no blemishes.
Celery	Trim, scrub and cut into 1-inch (2·5-cm) lengths, blanch, cool, drain.	3 minutes	Polythene containers	Use only for cooked dishes.
Celeriac	Wash, peel, slice, steam until almost tender.		Polythene containers	
Chestnuts	Wash, peel and blanch, or cook until tender and purée.	1–2 minutes	Polythene containers	Freeze some purée sweetened to use for mousses, soufflés or sauces.
Chicory	Remove outer leaves, add lemon to blanching water, drain and pack.	2 minutes	Polythene bags	Use in cooked dishes or with white sauce.
Corn-on-the-cob	Trim off leaves and silks, wash, blanch, cool and dry.	4–8 minutes depending on size	Individually in freezer film, then in polythene containers	Do not freeze late-season cobs, only the youngest, freshest ones.
Courgettes*	Wash, trim into $\frac{1}{4}$–$\frac{1}{2}$-inch ($\frac{1}{2}$–1-cm.) slices, blanch or sauté in butter.	1 minute	Polythene containers	Can also be halved, blanched 2 minutes and stuffed. Freeze whole unblanched. Store 3 months only.
Fennel	Trim, wash, cut into $\frac{1}{4}$–$\frac{1}{2}$-inch ($\frac{1}{2}$–1-cm.) slices, blanch, drain, pack.	3 minutes	Polythene bags	Not suitable to serve raw.
Marrow	Use small young ones, peel, cut into 1-inch (2·5-cm.) slices and remove seeds, blanch.	3 minutes	Polythene bags	
Mushrooms*	Use small button mushrooms, wash, sauté in butter for 1 minute, or slice larger ones.		Polythene containers	Freeze small button mushrooms unblanched; store 3 months only.
Onions*	Peel and chop.	2 minutes	Double polythene bags	Double throw-away wrapping to prevent contamination of other foods. Freeze chopped unblanched; store 3 months only.
Parsnips	Trim, peel, slice or dice, blanch, cool, drain.	2 minutes	Polythene bags	
Peas*	Shell, blanch and cool.	1 minute	Polythene bags	Freeze unblanched; store 3 months only.
Mange-tout	Trim ends, blanch and cool.	2–3 minutes	Polythene bags	Open freezing ensures free-flow packs.

Vegetable	Preparation	Blanching time	Packing	Extra points
Peppers*	Wash, remove stems, pips and pith, cut in halves or strips; blanch, drain and cool.	3 minutes	Polythene containers	Freeze sliced unblanched; store 3 months only.
Potatoes				
Chipped	Prepare chips and part fry. 2 minutes; cool.		Polythene bags	
New	Scrape, cook until just tender.			
Spinach*	Wash carefully, blanch, cool; drain well and pack.	2 minutes	Polythene containers	Wash and freeze in polythene bags, unblanched and tightly packed; store 3 months only.
Tomatoes*	Skin, simmer for 5 minutes, rub through sieve, pack.		Polythene containers	Wipe, remove stem and freeze whole in polythene bags; store 3 months only. Suitable grilling or frying only. Hold under running water to remove skin.
Turnips	Trim, peel and dice, blanch, cool; drain.	3 minutes	Polythene bags	
Herbs (Method 1)	Wash and chop herbs, pack into ice-cube trays and cover with water, freeze, turn out, pack and freeze.		Ice-cube trays, polythene bags	Store for 12 months and use individual cubes in soups, stews, sauces, etc.
Mint	For mint sauce, pour over sugar syrup instead of water; to serve, thaw and add a little vinegar.			Pick, before mint becomes woody, young shoots with full flavour.
Herbs (Method 2)	Remove any woody stalks, wash and shake dry, lay flat on sheet of freezer foil, fold in and crimp edges together. Crumble frozen leaves to save chopping them.			

* Remember that unblanched vegetables require the usual *full* cooking time, and although enzyme activity will not cause off flavours to develop during their short stay in the freezer, you will lose one desirable side-effect of blanching. That is, the reduction of the numbers of micro-organisms present, so that spoilage is more likely if the defrosted vegetables are not cooked and used up immediately. Blanched vegetables pack down conveniently into less space and are almost fully cooked – two points in favour of blanching.

Cooking frozen vegetables

Cook from the frozen state or only partially thawed.

To boil Add salt, very little water and cook covered for about half the usual cooking time.

To sauté Use a frying pan with a lid; melt a nut of butter, add partially thawed vegetables and toss over high heat for 1 minute. Cover and cook over lower heat until tender, seasoning to taste and stirring occasionally.

Conservative method Delicate vegetables (peas, broad beans, etc.) can be cooked in a covered pan over moderate heat without additional water, add seasoning and a nut of butter only, shake pan occasionally without removing lid.

Golden guide number

to fruit

Fruit is a marvellous 'freezable', has a long storage life (averaging 1 year) and needs no elaborate preparation. There are four easy ways to freeze fruit.

1. Open freezing

Especially good for berry fruits which are so delicate and easily damaged and require the quickest possible freezing because they have such a high water content, or fruits not necessarily required with sweetening. Wash fruits only if necessary, drain, hull *afterwards* to avoid water pockets, and spread out, so as not to touch, on clean baking trays or large Tupperware seals. Freeze at low temperature until hard (1–2 hours); remove from freezer and pack quickly before a bloom (caused by condensation) appears on the fruit. I use shallow polythene containers to distribute the weight, with foil dividers between layers. This prevents lower layers of fruit becoming crushed as they defrost. If packed in bags, I tip the fruit out onto a flat dish. To preserve a firm texture, serve while still chilled. As fruit is packed when frozen leave no additional head-space.

Suitable for strawberries, raspberries, cranberries, blackberries and loganberries.

2. Dry sugar pack

Very suitable for berry fruits with a skin, but also high juice content, and other juicy fruits. Prepare fruits as above, spoon alternate layers of fruit and sugar into polythene containers or bags; alternatively place all the prepared fruit and sugar together in a bowl and turn until lightly coated. Allow on average one-quarter weight of sugar in proportion to the weight of fruit, slightly more if fruit is sour. The smaller the individual fruit, e.g., currants, the larger headspace is necessary; allow on average a ½ inch (1 cm.). Use granulated or castor sugar, or icing sugar if a smooth syrup is liked when fruit defrosts. Dry packing of all these fruits is possible *without* sugar if they are likely to be required for recipes unsweetened.

Suitable for currants, gooseberries, rhubarb intended for cooking, pineapple, melon and citrus fruits.

3. Sugar syrup pack

Most suitable for stone fruits, and those which discolour easily. Prepare the fruit according to kind as shown in the following chart. Fruits which discolour easily (apples, pears, peaches, apricots) need special treatment either by sprinkling with lemon juice before freezing, or by adding ¼ teaspoon ascorbic acid crystals to each 1 pint (generous ½ litre) of hot sugar syrup. Make sure crystals are fully dissolved by the time syrup is cold. Chill in refrigerator before using.

Solution	Sugar	Water	Strength
10%	2 oz. (50 g.)	1 pint (generous ½ litre)	very thin
20%	4 oz. (100 g.)	1 pint (generous ½ litre)	thin
30%	7 oz. (200 g.)	1 pint (generous ½ litre)	medium thin
40%	11 oz. (300 g.)	1 pint (generous ½ litre)	medium heavy
50%	1 lb. (450 g.)	1 pint (generous ½ litre)	heavy
60%	1 lb. 9 oz. (700 g.)	1 pint (generous ½ litre)	very heavy

Pack prepared fruit into containers half-filled with chilled syrup, or pour over sufficient to just cover fruit. Leave a ½-inch (1-cm.) headspace and as fruit tends to rise in syrup, crumple a piece of foil a bit larger than the top of the container, press lightly down on the fruit, then put on the seal. The foil freezes into the syrup, but do not worry, when the pack is defrosted it can easily be removed.

Suitable for apples, pears, damsons, figs, citrus fruit, grapes, plums, melon, peaches and pineapple. The choice of strength for the syrup depends on two factors. One, whether the flavour of the fruit is very delicate and may be overwhelmed by a strong syrup. Two, whether the fruit is basically rather sour and may be improved by mellowing in a heavy syrup. A medium heavy syrup seems to suit most fruits.

4. Purée pack

Any fruit which can be stewed and made into a purée with or without sugar can be packed in this form. Delicate fruits can be packed puréed without cooking. Label unsweetened purées carefully to avoid confusion. Use the minimum of water, about 4 tablespoons to each 1 lb. (½ kg.) prepared fruit and cook to a pulp over gentle heat. Liquidise or sieve if necessary, cool quickly and pack into containers with ½-inch (1-cm.) headspace. Suitable for apples, plums, damsons, peaches, strawberries and raspberries.

Fruit	Preparation	Packing	Extra points
Apples	Peel, core and slice into cold salted water. Blanch 1 minute or steam blanch 2 minutes. Freeze dry in a dry sugar or syrup pack. Can be frozen as purée, sweetened if liked.	Polythene bags or polythene containers	Store for 12 months. Especially useful for apple sauce or baby food.
Apricots	Wash, halve and stone, syrup pack, add ascorbic acid.	Polythene containers	
Blackberries	Wash only if necessary and drain then remove hulls. Freeze dry in a dry sugar or syrup pack.	Polythene containers	Only use the best ripe, yet firm fruit.
Blackcurrants and redcurrants	Sprigs – wash and drain, open freeze then dry sugar pack. Or top and tail; dry sugar pack.	Polythene bags or polythene containers	
Cherries	Stalk, wash and drain. Open freeze then pack dry in a dry sugar or syrup pack.	Polythene bags	Red cherries freeze better than black, white ones best in syrup.
Coconut	Shred, add coconut milk. When using pour off milk. Can be toasted.	Polythene containers	
Cranberries	Stalk, wash and drain, open freeze and pack dry or in syrup, or freeze as sauce.	Polythene containers or polythene bags	
Damsons	Wash, halve and stone, syrup pack or as cooked purée.	Polythene containers	
Figs	Wash gently, leave whole, wrap individually, or peel and syrup pack.	Foil. Polythene containers	
Gooseberries	Wash and drain, dry sugar or syrup pack.	Polythene bags or polythene containers	Slightly under-ripe fruit freezes best.
Grapefruit	Wash, peel, segment or slice, dry sugar or syrup pack.	Polythene containers	
Grapes	Leave seedless grapes whole, or skin and pip, syrup pack.	Polythene containers	
Lemons and limes	Leave whole, slice or segment, dry sugar or syrup pack. Juice can be frozen and packed as ice cubes.	Polythene bags or polythene containers	
Loganberries	Wash only if necessary, drain and then hull, open freeze and pack dry, dry sugar pack or as purée.	Polythene containers	Best picked on a dry day when ripe but firm, Mildew develops within hours on wet fruit.
Melons	Cut flesh into balls, cubes or slices, dry sugar or syrup pack.	Polythene containers	

Fruit	Preparation	Packing	Extra points
Oranges	Juice can be frozen and packed as ice cubes.	Polythene bags	
	Grate, peel and mix with sugar.	Polythene containers	
	Segment or slice fresh, dry sugar or syrup pack.		
	Freeze Seville oranges whole.	Polythene bags	Freeze in season, make marmalade later.
Peaches	Skin, halve and stone, syrup pack with ascorbic acid. Can be frozen as purée with lemon juice and sugar.	Polythene containers	Marvellous to serve, for a special occasion, still slightly iced and flamed with brandy.
Pears	Slice, poach in boiling syrup $1\frac{1}{2}$ minutes, drain, cool and freeze with syrup or pack in cold syrup with ascorbic acid.	Polythene containers	Choose perfect, slightly under-ripe fruit.
Pineapple	Peel, core and slice. Dice or crush, layer with freezer film and sprinkle with sugar, or syrup pack.	Polythene containers	
Plums	Wash, halve and stone, dry sugar or syrup pack. Can be frozen as cooked purée.	Polythene containers	
Raspberries	Wash only if necessary, drain and then hull, open freeze and pack dry, dry sugar pack or as purée.	Polythene containers	Best picked on a dry day when ripe but firm. Mildew develops within hours on wet fruit.
Rhubarb	Wash, trim and cut into 1-inch (2·5-cm.) lengths, blanch in boiling water 1 minute, cool, pack in syrup.		Blanching is necessary because rhubarb is the stem of the plant and not the fruit.
Strawberries	Wash only if necessary, drain and then hull, open freeze and pack dry sugar or syrup pack. Can be frozen as purée if preferred.	Polythene containers	

Golden guide number 

to fish and shellfish

All fish must be frozen as soon as possible after catching, at least within 24 hours. If you are not sure how long it is since the fish was caught, it is better not to freeze it. If there is a fisherman in the family, equip him with large polythene bags and an insulated carrying bag so that the fish is not spoilt on its journey home. Remind him tactfully to avoid bruising the fish by letting it bang about in the boat or on the bank. It is an advantage if he will gut the fish and remove the gills immediately.

Lean fish

Prepare fish for freezing by removing fins, tail and loose scales then gut if necessary and wash well in cold water.

Steaks or fillets If cut into portions, it is not necessary to glaze the fish. Dip steaks or small fillets into a salt solution (1 oz. (25 g.) salt dissolved in 1 pint (generous ½ litre) cold water). Drain well, wrap closely and freeze. Separate fish portions can be moulded in foil, or a number of portions frozen together with foil dividers between them, in polythene bags.

Whole fish Fish small enough to be frozen whole or divided into two large fillets only have a slightly longer storage life if glazed. Open freeze on baking trays until solid, about 2 hours. Have ready a bowl of iced water and dip the frozen fish up and down until a thin coating of ice forms over it. Place on a wire tray over the baking tray and return to the freezer. When the glaze is quite solid, about 30 minutes, repeat the dipping process until glaze is ¼-inch (½-cm.) thick. Wrap in freezer foil or polythene bag.

Oily fish

These include salmon, salmon trout and mackerel. Prepare as for lean fish. If salmon are large enough to cut into steaks, dip the portions into a strong ascorbic acid solution (1 teaspoon crystals to 1 pint (generous ½ litre) water). Drain, wrap as for lean fish and freeze. Whole fish should be treated by glazing, as for lean fish but it is difficult to accommodate a large whole salmon in an upright freezer without bending it. Oily fish tend to dry out and become freezer burnt unless carefully protected.

Cooking frozen fish

Steaks and small fillets can be cooked from the frozen state. Small whole fish moulded in foil can be placed on a baking tray, with a little water poured round, and baked in a moderate oven for 1–1¼ hours, or roughly twice as long as you would allow for the same weight of fresh fish. To prevent dehydration of the gap left by gutting while the fish is frozen, pack it with crumpled foil or stuffing, or a knob of butter. Fish fillets which have been coated in egg and breadcrumbs before freezing should be partially thawed (1 hour at room temperature) or the coating tends to separate from the fish. It is possible to freeze fish fillets coated in batter and cooked until just set but not coloured. Deep fry these from the frozen state.

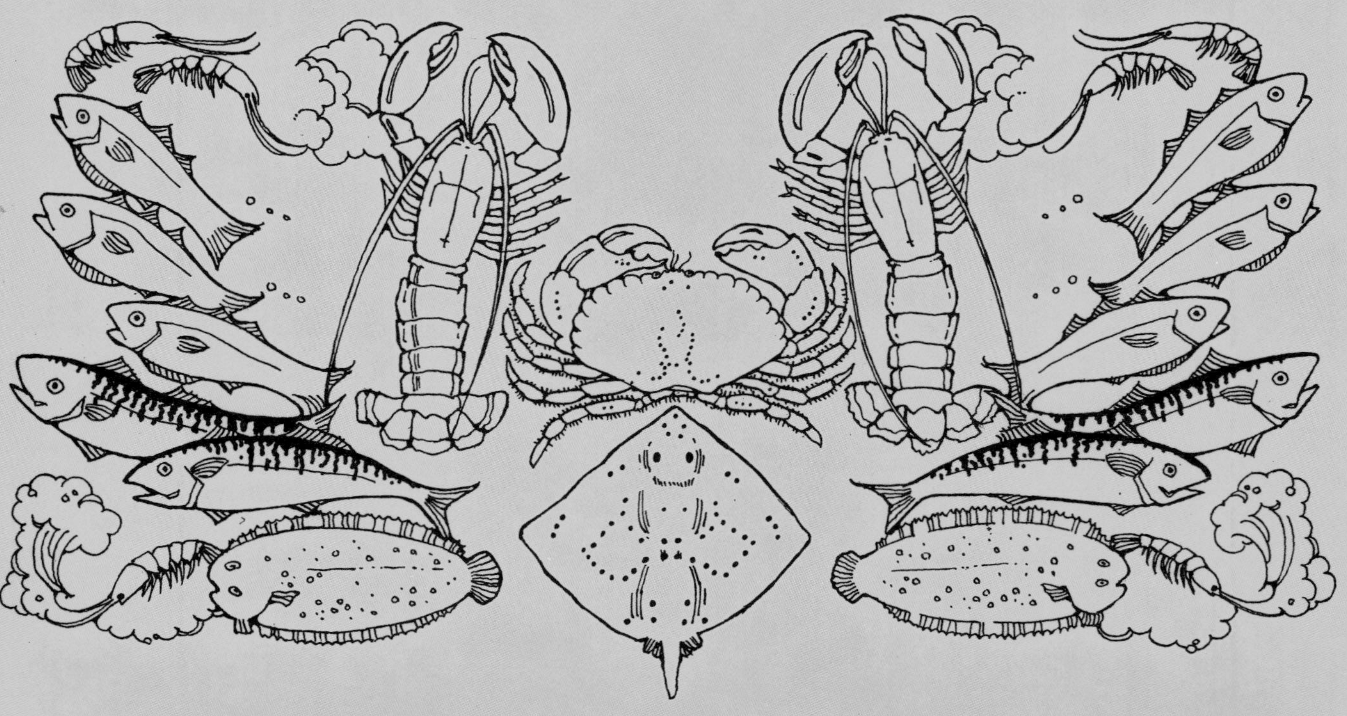

Fish calendar

Many fish are available all the year round, others are at their best during a shorter season of the year.

January
February
March
April
May
June
July
August
September
October
November
December

Mussels
Sprats
Scallops

Lobsters
Crabs
Prawns
Trout

Scallops
Mackerel
Sprats

Whiting
Haddock
Skate

Whitebait
Salmon
Salmon Trout

Whiting
Mussels
Oysters

Mackerel
Oysters

Haddock
Skate
Hake

January
February
March
April
May
June
July
August
September
October
November
December

Available all the year round:
Cod, Halibut, Turbot, Plaice, Sole
Herring
Shrimps

Bear in mind that quality naturally declines during the spawning season and fish should then be eaten fresh and not frozen.

Fish	Preparation	Packing	Storage	Calendar
Whole fish	Remove head, fins, tail and scales, gut and wash. Glaze.	Wrap tightly. Foil or polythene bags	3 months	All year round
Fillets or steaks	Skin and dip oily fish in ascorbic acid solution, lean fish in salt water.	Layer individually with foil dividers in polythene bags	3 months	All year round
Crab	Kill, bring slowly to boil in salt water, boil 15 minutes per 1 lb. ($\frac{1}{2}$ kg.) Drain, open and remove meat.	Polythene containers or bags	2–3 months	Midsummer
Lobster	Kill, bring to boil, cook 15 minutes per 1 lb. ($\frac{1}{2}$ kg.). Remove meat from shell, pack.	Polythene containers or bags	2–3 months	Midsummer
Prawns and shrimps	If raw, add to boiling salted water cook for 3–5 minutes, cool quickly, peel. Pot shrimps with less seasoning.	Polythene containers or bags	1 month	All year round
Oysters	Remove from shells, dip in salt water, pack with shell liquid.	Polythene containers	2–3 months	September–April
Mussels	Prepare in the normal way, freeze in the liquid in which they were cooked, use for soups, etc.	Polythene containers	3–4 months	September–April

Note Cross contamination of other foods in the freezer by fishy smells or flavours can only be avoided by extra care in packing, and if possible, by enclosing all your fish and shellfish packs in a large polythene container or polythene batching bag.

Golden guide number 4

to meat

Packing meat for the freezer is probably the biggest operation of its kind you will undertake. At first glance the quantities involved seem enormous, as to get the big saving associated with bulk buying you do have to think in terms of a whole lamb, half a pig, a forequarter or even hindquarter of beef. Weights vary considerably, as you might be offered a tiny New Zealand lamb weighing in at 23 lb. (11½ kg.) or a fully-grown English lamb of a robust 45 lb. (22½ kg.). Meat has every quality to entitle it to a big proportion of freezer space. 1. Buying it in bulk represents a substantial saving. 2. Good quality meat in peak condition, properly packed, seems to benefit rather than lose quality while stored. 3. Has an average storage life of around a year – I have eaten roast beef from my freezer, 18 months old, that was superb.

Here's what you can expect for your money.

Beef

Forequarter (100–160 lb., 50–80 kg.) Back, top and forerib, feather steak, leg of mutton cut (shoulder), flank and brisket, clod and sticking, shin, chuck and bladebone.

Hindquarter (150–180 lb., 75–90 kg.) Sirloin (including wing rib), fillet, rump, topside, silverside, top rump, leg, skirt, flank, ox kidney. You must expect 24–30 lb. (12–15 kg.) of the purchase to be bone and 12–20 lb. (6–10 kg.) to be fat. The bones can be used to make stock and the fat reduced to make dripping. The hindquarter supplies several pounds of kidney suet for puddings and has a relatively low proportion of bone to meat, but a greater proportion of fat to meat than in the forequarter.

Note Veal is not a favourite of mine for home freezing as the meat is exceptionally delicate in flavour and does seem rather tasteless after even a few weeks' storage. Half a calf weighs in the region of 100 lb. (50 kg.).

Lamb

(23–45 lb., 11½–22½ kg.) Two of each cut – leg, loin shoulder, best end of neck, breast, middle neck and scrag, kidney (if home killed). Nothing need be discounted for additional bone and fat content. Although it is possible to buy half a lamb, this is really unnecessarily timid as a whole jointed lamb, uncooked, can be stored in less than 1½ cubic feet of space.

Pork

Side (40–60 lb., 20–30 kg.) Leg, loin (chops), belly, hand and spring, spare ribs, bladebone, half a head. If you particularly require the trotters or offal, ask for them as they are not usually supplied. A whole pig is only a viable purchase for large families, as pork does have a shorter storage life than other meats.

Basic rules

Meat must come from the butcher, freshly slaughtered, and chilled in the case of pork and veal, or hung for 6–12 days in the case of lamb and beef. New Zealand lamb, has been hung before freezing for transport from New Zealand and if delivered to you in the frozen state by the butcher, has as long a potential storage life as home-killed meat. Good hanging not only preserves flavour and texture, but reduces the amount of drip when meat is thawed.

Since a bulk purchase of beef, for example, may yield well over 100 lb. (50 kg.) of edible meat to be frozen, this represents at least a three-day operation and with a modern efficient freezer I have added 35 lb. (17½ kg.) each day to a 12 cubic foot freezer without overwhelming the machine's freezing capacity. The secret is to select, firstly, the cuts you want to keep the longest, for early freezing down, and secondly, the cuts which will benefit from the further tenderising action of storage, in the

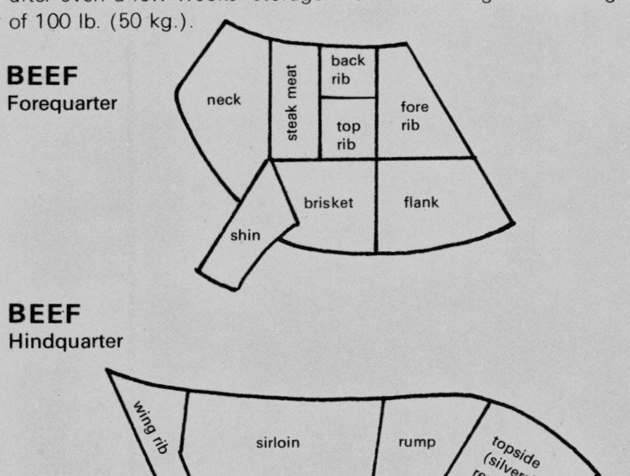

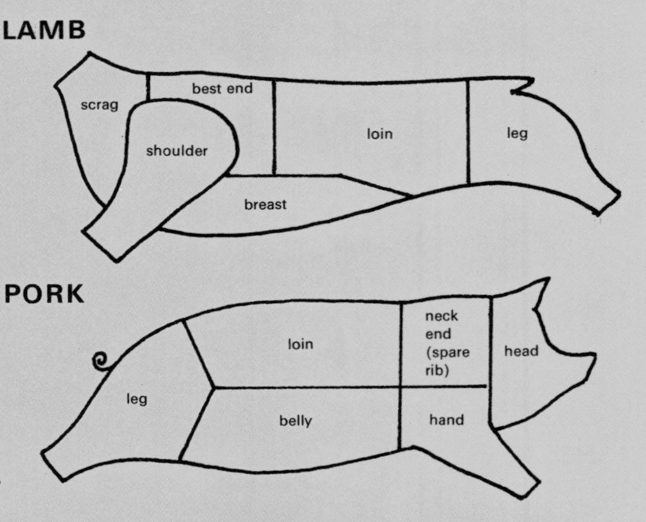

1 Wrapping a defrosted leg of lamb in roasting wrap and folding it down closely to the meat by the druggist's wrap method.

2 Twisting and turning the opaque foil borders up to ensure a good seal and to keep the meat juices in during cooking.

3 Inserting a meat thermometer into the top of the meat, through the roasting wrap, shortly before the end of cooking time, to see whether it has reached the required temperature.

1

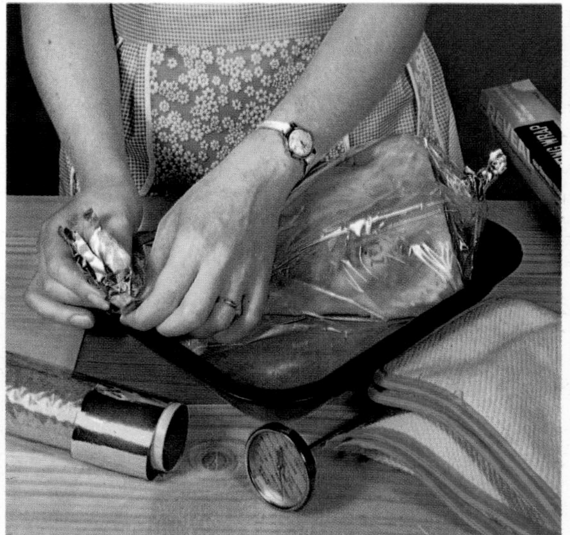
2

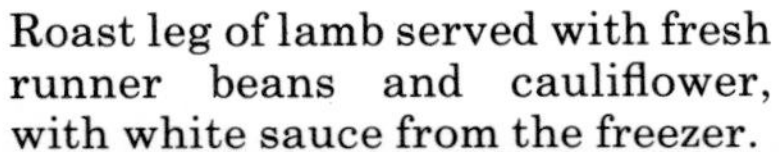
Roast leg of lamb served with fresh runner beans and cauliflower, with white sauce from the freezer.

3

1 Fork-mixing pastry in a quantity large enough for a batch of two double-crust pies and for four pastry circles.

2 Filling shaped foil pie plates lined with pastry. Pastry circles are interleaved with circles of foil.

3 Trimming a veal and pimento pie ready for fluting. The remaining pastry is closely wrapped in foil for rolling out when defrosted.

1

2

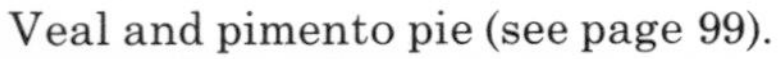
Veal and pimento pie (see page 99).

3

refrigerator or at room temperature, to be frozen down last. Minced and diced meat should of course be frozen first as so many cut surfaces exposed to the air invite rapid deterioration.

Packing for freezing

Since the oxygen in air coming into contact with meat surfaces causes oxidation, and meat expands so little on freezing, the closer the wrap the better. Large cuts, if a convenient shape, can be enclosed in sheet wrapping and sealed.

1. Butcher's wrap Use a *square* sheet of polythene, place meat diagonally across one corner. Bring corner over and tuck under meat, then roll forward, bringing the corners on either side to the centre. Continue rolling, and ensure that the folded wrapping narrows towards the opposite corner. Seal the point with freezer tape.

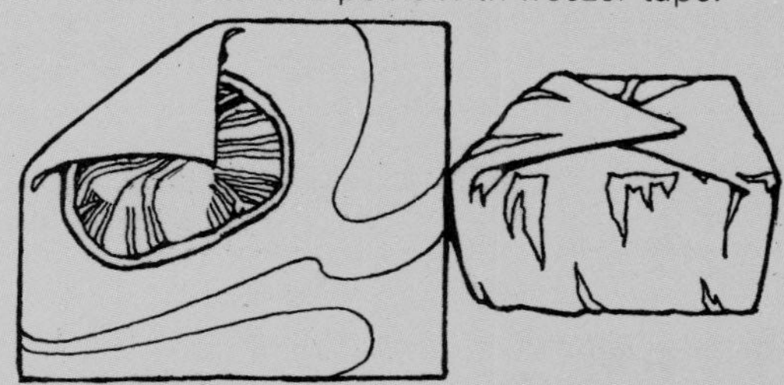

2. Druggist's wrap Use a *rectangular* sheet of polythene, place meat in centre and bring long sides together over the top, folding down until tight against the meat. Fold short sides in to form points, bring in towards centre and seal points with freezer tape.

Large joints are not suitable for either of these wraps; a leg of lamb should be moulded in freezer foil. If you use polythene bags, it is essential to withdraw surplus air before sealing.

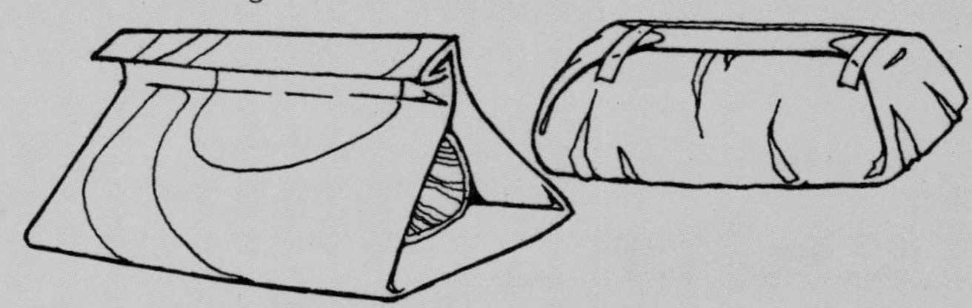

Small cuts and offal Pack small cuts intended for stewing, ready-diced if necessary, in quantities sufficient to cook at one time. Small cuts intended for frying or grilling should be packed with foil dividers so that steaks or chops can be easily separated without defrosting. Never pack so that you need to defrost part of the contents, then refreeze the remainder. Steaks can easily be piled up, with dividers, one on top of another, but chops are easier to pack side by side, skin side down.

Another way to pack a selection of small cuts to cover your family requirements for, say, a fortnight, is to mould each one individually in kitchen foil then pack them altogether in a large polythene container (as for poultry). Pad the bones; meat bones are almost incredibly sharp and so easily puncture the wrapping when frozen that it is worth padding them with foil. If the joint has a bony piece which yields practically no meat when roasted, cut it off before wrapping. If you use a serrated knife to cut bones, never cut so that if the knife slips it might cut your hand, and provide it with a holster or keep it in the original box as it is quite a dangerous weapon.

Over-wrap, unless the cuts are enclosed in a polythene container; over-wrapping is more necessary with meat than with any other food as the pack is so much more likely to be punctured.

Bacon

Bacon being a cured meat reacts differently to freezing than does fresh meat. This is due to salt and other ingredients in the curing brine. Vacuum-packed joints, steaks and rashers can be stored up to 10 weeks in the frozen state; home-wrapped bacon joints of smoked bacon up to 8 weeks; and unsmoked bacon up to 5 weeks; home-wrapped smoked rashers, chops and steaks up to 6 weeks; unsmoked up to 4 weeks. A special freezer pack of either smoked or unsmoked rashers by Danepak Limited is available at freezer food centres. It contains five 8-oz. (225-g.) vacuum packs sealed together in one bag. Providing the frozen individual packs are not punctured and they are kept at a temperature of 0°F (−18°C) or below, all the packs will remain in perfect condition for 20 weeks after the expiry date printed on the pack. Packs accidentally punctured lose their rigidity and should be used up as soon as possible.

Buying bacon for freezing

If you freeze down bacon at home, remember that it must be completely fresh so order it from your retailer in such a way that you can collect your supplies on the day he receives his delivery. Smoked bacon can be stored for longer than unsmoked bacon. It is virtually impossible to freeze bacon joints weighing more than 5 lb. (2½ kg.) sufficiently quickly to give good results.

Preparing bacon for freezing

Joints Mould in freezer foil excluding as much air as possible. Place the moulded joint in a heavy gauge polythene bag (the thicker the bag the better the result), withdraw air and seal.

Rashers, chops and steaks Small cuts have a much shorter storage life because far more of the meat and fat surface has been exposed to air, with an increased risk of rancidity developing. Use foil dividers between layers and pack as for joints.

Vacuum packs This commercial process which cannot be achieved at home delays rancidity because air has been entirely withdrawn. Some other types of wrapping resemble the vacuum pack so when buying for your freezer inspect carefully and make sure you buy genuine vacuum packs and that the vacuum has not been damaged, i.e. the bacon slices loose in the packet. Overwrap with another polythene bag for protection.

Defrosting and cooking

Defrost joints for 24 hours in refrigerator, or overnight at room temperature, loosely wrapped. Small cuts can be defrosted overnight in the refrigerator or in 3 hours at room temperature. Vacuum-packed joints and rashers can be thawed in the bag, and small cuts can be defrosted rapidly by placing the pack in hot water for a few minutes. Spread rashers on kitchen paper to absorb surplus moisture before cooking. To give the best flavour, bacon should be cooked by any of the usual methods as soon as it is thawed.

Golden guide number 

to pasta and rice

Pasta

Good quality pasta, made from durum wheat with a protein content of 12 % or above in its uncooked form, can be successfully frozen. It should be cooked (but not over-cooked) in the usual way. Rinse immediately after draining to prevent it from continuing to cook in its own heat. Stir in sufficient oil to coat the strands, divide into serving portions and freeze in polythene containers ($2\frac{1}{2}$ oz. (65 g.) uncooked dry weight of pasta is normally considered one adult portion).

To defrost, the pasta can be taken straight from the freezer and added to a pan of fast boiling water. Leave just long enough over heat for the water to return to the boil. Do not continue to cook as the pasta will be defrosted and reheated during the time that the water takes to come back to the boil. The larger the quantity of pasta the greater the volume of water required for quick reheating. A two-portion pack will require about 1 pint (generous $\frac{1}{2}$ litre) of water.

Pasta is also excellent for freezing as part of a complete made-up dish ready to be placed in the oven to defrost and reheat. Foil containers are the most practical as the prepared dish can be taken straight from freezer to oven.

Layered dishes Lasagne may be reheated, uncovered, providing the top layer of lasagne is completely covered with sauce. If the covering of sauce exposes some of the pasta, it is better to cook the dish covered and then brown off under a hot grill before serving, otherwise the pasta becomes leathery.

Composite dishes Pasta mixed with a rich sauce must be reheated covered. The way the sauce and pasta are combined for such dishes is important when they are to be frozen. The general practice is for the pasta to be put in the dish and a well made in the middle into which the sauce is poured. When such dishes are to be frozen, it is essential that the pasta is put in the middle and the sauce poured around the outside. There is a good reason for this. The sauce takes longer to reheat than the pasta. Being around the outside of the dish it heats first. Additionally, the moisture from the sauce keeps the pasta from drying out. If the reverse presentation is used, with the sauce in the middle, the pasta near the edges tends to over-cook and burn before the sauce is hot.

To freeze single portions Use shaped foil trays. Put the pasta in first, in one diagonal half of the dish, and the sauce in the other half. When the lid is removed for serving the triangle of pasta and a similar triangle of sauce look most attractive. If lids are not available cover with foil, crimping under the edges.

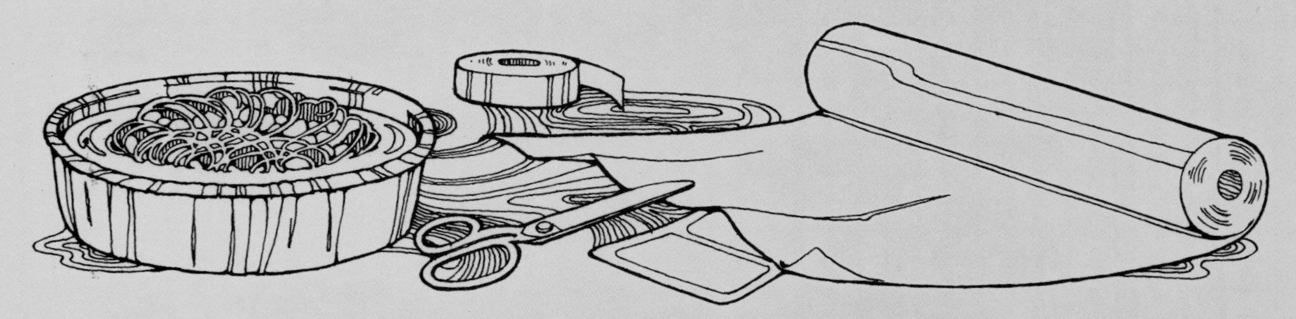

Garlic bolognaise sauce

Makes 6 pints ($3\frac{1}{4}$ litres) sauce.

To freeze Pour into polythene containers, leaving 1-inch (2·5-cm.) headspace, or line containers with polythene bags, pour in sauce and freeze until solid. Remove bag from container. Seal and label.

To serve Defrost and reheat gently in a saucepan. Simmer for 2 minutes.

Storage time 3 months.

Imperial	Metric	American
4 tablespoons oil	4 tablespoons oil	$\frac{1}{3}$ cup oil
4 large onions, chopped	4 large onions, chopped	4 large onions, chopped
3 cloves garlic, crushed	3 cloves garlic, crushed	3 cloves garlic, crushed
4 sticks celery, chopped	4 sticks celery, chopped	4 stalks celery, chopped
4 lb. minced beef	2 kg. minced beef	4 lb. ground beef
3 teaspoons salt	3 teaspoons salt	3 teaspoons salt
$\frac{1}{2}$ teaspoon black pepper	$\frac{1}{2}$ teaspoon black pepper	$\frac{1}{2}$ teaspoon black pepper
$\frac{1}{2}$ teaspoon paprika pepper	$\frac{1}{2}$ teaspoon paprika pepper	$\frac{1}{2}$ teaspoon paprika pepper
4 tablespoons Worcestershire sauce	4 tablespoons Worcestershire sauce	$\frac{1}{3}$ cup Worcestershire sauce
1 small can tomato purée	1 small can tomato purée	$\frac{1}{4}$ cup tomato paste
1 16-oz. can tomato juice	1 450-g. can tomato juice	1 16-oz. can tomato juice

Heat the oil in a large saucepan and use to fry the onion, garlic and celery until softened but not browned. Add the meat and stir until it changes colour then add the remaining ingredients. Bring up to the boil, cover and simmer for about 20 minutes. Cool.

Stuffing for pasta shapes Leftover meat, minced and moistened with thick gravy, forms the basis of a savoury stuffing for large hollow pasta such as cannelloni. The filling can have additional flavourings, such as a few drops of Tabasco or Worcestershire sauce, tomato purée and paprika, or finely chopped mushrooms. It should be fairly highly seasoned and then used to stuff the cooked pasta. Pack in foil containers, brush a thin layer of oil over the exposed surface of the pasta to prevent it drying out, and cover surface closely with freezer film, then seal with lid or foil. To serve, remove seal, sprinkle thickly with grated Parmesan cheese and lay a sheet of kitchen foil lightly on top. Defrost and reheat in a moderately hot oven (400° F, 200°C, Gas Mark 6) for 25–30 minutes, removing the foil after 15 minutes.

Rice

There are several ways of cooking long-grain white rice and probably the most used method is to cook it in a large quantity of boiling salted water. However, I find the best way is the American one-two-one way, as follows: using a large saucepan, measure in 1 cup long-grain rice, 2 cups water and 1 teaspoon salt. The size of cup is not important but the same one must be used for rice and water. Bring to the boil and stir once. Lower heat, cover pan and simmer for about 15 minutes, without removing the lid or stirring. Take out a few grains, cool and bite them. If not quite tender, or the liquid not completely absorbed, replace the lid and cook for a few minutes more. Remove from the heat; if serving immediately, fluff the rice lightly with a fork. Alternatively, you may find it more convenient to cook the rice in the oven. Using exactly the same proportions as before, put the rice and salt into an ovenproof casserole. Add *boiling* water, cover and cook in a moderate oven (350°F, 180°C, Gas Mark 4) for about 40 minutes. Test in the same way and fluff with a fork before serving. The rice will be plump and fluffy with every grain separate. Brown rice, which is available from health stores, is the whole unpolished grain of rice with only the outer hull and a small quantity of bran removed, so that it retains more natural vitamins and minerals than white rice. It requires longer slower cooking over a very low heat.

To freeze cooked rice Cool quickly and pack loosely into polythene bags. Seal with a twist tie and partially freeze. Remove from the freezer and crumple the bag with both hands to separate the grains, then replace in the freezer. Storage life is 4–6 months. When required to serve, allow to defrost at room temperature for about 1 hour. Spread out on an ovenproof dish or baking tray, dot with butter, cover with foil and reheat in a moderate oven (325°F, 170°C, Gas Mark 3) for 20 minutes.

How much to cook It is usual to allow 2 oz. (50 g.) rice per person. For two people, use 4 oz. (100 g.) rice and ½ pint (3 dl.) liquid. The cooked yield will be three times the original weight of rice, so from 4 oz. (100 g.) you will get 12 oz. (350 g.) cooked rice. It is useful to remember that one average teacup holds 6 oz. (175 g.) rice, one breakfastcup 8 oz. (225 g.) rice or ½ pint (3 dl.) liquid. Bag in quantities to serve two or four people, or in convenient family amounts.

Rice fillings and stuffings These usually have as long a freezer life as any other cooked dish. A well-flavoured savoury rice filling, based on 8 oz. (225 g.) uncooked long-grain rice would be suitable to use as a stuffing for 8 medium-sized green peppers or large tomatoes. The cooked rice mixture is packed inside the raw vegetable (blanched if necessary) and frozen; when required to serve it is baked after partial defrosting. Made-up cooked rice stuffing for poultry should be frozen in polythene bags until required as complete defrosting is a lengthy process with stuffed poultry.

Buttered rice Add a thick slice of lemon to the rice while cooking. When cooked, place in serving dish and fork in 1–2 oz. (25–50 g.) butter.

Golden rice Add ⅛ teaspoon powdered saffron to each 8 oz. (225 g.) rice when cooking. Fork in butter as above when serving, together with 2 oz. (50 g.) chopped toasted almonds.

Curried rice Melt 1 oz. (25 g.) butter in a large saucepan and use to fry 1 tablespoon finely chopped onion until soft. Stir in 1½ teaspoons curry powder and add water and rice as usual. When cooked, turn into serving dish and fork in 1 oz. (25 g.) butter.

All these savoury rice dishes, if prepared for freezing, have a better consistency if the butter is added after defrosting while the rice is reheating.

Sweet rice dishes Sweetened rice cooked in milk can be combined with fresh berry fruits, stiffly beaten egg whites, cream and sufficient dissolved gelatine to set. Freeze in a pretty serving dish or mould, or for everyday eating, in a polythene container. A cornflour custard mixed with cooked rice, egg yolks and fruit can be topped with a stiff meringue made with the egg whites, lightly baked and frozen. Can be served cold, or reheated.

Golden guide number 

to poultry and game.

Poultry

How birds are prepared for freezing depends on how you intend to cook them. Since whole birds and portions are available from freezer food centres you will probably prepare them for freezing only when you have the chance of buying a number of chickens at a bargain price.

To freeze whole for roasting Remove giblets, clean carefully, pad leg bones with foil and fasten together with a rubber band. This prevents them from piercing the wrapping. Place in a gussetted polythene bag, chosen to fit (a bag 6 inches (15 cm.) by 8 inches (20 cm.) fits a 3-lb. ($1\frac{1}{2}$-kg.) chicken, withdraw air by straw or pump and secure with twist tie. Use freezer tape to fix the corners in closely. The same method would apply to a fowl for boiling. Freeze giblets separately, as these have a shorter storage life than the bird. Do not freeze poultry stuffed, as this has also a shorter freezer life, and causes such a delay in defrosting that bacteria may have time to get lively again.

To freeze in portions Trim off winglets, mould portions in kitchen foil, then pack closely together in a polythene container. The Tupperware flavour saver takes a whole chicken and at least four portions comfortably, and this would be an average family requirement for 2 weeks. Coat small, tender portions in seasoned flour, egg beaten with a little water, and toasted breadcrumbs, ready for frying.

To joint a whole chicken Cut in half, through and along the breast bone. Open the bird out and cut along the length of the backbone, or either side of it, and lift it out. If you use a sharp knife, tap the back hard to force it through the bones. Secateurs make the job easier. Lay the two halves of the bird skin side up, and cut each diagonally across between wing and thigh, allotting the breast meat to the wing portion. All discarded trimmings and any available giblets should be used to make stock (reserve the liver). To make stock, cover bones with water, add bouquet garni, salt, pepper, sliced onion and carrot. Simmer, covered, for 2 hours. Reduce the stock and season well. The trimmings and giblets from 2 lb. (1 kg.) chicken bones and giblets should produce about 1 pint (generous $\frac{1}{2}$ litre) strong stock. Cool quickly, skim off excessive fat and freeze in ice-cube trays or small containers (2-oz. (50-g.) Tupperware tumblers are ideal). The tiny top-hat of chicken fat makes the stock invaluable for improving sauces and soups.

Many favourite recipes from classic cook books can be made up and stored using chicken portions – chicken Kiev, chicken à la king, chicken curry, Chinese-style chicken.

Turkey and duck The same rules apply as for chicken, except that turkey portions are too large to be very manageable in the raw state. The best way to split up a turkey is to carve off thin slices of the breast and thigh for escalopes, coat as for veal and freeze. Cook the remaining turkey, padding the breast with stuffing, and serve hot. Roughly chop the leftovers for made-up dishes. Large turkeys are often very cheap and provide a lot of meat which does not become boring if used in this way.

Game

Game birds should be bled and kept in insulated bags until ready to be prepared. Game animals should be beheaded and bled as soon as possible after killing. All game should be fully hung before freezing as it goes bad very quickly after thawing. Allow for 1 days' extra maturing during the thawing process if you do not like too gamy a flavour. If possible, pluck or skin and draw game before freezing, as it takes less space in the freezer and these jobs are not easy to do after thawing.

Rabbits and hares Unless required for roasting, joint and wipe the joints with a damp cloth. Do not wash them. Wrap each joint, either in freezer film or by moulding in foil, then together in one large polythene container or bag. With hare, the saddle may be packed separately for roasting, and the other joints together for jugged hare.

Game birds Pluck, draw, take out shot, wipe with a damp cloth, and pack as for rabbit and hare. Some connoisseurs prefer to hang them undrawn, and draw, removing shot and the oil sac from the base of the tail, just before freezing. Remember to keep tail feathers of pheasant if required for serving.

Venison This is prepared as for beef, and the meat is usually chilled, then hung for 5–6 days to mature before jointing and freezing. Unhung venison is apt to be rather dry and tough after freezing.

Seasons when game is available February to June is a close season for all game. Most game birds are in season from mid-August to the end of January. Venison (buck) is in season from July to September, and venison (doe) from October to December.

Glossary of freezer terms

Aluminium foil Heavy gauge foil which is made of aluminium therefore moisture-vapour-proof; used in sheet form or as shaped containers. Not suitable for foods containing a high proportion of vinegar, or citrus fruit juices.

Ambient atmosphere Temperature of the air immediately surrounding the freezer. The higher it is the harder the freezer works to maintain a desirable low temperature inside the cabinet. If humid, can cause condensation and rust.

Anti-oxidant Chemical agent (usually ascorbic acid) added to sugar syrup to prevent discolouration of fruit or to water as a dip for oily fish.

Ascorbic acid Synthetic vitamin C product available from chemists in crystal or tablet form. Crystals are preferable because it is easier to make a solution with them, and they are also cheaper than tablets.

Blanching Immersion of fresh vegetables in boiling water before freezing to halt enzyme action which otherwise continues in the frozen state and causes deterioration.

Butcher's wrap Method of using sheet wrapping to protect food for freezing. Food is placed near one corner of a square sheet and folded diagonally to opposite corner, bringing in the points from the sides.

Conservator An appliance capable of maintaining pre-frozen food at a temperature of 0°F (−18°C).

Dividers Small pieces of foil or freezer film used to interleave layers of food which make them easier to separate for defrosting.

Drip loss The loss of any food's natural juices during defrosting, particularly of meat which has not been quick frozen.

Druggist's wrap Method of using sheet wrapping to protect food for freezing. Food is placed in centre of rectangular sheet, and the two long sides of the sheet are brought together in the centre, and folded over and over until tight against the food. Seal, then tightly fold in the two end pieces and seal.

Dry ice Solid carbon dioxide, often used to keep frozen food solid during transport. It should never be allowed to come into contact with the hands, with food or the lining of the freezer cabinet.

Dry pack — To pack foods, especially fruit, without adding liquid or sugar.

Dry sugar pack — To pack fruits in dry sugar.

Enzymes — Natural substances present in all foods, which cause noticeable deterioration in vegetables and are inactivated by blanching.

Freezer burn — Dehydration of inadequately wrapped food resulting in oxidation.

Freezer log — A notebook or index file which records the date of addition, type of food and container, and position in the freezer.

Glaze — A thin coating of ice formed by dipping frozen fish into ice water. The fish should be refrozen and dipped a second time.

Headspace — Space left between food and top of container to allow for expansion of water content on freezing.

Heat seal — To seal a polythene bag by pressure between the teflon coated bars of a heat sealer, or with a warm iron between layers of brown paper.

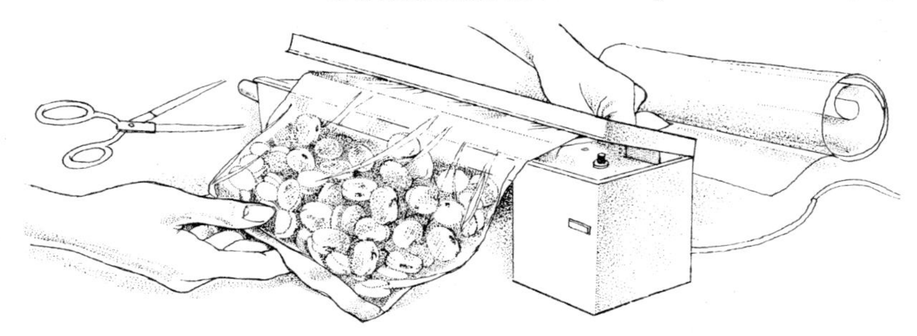

Microwave — Type of cooking which works on the principle of high frequency rather than heat. Food can be frozen and cooked, in a suitable polythene container, in a microwave oven very rapidly.

Moisture-vapour-proof — Packing materials not permeable by moisture vapour, thus preventing dehydration.

Open freezing — Food spread out (without touching) on trays to freeze so that it is free flowing when packed. Also used for delicate items which require especially quick freezing, or for decorated cakes to prevent damage.

Syrup pack — To pack fruits in sugar syrup of varying strengths.

Temperature gradient — A difference in temperature between one part of the freezer cabinet and another, causing air circulation, and dehydration of inadequately protected foods in the form of moisture vapour which is redeposited on the inside walls as frost.

To freeze or not to freeze

The more experiments I have seen carried out, the more optimistic I become about the freezing possibilities of most foods. Jelly, for example; after a week or two, an ordinary fruit jelly made average strength tastes grainy when partly defrosted but is quite normal in texture and flavour when fully defrosted. For long term storage, make it a little stronger, that is make up to $\frac{3}{4}$ pint ($4\frac{1}{2}$ dl.) with water rather than 1 pint (generous $\frac{1}{2}$ litre). Milk jellies, made with evaporated milk, defrost perfectly, and so do small savouries set in aspic jelly. Custard is another matter. Cornflour-based custard can 'weep', and one is left with rather a solid mass, and a lot of whey. Canned custard freezes best, or custard made with fresh milk. If made up into trifles, keep the sponge and fruit base rather dry, so that any fluid which separates sinks into it with the brandy or sherry. Egg custards also 'part company', but this can be overcome by freezing uncooked. Mayonnaise tends to curdle, and must be rewhisked. I have had the best success in freezing blender mayonnaise where the whole egg is used rather than two yolks, and the beating gives a fully homogenised texture which counteracts the tendency to separate. Cream with less than 40% butter fat content separates, and sometimes no amount of beating will bring it back; technically, it is due to the nasty habit of the fat globules separating as the water content freezes. They then agglomerate and stay that way. Absolute unfreezables are whole eggs. If frozen in the shell, they crack as the water content expands; if hard-boiled, the white goes rubbery. Separated and whisked lightly, both whites and yolks can be frozen in small amounts for cooking. Garlic is another temperamental customer. It does seem to impart a musty off flavour to savoury dishes after about six weeks. Although we have tested carefully, made-up dishes thickened with flour seem little different to those thickened with cornflour. However, if you are timid about this reduce the thickening agent to half the quantity recommended and use beurre manié to thicken further at the reheating stage. All salad greens from lettuce to cucumber are out if intended to be eaten as salads as they will *not* stay crisp, but can be frozen cooked, as in lettuce or cucumber soup. If in doubt, the motto is 'try it and see'. You have nothing to lose but the subject of the experiment, should it prove disappointing.

When is it safe to refreeze?

This is largely a matter of common sense. On one hand, there is no black magic which transforms food that has been once frozen, then defrosted and refrozen, into a dangerous substance when defrosted for the second time. On the other hand, there are changes at each freezing in the texture, colour, aroma, flavour and even nutritive value which vary enormously from one food to another. Also, there is the danger of a build-up of contamination by harmful organisms which multiply in favourable conditions and are not destroyed by freezing.

Obviously, frozen meat which has been thawed and cooked is perfectly safe to refreeze in its cooked form. Only very small changes take place in the cell structure of meat each time it is frozen and the cooking process should destroy any dangerous organisms. The made-up dish will be as palatable, and as safe, as one made from fresh meat for freezing. Fruit, such as strawberries, and other foods with a more delicate cell structure than meat, are sad evidence of the fact that each refreezing would have a seriously detrimental effect. This applies however to the quality of the food, not its safety. Spoilage still remains the vital factor in whether or not it is safe to refreeze.

All food, however carefully handled, is exposed both beforehand and during preparation for freezing, to contamination. Enzyme activity, responsible for growth, also causes spoilage. In some foods it is insufficiently arrested by freezing unless the food has been cooked sufficiently to destroy the enzymes, or, as in the case of blanching vegetables, to inactivate them. As well as causing deterioration and loss of quality, some micro-organisms are capable of causing serious food poisoning. These, like enzymes, are not destroyed by freezing, and in the process of restoring food to the temperature at which you eat it, you allow food to pass through the temperature range most likely to encourage growth of bacteria, yeasts and moulds, all of which could be harmful. Each time food is defrosted and exposed to conditions favourable to this growth, the danger rises. So you have to consider the loss of quality on refreezing and the possible presence of harmful organisms which may multiply when the food is again thawed out. Food which is defrosted while still wrapped, in the refrigerator, and still retains some ice crystals, is therefore infinitely safer to refreeze than food which has been unwrapped and allowed to stand exposed in a warm kitchen for hours. The more hygienically the food has been handled at every stage, the more confidence you can have in eating twice-frozen food.

My five favourite rules

1. Choose only the freshest and best food for freezing.
2. Keep food, hands, utensils, packing materials scrupulously clean.
3. Cook, cool, wrap, freeze fast. Hurry, hurry, from shopping basket to freezer.
4. Use moisture-vapour-proof materials, but don't ruin it all by forgetting the airtight seal, or the exclusion of unnecessary pockets of air.
5. Label clearly, record in your freezer log, and use up in good time. Forgotten food in the freezer is a waste of money.

Freezer burn and oxidation

Freezer burn is apparent because the surface of the food is dark, dehydrated and disfigured by greyish patches on the surface of meat, fish or poultry. If not too badly affected, exposing the food to moist conditions, such as in cooking, may make these changes undetectable. But if the food has been exposed to dry air, without proper protection for a long period, the surface may be completely desiccated. This permits oxygen from the surrounding air to penetrate the food tissues and the outer layer of fat, on meat especially, then becomes oxidised and rancid. This is why meat with a high fat content has a limited storage life.

Safe storage and defrosting

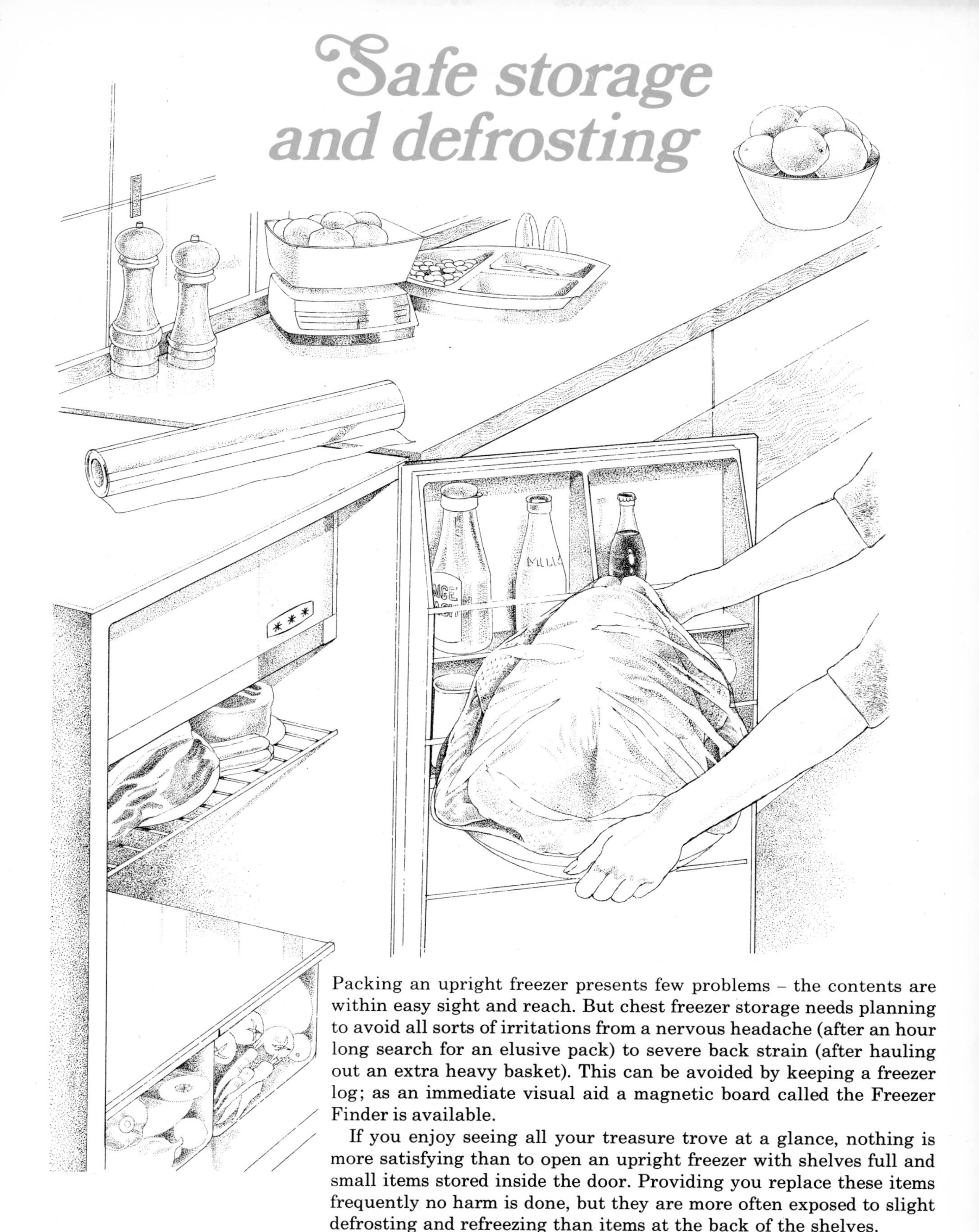

Packing an upright freezer presents few problems – the contents are within easy sight and reach. But chest freezer storage needs planning to avoid all sorts of irritations from a nervous headache (after an hour long search for an elusive pack) to severe back strain (after hauling out an extra heavy basket). This can be avoided by keeping a freezer log; as an immediate visual aid a magnetic board called the Freezer Finder is available.

If you enjoy seeing all your treasure trove at a glance, nothing is more satisfying than to open an upright freezer with shelves full and small items stored inside the door. Providing you replace these items frequently no harm is done, but they are more often exposed to slight defrosting and refreezing than items at the back of the shelves.

Storage expertise

1. **Upright** Leave the least accessible shelf for reserve supplies. Resist the temptation to pack too tightly because it is so easy. Many packs stack well; but leave finger spaces to facilitate removal.

2. **Chest** Use the three-layer system. Bottom layer – coloured batching bags, polythene shopping bags or cardboard cartons clearly labelled on *top* with contents which are not soon to be withdrawn. For instance, as a whole lamb gives you 'two of everything' pack half in an upper basket and its exact twin in a bag to go at the bottom of the chest. Middle layer – numbered baskets (colour identify contents and match up with labels on inside lid), for example 'basket 2 – bread and cakes'. These baskets should be for staple foods – your month's supply of fresh meat and cooked dishes, for instance. Top layer – numbered sliding baskets with a space you can peer through to see second layer. These are for quick turnover or short storage items – vegetables, sandwiches, shellfish.

Catering packs are much easier to handle if divided and packed in meal-sized portions. It is easy to remove exactly the quantity you need from a pack of flow-frozen vegetables or fruit – and seal the bag down again close to the food – in theory. But in practice, more than you want tips out, and the bag tends to get damaged by constant handling.

NOTE Always label packs on the side for upright freezers, on the top for chests. All bag packs should be shaped to as neat a square as possible, as awkward shapes are hard to keep sorted and to stack. Pack foil containers in pairs, taped face to face so that only one pack needs a lid.

How long is too long?

A carefully kept freezer log or index file ensures that no pack escapes your notice and is stored too long. Some housewives who freeze large quantities of home grown produce find it worthwhile to colour identify early and late season's crop of, for instance, green beans, and take specimen packs, at intervals, from each batch to control using it up during the period of maximum quality. If expensive ingredients have been used for party dishes and heavily seasoned with spices, herbs and garlic, freeze a test container of the dish in, for example, a 2-oz. (50-g.) Tupperware tumbler. If you are running near the storage limit and want to hang on for an important party still two weeks ahead, you can taste to make sure undesirable changes in flavour are not taking place.

The maximum storage time recommended by most experts on home freezing is, in my opinion, too pessimistic. I have experimented with extending even my own recommended storage times and had results which were almost incredibly successful. For example, a ratatouille which was made according to the recipe given in *Entertaining from your Freezer* (but without the garlic) and photographed for the cover of the book, was then frozen down in small packs (16-oz. (450-g.) Tupperware square rounds). The last of these packs was eaten by myself and Christine Curphey after 18 months storage and was quite palatable. The texture of the aubergines was rather spongy, but the general flavour was excellent.

Golden guide number 7

to storage and defrosting

Safe storage life To exploit the use of your freezer to the full, keep the contents on the move and use all your packs long before the recommended time limit. But this does not mean that food stored longer is unsafe, or uneatable, merely that it may have suffered some loss of flavour, colour and texture.

Recommended storage time table

Food	Storage time
Fruit packed in sugar or syrup (pineapple, 3–4 months only)	9–12 months
Fruit packed dry without sugar	6–8 months
Fruit purées	6–8 months
Fruit juices	4–6 months
Most vegetables (mushrooms, 6 months only)	10–12 months
Vegetable purées	6–8 months
White fish, such as cod	6 months
Oily fish, such as salmon	3–4 months
Fish portions coated in breadcrumbs or batter	3 months
Crab and lobster	3 months
Oysters and scallops	1–2 months
Prawns and shrimps (raw)	3 months
Prawns and shrimps (cooked)	1 month
Chicken and turkey	12 months
Duck and goose	4–6 months
Giblets	3 months
Venison	12 months
Rabbit and hare	6 months
Beef and lamb	9–12 months
Veal and pork	4–6 months
Sausage meat, minced meat (unseasoned)	3 months
Sausage meat, minced meat (seasoned)	1 month
Offal such as liver, kidney	2–3 months
Offal such as sweetbreads, tripe	3–4 months
Cured meats such as bacon	1 month
Milk	3 months
Cream (minimum 40% butterfat)	12 months
Fruit-flavoured yogurt	3 months
Butter, unsalted	6 months
Butter, salted	3 months
Eggs (separated and beaten)	9 months
Cheese, hard	6 months
Cheese, soft or cottage	3 months
Ice cream	3–6 months
Risen bread dough	3 weeks
Unrisen plain white dough	8 weeks
Unrisen enriched white dough (i.e. made with added eggs and sugar)	5 weeks
Baked or partly baked bread	4 weeks
Enriched bread and soft rolls, baked	6 weeks
Teabreads and decorated cakes	3 months
Yeast pastries	2–3 weeks
Yeast (packed in 1-oz. (25-g.) cubes)	12 months
Sandwiches, various fillings	2 months

Other foods

Pasta Used as an ingredient in cooked dishes can be frozen successfully if it contains a high percentage of protein, and is manufactured from the finest durum semolina – as is most of the pasta made in this country today. Some pastas are enriched with egg; these freeze very well indeed. Storage life, 2–4 months.

Pancakes If made from a specially enriched batter, freeze well. Add 1 tablespoon corn oil and 1 extra egg yolk to usual ingredients. Freeze in piles, with dividers. Storage life, 2–4 months.

Stuffings If made with breadcrumbs, dried herbs and egg (also with fat if liked) should be packed separately because of their relatively short life in store compared with meat and poultry. Storage life, 4 weeks.

Defrosting techniques

Watertight packs Under running water, preferably cold, but warm if haste is essential. To judge when food will slide out easily, turn pack over. Allow 20–45 minutes for crisp foods, i.e., fresh fruits; 15–20 minutes in warm water for foods which will not lose texture, i.e., stews. For the latter, the food only needs to be defrosted sufficiently to allow it to slide out of the container into a heavy-based saucepan, double boiler or ovenproof casserole for reheating or cooking.

Room temperature defrosting Suitable for foods which defrost in 3 hours or less. Large packs require many hours even in an ambient temperature of 70°F (21°C), which means the outer layer of food will exceed the temperature at which bacteria again become active some time before the centre is defrosted. If you cannot thaw large packs in a refrigerator allow them to defrost, lightly covered, in a cool place.

Refrigerator defrosting Suitable for all foods. It takes longer and the food must be covered as inside the enclosed refrigerator cabinet cross flavours are so easily absorbed.

Note Vegetables and commercial packs with instructions to cook from the frozen state should not be thawed.

Defrosting time table

Food	Approximate time in refrigerator	Approximate time at room temperature	Notes
Soft fruits (skinless)	6–7 hours	2–3 hours	
Soft fruits (with tough skins)	7–8 hours	3–4 hours	
Crisp fruits	7–8 hours	3½–4 hours	
Fruit purées	6–8 hours	2–4 hours	
Beetroot	9–10 hours	2–3 hours	
Mushrooms	6–8 hours	2–3 hours	
Peppers (sliced)	24 hours	1½ hours	Plunge, frozen, in boiling salted water for 5 minutes.
Corn-on-the-cob	9–10 hours	2–3 hours	If cooked from frozen state cob will not heat through.
White fish (whole)	6–8 hours per lb. (½ kg.)	3–4 hours per lb. (½ kg.)	Can be cooked, wrapped, from the frozen state.
White fish (steaks and fillets)	4–6 hours per lb. (½ kg.)	2–3 hours per lb. (½ kg.)	Interleave and spread out to defrost. Can be cooked from the frozen state.
Oily fish (whole)	6–10 hours per lb. (½ kg.)	3–5 hours per lb. (½ kg.)	Can be cooked, wrapped, from the frozen state.
Oily fish (steaks and fillets)	4–8 hours per lb. (½ kg.)	3–4 hours per lb. (½ kg.)	Interleave and spread out to defrost. Can be cooked from the frozen state.
Shellfish	10–12 hours per lb. (½ kg.)	3 hours per lb. (½ kg.)	
Smoked fish	5 hours per lb. (½ kg.)	2–2½ hours per lb. (½ kg.)	
Chicken (over 4 lb., 2 kg.)	1–1½ days		
Chicken (4 lb., (2 kg.) and under)	12–16 hours	1–1½ days, covered, in a cool place.	Make sure pack of giblets has been removed and that no ice crystals linger in the cavity, since quick roasting may not raise the temperature of the bony structure of the bird to a high enough temperature to destroy all harmful micro-organisms.
Duck (3–5 lb., 1½–2½ kg.)	1–1½ days		
Goose (4–14 lb., 2–7 kg.)	1–2 days		
Turkey (over 16 lb., 8 kg.)	2–3 days		
Turkey (16 lb., (8 kg.) and under)	1–2 days		
Venison	5 hours per lb. (½ kg.)		After fully defrosting, wash in 1 part vinegar to 2 parts water before cooking.
Rabbit and hare (over 4 lb., 2 kg.)	1–1½ days		

Food	Approximate time in refrigerator	Approximate time at room temperature	Notes
Rabbit and hare (4 lb. (2 kg.) and under)	12–16 hours		
Game birds	12–16 hours		
Poultry portions	5–6 hours per lb. ($\frac{1}{2}$ kg.)	1 hour per lb. ($\frac{1}{2}$ kg.)	Can be poached in boiling stock from frozen state.
Meat joints	5 hours per lb. ($\frac{1}{2}$ kg.)	2 hours per lb. ($\frac{1}{2}$ kg.)	Can be slow roasted from frozen state, covered, allowing 1 hour per lb. ($\frac{1}{2}$ kg.).
Chops, steaks, sausages	6 hours	2 hours	Interleave and spread out to defrost, or cook from frozen state; takes approximately twice as long.
Minced meat and offal	10–12 hours per lb. ($\frac{1}{2}$ kg.)	1–1$\frac{1}{2}$ hours per lb. ($\frac{1}{2}$ kg.)	
Cakes (large) iced		4–6 hours	Treble for very rich fruit cakes.
Cakes (large) plain		2 hours	
Cakes (small) iced and wedges		1–1$\frac{1}{2}$ hours	Spread out
Cakes (small) plain		30 minutes–1 hour	Spread out
Biscuits		15–30 minutes	Spread out
Bread (large loaf)		2 hours	Or reheat in oven (see Golden Guide to Bread); toast slices from frozen state.
Sandwiches		2–4 hours	Spread out 1 hour
Cheese	12 hours per lb. ($\frac{1}{2}$ kg.)	4 hours per lb. ($\frac{1}{2}$ kg.)	Must be wrapped. Grated can be used from frozen state.
Cream	12 hours per pint (generous $\frac{1}{2}$ litre)	2 hours per pint (generous $\frac{1}{2}$ litre)	Use rosettes from frozen state.
Milk	18–20 hours per pint (generous $\frac{1}{2}$ litre)	1$\frac{1}{2}$ hours per pint (generous $\frac{1}{2}$ litre)	Or reheat from frozen state.

Note Cooked made-up dishes should be thawed in the refrigerator overnight, or thawed and cooked (or reheated) direct from the frozen state. However, since refrigerator space is limited and speed may not be essential, dishes intended to be served cold after defrosting thaw in about one-quarter of the time at room temperature.

Golden guide number

to bread, sandwiches and cakes

Baking day favourites are also my freezer favourites. They are easy to pack and store, and always seem to emerge in perfect condition.

Bread

Both yeast and baking powder breads should be frozen as soon as they are cooled, and while still very fresh. Mould in freezer foil, wrap in sheet polythene or pack in shaped polythene containers, filling what would otherwise be unnecessary air spaces, with bread rolls. Loaves wrapped in foil can be placed, frozen, in a moderately hot oven (400°F, 200°C, Gas Mark 6) for 35–45 minutes, according to the size of the loaf. Sliced bread can be toasted while frozen. Rolls can be defrosted and reheated in a moderately hot oven for 15 minutes if wrapped in foil. Storage times: white and brown bread, 4 weeks. Enriched bread and rolls, 6 weeks. Bread with a crisp crust has the shortest storage time as the crust begins to 'shell off' after 1 week.

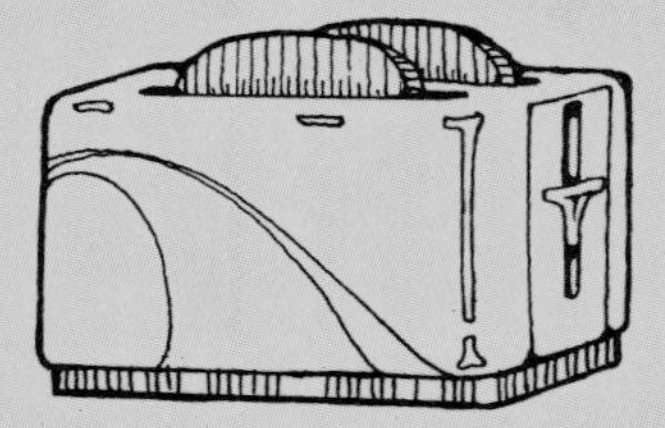

Croissants and brioches

These richer baked goods have a longer storage life than breads described as 'enriched', because of their high fat content. They also have the advantage that they can be defrosted and warmed, unwrapped, without drying out, and can be taken straight from the bag or polythene container and popped into a moderately hot oven (400°F, 200°C, Gas Mark 6) for about 15 minutes. Do not attempt to hasten defrosting by putting into a *hot* oven to hurry up breakfast, or the outside may be deceptively hot and a cube of frozen dough may still nestle in the centre.

Bread doughs

Uncooked bread dough gives a better result if the proportion of yeast is slightly increased.

Unrisen dough, weigh and wrap in quantities suitable for baking, i.e., 1 lb. 2 oz. (500 g.) dough for a 1-lb. ($\frac{1}{2}$-kg.) loaf tin. Freeze in lightly greased polythene bags, leaving just sufficient headspace above dough for it to rise during the freezing process. Storage times: plain white and brown dough, 8 weeks. Enriched dough, 5 weeks.

Risen dough, after rising, knock back, shape as required and seal in bags tightly and freeze at once. Storage time: 3 weeks. To thaw unrisen and risen dough: unseal polythene bag, tie again loosely to allow space for rising. Leave for 5–6 hours at room temperature or overnight in the refrigerator. Unrisen dough, knock back, shape, prove and bake. Risen dough, prove and bake.

Part-baked bread and rolls Slip the pack into a polythene bag and seal. Defrost and then follow instructions on pack for baking. Storage time: 4 weeks.

Everyday sandwiches

For everyday supplies, make up sandwiches by the assembly line method. Spread slices with softened butter or margarine to cover completely, to prevent filling sinking into the bread. Fill, put together and wrap as they will be used, individually or in quantities sufficient for a sandwich meal. Pack some to fit a Tupperware double diner so that you can put sandwiches in one end and biscuits, cake or fruit in the other. When packing assorted sandwiches, wrap those with strong smelling fillings in freezer film to prevent cross flavours within the pack. Storage time: 2 months. Defrosting: $2\frac{1}{2}$–3 hours at room temperature. Spread out on plate: 1 hour.

Suitable fillings

Peanut butter with marmalade, chopped ham or crumbled cooked bacon.
Liver sausage with chopped gherkins, stuffed olives or mashed hard-boiled egg yolk.
Cream cheese with chopped dates and oranges, chopped nuts or chopped gherkins.
Canned sardines, salmon or tuna.
Sliced roast meats – beef, pork, lamb or corned beef.
Chicken, finely chopped and mixed with very little mayonnaise or cream cheese.
Ham or grated cheese with sweet brown pickle or mango chutney.
Chopped prawns with very little mayonnaise or cream cheese and curry powder.

The variations are endless but just be wary of using salad vegetables, chopped whole cooked egg, tomatoes and any filling liable to turn the bread soggy during defrosting.

Teatime entertaining Trim off crusts and pack in closely moulded foil using druggist's wrap, or in shaped polythene containers.

Open sandwiches and canapés Freeze in a single layer in a shallow polythene container or shaped foil tray. Arrange on serving dish in the frozen state. Storage time: 1 week.

Pinwheel, club and ribbon sandwiches Freeze uncut and slice when partially defrosted. Storage time: 2 months.

Rolled sandwiches Pack in shallow foil trays close together and seal with foil. Arrange on plates when partially defrosted. Storage time: 2 months.

Toasted sandwiches Toast one side of bread slices, spread the other side with cheese rarebit mixture or flaked canned fish mixed with thick white sauce. Cool, pack with foil dividers and freeze. Spread out while still frozen (cut into fingers for a buffet party) on the grid and grill until golden brown. Storage time: 2 months.

Fried sandwiches Put buttered slices of white bread together with sliced processed or Gouda cheese in the middle. Wrap with foil dividers and freeze. Shallow fry on both sides in butter until golden brown and cheese begins to melt; drain. Storage time: 2 months.

Savoury bread loaves Make cuts 1-inch (2·5-cm.) apart along a French or Vienna loaf to within ½ inch (1 cm.) of the base. Spread the cuts generously with butter, flavoured with crushed garlic, seasoned grated cheese, or chopped mixed fresh herbs. Mould closely in freezer foil. To serve, place the foil-wrapped loaves in a moderately hot oven (400°F, 200°C, Gas Mark 6) for 30–40 minutes, according to size and shape. Storage time: 1 week.

Breadcrumbs Fresh breadcrumbs remain quite separate when frozen in polythene bags or containers and need not be defrosted before using in cooking. Storage time: 3 months.

Fried bread croûtons Cut out fancy shapes from thin slices of stale white bread and fry in a mixture of oil and butter until crisp and golden brown. Drain well, cool and pack as for breadcrumbs. Spread out on baking tray and place in a moderately hot oven (400°F, 200°C, Gas Mark 6) to defrost and reheat. Storage time: 1 month.

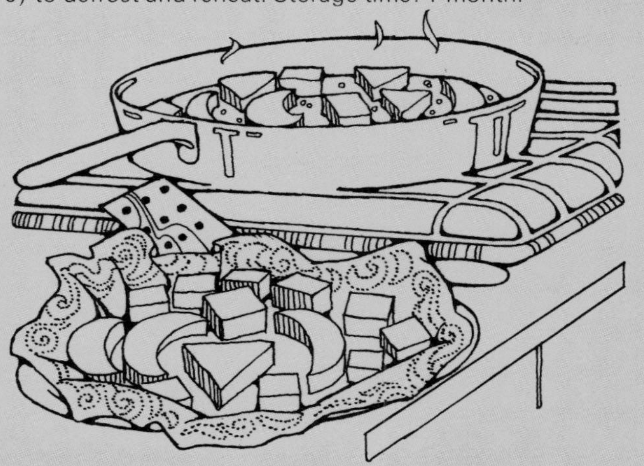

Yeast Fresh yeast is not always available and many cooks prefer to use it rather than dried yeast granules. Because yeast cells are destroyed by high temperatures, it is generally supposed that freezing will also kill them. Fortunately, this is just an old wives' tale and yeast keeps perfectly for up to 1 year. Pack in ½-oz. (15-g.) or 1-oz. (25-g.) cubes, individually wrapped. Put the cubes all together in one container. It will thaw at room temperature in about 30 minutes but if urgently needed can be grated, from the frozen state, on a coarse grater.

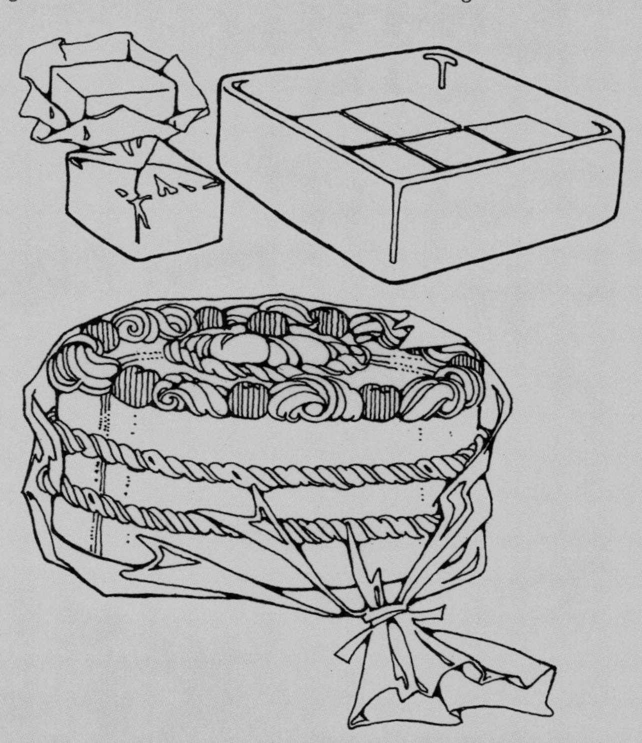

Baked cakes and scones

Plain family cakes and scones can be packed in polythene bags or wrapped in sheet polythene, as they are not likely to be damaged. Storage time: 6 months. Scones defrost at room temperature, in the pack, in 1–1½ hours, or spread out on a plate, in 1 hour. Scones can be defrosted and reheated, from the frozen state, in a moderately hot oven. Spread out on a baking tray, cover with foil and place in the oven for 10 minutes. Storage time: 6 months.

Undecorated cakes Defrost in the pack at room temperature. Storage time: 3 months.

Decorated cakes If close wrapped, the cake must be open frozen first to prevent damage and the wrapping removed while the cake is still in the frozen state. Defrost at room temperature on serving dish, with large polythene container inverted over the top. Storage time: 3 months.

Unbaked cake mixture Mixtures such as a Victoria sandwich cake mixture freeze well, uncooked. A suitable amount for one cake layer just fills a Tupperware cereal bowl, leaving the necessary headspace. The whisked type of sponge mixture does not freeze satisfactorily. Defrost containers at room temperature for 3 hours, then turn into prepared tins and bake. Storage time: 2 months.

Golden guide number 

to dairy foods

Milk
Pasteurised homogenised milk can be frozen in cartons. It takes 1 hour to defrost at room temperature. It is unsafe to freeze milk in narrow-necked pint bottles. Since it is usually wanted in a hurry, ½-pint (3-dl.) cartons or Tupperware tumblers are more convenient when returning from holiday or if you have unexpected guests. Shake the container gently once or twice during defrosting. It can be reheated from the frozen state. Storage time: 3 months.

Yogurt
Fruit-flavoured yogurt is a useful stand-by to have in the freezer although the saving on buying in bulk is very small. Store in the cartons in which it is purchased. Storage time: 3 months.

Cream
Double cream (pasteurised) or whipping cream (providing it has a minimum 40% butter fat content) can be frozen, lightly whipped with a pinch of sugar as a stabiliser. Single cream separates, and can rarely be reconstituted. Sweetened whipped cream should not be made too stiff as it is beaten again to restore the smooth consistency when thawing, and this may turn it buttery. Freeze in bags or polythene containers. I prefer the latter, as the cream can be beaten in the container. Pipe rosettes of whipped cream onto foil-lined trays, open freeze and transfer to polythene containers, or pipe straight into shallow foil trays. Cover before freezing. These can be removed with a warm palette knife onto cakes and jellies in the frozen state. Storage time: 6 months. Clotted cream, which has been pasteurised, stores well for up to 1 year.

Ice cream
Home-made or bought ice creams, and sorbets, can be stored in the freezer. Over-wrap commercial carton packs with polythene, or pack all together in a large polythene container. The large polythene boxes now being used for litre and gallon packs are sufficient protection. Transfer to the freezing compartment of the refrigerator to soften for 3–4 hours before serving. An ice cream scoop, dipped in very hot water, will remove portions from very solid ice cream. Cover surface of ice cream exposed to air, in large containers, with freezer film. Home-made ice cream in fancy moulds is better taken straight from the freezer, and dipped quickly in very hot water to turn out, rather than softened in the refrigerator, to hold its shape. Tupperware jellettes and Jel'n'serve make this extra easy because the removal of the fancy seal releases the vacuum. Storage time: 3–6 months.

Butter
If buying from a shop, make sure it is a fresh stock, if from a farm, that butter has been made from pasteurised cream. Freeze, over-wrapped in freezer foil, in 8-oz. (225-g.) packs. Storage time: unsalted, 6 months; salted, 3 months. Thaw out, wrapped, in the refrigerator. This takes about 4 hours for a small pack.

Cheese
Most hard cheeses freeze perfectly and defrost in prime condition for a ploughman's lunch or to serve with biscuits. Those cheeses which are naturally crumbly in texture might be slightly more so, and where a protective coating has been applied (like the lovely red overcoat of a Dutch Edam) there may be some tendency for the coating to separate from the cheese. But the texture, flavour and appetising appearance of the cheese itself are not in any way impaired. Storage time: 6 months.

A whole cheese, or a very large wedge, weighing as much as 10 lb. (5 kg.) may therefore be a good buy, since it is so easy to divide the cheese and pack in family-sized portions (weighing 1–2 lb., ½–1 kg.). I always reserve one extra large portion for a cheese and wine party, as I usually give one every few months. Particularly good are Dutch Edam and Gouda with their close, creamy texture.

1 Chopping pineapple to mix with lemon juice and grated zest and add to the sieved cottage cheese.

2 Sprinkling the crumb mixture over the cheesecake in a well-oiled 7-inch (18-cm.) cake tin. This is chilled until firm enough to unmould before freezing.

3 The undecorated cheesecake, frozen on serving dish, being taken out of polythene bag while still frozen to avoid damage.

1

2

Pineapple cheesecake (see page 208).

3

Blue cheeses, especially Stilton, freeze well but foreign ones are inclined to crumble when thawed, but this makes them all the more useful for salads, dips or a rather exotic salad dressing. Cream, curd and even cottage cheeses have a place in the freezer, especially as they can be beaten or sieved to improve the texture after defrosting. Personally I have not been so satisfied with the very delicate flavour of such French cheeses as Brie and Camembert, but this may be because it is extremely hard to judge the exact peak of maturity of these cheeses. All cheese is best if frozen when fully mature, as it will then have developed the maximum flavour. Thaw out, in the wrapping, overnight in the refrigerator, and then allow it to come up to room temperature for serving. Semi-frozen or too quickly defrosted cheese does tend to be tasteless and soapy in texture.

Three ways to freeze

1. For the cheeseboard The cheese should be slightly chilled, cut into portions of the size required and moulded in freezer foil or double thickness kitchen foil. Single thickness is sufficient protection if the wedges are then placed in a polythene bag. My secretary who has carried out a number of tests for me, with all the cheeses mentioned above, finds that a heavy gauge polythene bag alone is sufficient protection if all surplus air is carefully pressed out of the bag before it is sealed, so that as far as possible the polythene adheres to all the surfaces of the cheese.

2. For cooking For all cooking purposes 4-oz. (100-g.) and 8-oz. (225-g.) bags of coarsely grated cheese are invaluable. You can use the cheese almost from the frozen state if you crumble it in the bag between your fingers, or it can be defrosted at room temperature. Spread out on a plate this takes only about 1 hour. A pleasant alternative is to freeze a basic cheese sauce. If the container of sauce is placed under warm water for a few minutes, the block of frozen sauce can be turned out and reheated gently in a saucepan.

3. In made-up dishes Very many made-up dishes for freezing include cheese in the recipe. I have only one piece of special advice here, once fresh cheese has been added to a dish (not cheese sauce) and cooked quickly, it does tend to get both stringy and oily, and reheating after defrosting will accentuate this. Where the dish has a topping of grated cheese, or possibly cheese mixed with breadcrumbs, it is much better to add this after defrosting, at the reheating stage. A typical example of a successful achievement in freezing foods which many people regard as temperamental, is in Italian dishes made with pasta, meat sauce, béchamel sauce and cheese, commercial versions of which are now sold with great success.

Eggs

Packed beaten egg can be quickly thawed under cold running water, but take care not to use warm water or the egg may go lumpy. Eggs for freezing must be really fresh, and raw. Cooked whole eggs go rubbery, and eggs in the shell burst as the water content freezes and expands.

To freeze whole eggs Wash eggs carefully in cold water so that the liquid egg is not contaminated by dirt from the shell. Break them, one at a time, into a cup and add to a bowl, checking that each one smells entirely fresh. Beat very lightly, but do not whip in air. Add 2 teaspoons salt or 1 tablespoon sugar to each 1 pint (generous $\frac{1}{2}$ litre) of mixture (about 10 eggs make a pint) to act as a stabiliser and prevent it from hardening when frozen. Pack as for any liquid, allowing a headspace, and be sure to label the pack whether sweet or savoury. Small amounts, frozen in ice-cube trays, are useful to add to enrich puddings and sauces. Defrost at room temperature, from 30 minutes to 2 hours, according to quantity.

To freeze egg whites Separate the whites from the yolks and freeze without beating or any addition. Defrost completely before beating; frozen egg whites make excellent meringues.

To freeze egg yolks Separate the yolks from the whites and beat lightly, but do not whip in any air, together with salt or sugar in the same proportions as for whole eggs. Label whether intended for sweet or savoury use. Storage time: 9 months.

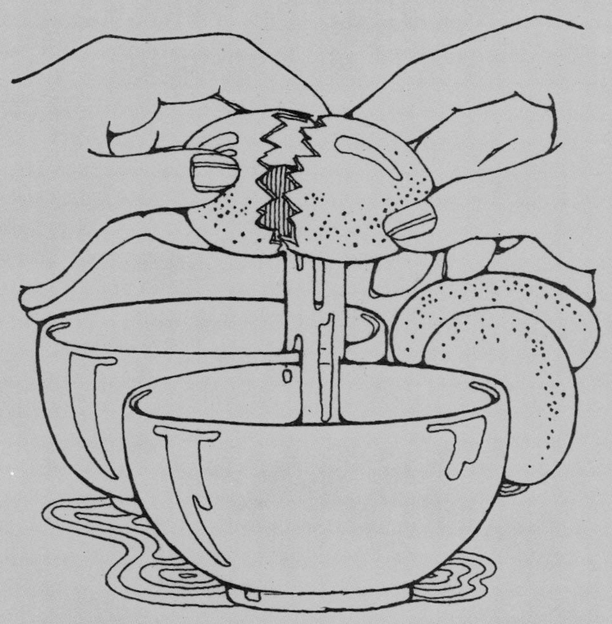

Cauliflower quickie (see page 150),
Edam cheese salad.

Allocating freezer space for your needs

Family Smith A young couple both out at work, have a freezer fridge, with about 6 cubic feet frozen storage space. They keep small portions of sauces and creamy individual sweets in the door, one shelf packed with 20 four-portion packs and 20 two-portion packs of meat. Another shelf holds poultry, fish and selection of gourmet made-up dishes in the same handy-size packs. The third shelf holds about 30 two-portion packs of assorted vegetables, and a dozen or so 16-oz. (450-g.) square round Tupperware containers of fruit, home-packed by Mrs. Smith. The bottom shelf holds bread and a pile of sandwich layers for cakes packed ready for filling and spreading as required, with the necessary 2-oz. (50-g.) tumblers of freezer jam and whipped cream – just the amount they reckon to fill a 6-inch (15-cm.) Victoria sandwich cake. They entertain infrequently.

Family Jones A couple in the thirties, with three young children, and Mrs. Jones stays home to look after them. They have a 16 cubic feet chest freezer, and apart from the fast freeze compartment which she tries to keep clear for new additions, they divide the storage space into four. One set of baskets, above and below – bread and sandwich packs, simple cakes, biscuit doughs ready to slice and bake. One set of baskets for meat, with a Tupperware square keeper full of meat and poultry cuts for a week's meals right on top. One set of baskets for packs of fruit and vegetables, most of which come from the allotment. The last set contains soups, sauces, freezer jam, ice cream, (bought in litre containers), and a catering size pack of potato croquettes, a family weakness. They hardly ever entertain.

Family Robinson A couple in the forties, children grown up and married, except for one son who lives at home; Mrs. Robinson does a part-time job. They entertain a lot, and have an 11 cubic foot upright freezer. The allocation of space is half to bought gourmet dishes, cheesecakes, Danish pastries, exotic vegetables like courgettes, and catering packs of mixed root vegetables for stews. Mrs. Robinson reckons saving her time is in fact saving money. The other half of the freezer holds two 10-lb. (5-kg.) packs, one of rump steaks and one of chicken portions, and always part of a whole lamb, or half a pig, which they reckon to demolish within six weeks of purchase. Made at home are mousses, Bavarian creams and layered ice creams, because Mrs. Robinson likes making them, and packs them straight into Tupperware dessert and parfit dishes, to take out, pipe with cream, and serve. Usually she crams in a few loaves, rolls and croissants for Sunday breakfast. Door storage holds fruit syrups and purées, and cooked mushrooms in small containers to liven up stews.

Buying food in bulk

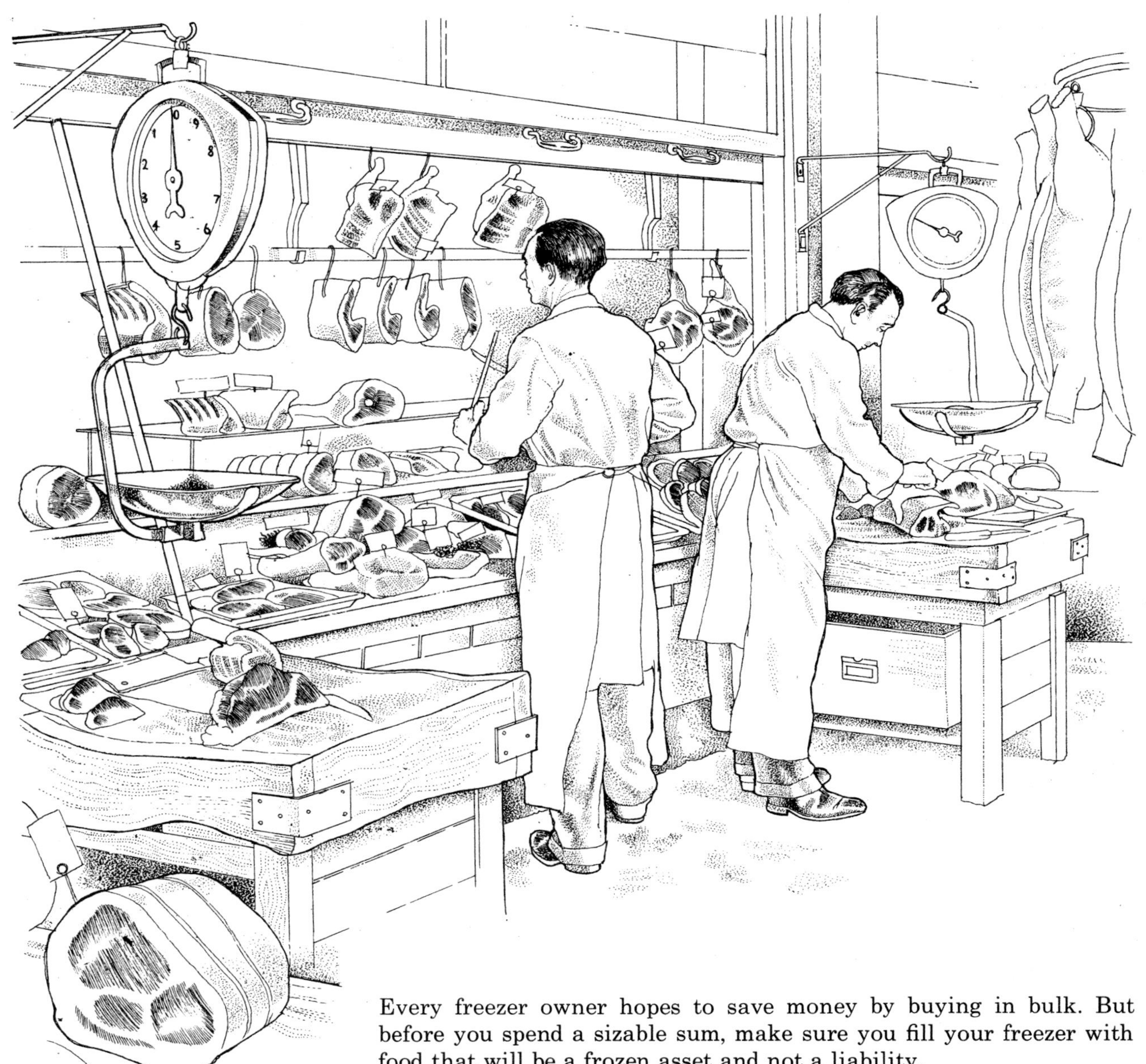

Every freezer owner hopes to save money by buying in bulk. But before you spend a sizable sum, make sure you fill your freezer with food that will be a frozen asset and not a liability.

The primary aim of bulk buying is to save money, and the secondary aim to save time spent on unnecessary shopping. Some owners may feel that saving time is even more important than money. Not only avoiding time-wasting shopping trips, but by buying ready-prepared meals and saving many hours that would have been spent in the kitchen.

Before you buy in bulk, get out pencil and paper and consider your answers to such questions as these.

1. Would you prefer to save *time* making everyday dishes such as fish

cakes, sausage rolls, steak and kidney pies – even if these made-up dishes cost more than the home-cooked equivalent?
2. Will your family enjoy a rather repetitive diet of, perhaps, peas and fish fingers if you take full advantage of the price reduction by buying a 20-lb. (10-kg.) bag of peas and a huge pack of fish fingers?
3. If you buy a forequarter of beef at a favourable price, will your family be happy to eat stewed meat at frequent meals? From a forequarter weighing 125 lb. ($62\frac{1}{2}$ kg.) you could expect to obtain about 85 lb. ($42\frac{1}{2}$ kg.) of edible meat and of that about 60 lb. (30 kg.) would be in stewing and braising cuts!
4. Is an attractive bulk purchase going to take up more freezer space than you can comfortably afford for any one type of food? Man does not live by ice cream alone!

Having decided what purchases will really justify their stay in your freezer, it then pays to shop around and discover where you can get the best value. You may not have noticed how much prices vary from one shop to another when you used to buy in penny numbers, but when you are buying 10 or 20 lb. (5 or 10 kg.) weight of food, you will notice how these odd pence mount up and the difference between two bulk offers might pay for one full family meal. Let us look at some comparative prices and how they may be influenced by the quality of the food.

Vegetables

There are various grades of peas, and of course you will enjoy flavourful top grade peas much more than standard. Do not be influenced too much by such descriptions as 'choice', 'extra fancy' and so on; by and large the most expensive pack on the price list will be the best quality. If you find the best bargain per lb. ($\frac{1}{2}$ kg.) weight from one outlet has poor eating quality, you may be well advised to reduce your overall saving and buy better quality peas from another outlet.

Some vegetables may cost half as much again per lb. ($\frac{1}{2}$ kg.) for different grades within the same store. Again, vegetables of the same high quality may cost less at a branch of a multiple frozen food suppliers. Before you opt for the delivery service, remember that frozen food requires special care in handling and transport and you are bound to pay something for this. There might be a saving of 5p on every dozen individual chicken pies you collect yourself, against the price if they are delivered to you.

Fish and shellfish

Some readers have been disappointed with large packs of frozen prawns, which seemed a bargain until they were found to be flabby and almost tasteless. If this is not indicated on the pack, enquire whether they come from Scandinavia or the Pacific Ocean. Prawns caught in cold water areas are much superior to those caught in warm waters. Then there is the problem of the large pack of fish fillets. Although vegetables are either free flowing or easy to cut into portions from a block, a big pack of fish fillets which have been packed without dividers is a nightmare. It may be impossible to remove any complete fillets without defrosting the whole pack, and slices cut off the frozen block with a freezer knife would defrost into strips of fish which look extremely odd on the plate. It might be worth a visit, when the catch comes in, to a port or fishing village. This way you can be sure the fish is fresh enough to be worth freezing, but have it gutted and filleted if you do not wish to carry out these messy processes yourself.

Fruit

The best way to buy fruit in bulk is to pick your own on the farm. Take your insulated bags with you or bring the fruit home in shallow trays so that it is not bruised and damaged before freezing. All sorts of freezer bargains can be found if you keep your eyes open when driving in the country. Vegetables as well as fruit are often offered to passers-by, when the price in the market would not show a profit to the grower.

Poultry

An advertisement in your local paper of chickens by the dozen straight from the farm sounds very economical. Check carefully whether this refers to plucked and drawn birds. Remember that the head and feet represent quite a proportion of the total weight of the dressed bird, and drawing chickens may not be your idea of an afternoon's entertainment! If the supplier explains exactly what you are getting for your money and you are still happy that it is a bargain, there is a lot of satisfaction to be gained from chicken chain cooking.

Meat

Check that you have room available to accommodate the meat. For example, a forequarter of beef weighing 125 lb. (62½ kg.) will take up at least 4 cubic feet of freezer space (meat 85 lb. (42½ kg.), fat 25 lb. (12½ kg.) and bones used to make 4 pints (2¼ litres) stock).

1. Go for good quality meat, even if you pay a penny per lb. (½ kg.) more than the lowest price quoted. Insist on beef being properly hung before delivery or it may be tough. Inform the butcher if you like your beef particularly well hung. Remember that a hindquarter includes the choicest cuts, whereas a forequarter is a more manageable weight but includes many cuts only suitable for stews and mince.
2. Especially with regard to beef, make a list of the cuts included in your order and note the purposes for which each cut can be used. Ask the butcher to advise you what proportion of meat you will receive that is suitable for roasting, braising and so on before you order. The first time, ask him to bag the meat in convenient quantities and mark each pack with a description of the cut and purpose for which it is best suited.
3. If you have a preference for meat sliced for braising, mention this and ask the butcher to do this job for you, or he may dice it all, or worse still leave it in large pieces. If he advises mincing the coarser meat and you want to freeze it uncooked, ensure that he does not add too much of the waste fat, as a high proportion of fat shortens the storage life. If you have a pan big enough, ask for all the bones and make stock with these. Surplus beef fat can be rendered down to make excellent dripping, but the butcher will probably not send you either the bones or the fat unless you specially ask for them. He may also be prepared to mince the suet for you which again saves time and trouble in your own kitchen.

Lamb The cuts are all well-known and easily recognisable. The only special advice needed is to decide beforehand whether you wish the best end of neck left as a whole joint for roasting, divided into cutlets, or boned and divided into noisettes. There is not a large proportion of stewing meat.

Pork A whole pig represents a lot of meat, but a half pig is a good purchase for the beginner. Unless you specialise in old country recipes, relinquish your rights to the head and trotters to the purchaser of the other half of the pig. The cuts are mainly for roasting and frying and the only ones likely to be unfamiliar are the belly and spare ribs. Since pork is a very fat meat, and therefore has a limited storage life, do not buy more than you can use up within six months.

Beef Because it is a large animal, you cannot avoid buying a lot of meat unsuitable for roasting and frying. If, for example, your family will eat nothing but roasts and grills, you will have on your hands a problem rather than a bargain. You may also find yourself landed with far more unaccustomed cuts than you knew even existed, such as the following: brisket, silverside and thick flank – suitable for very slow roasting, pot roasting or braising; chuck steak and skirt – suitable for braising; leg, shin and neck (or clod) – suitable for stewing only. Skirt can be sliced thinly across the grain while still partially frozen and fried to make economical minute steaks.

ROASTING MEAT FROM THE FROZEN STATE

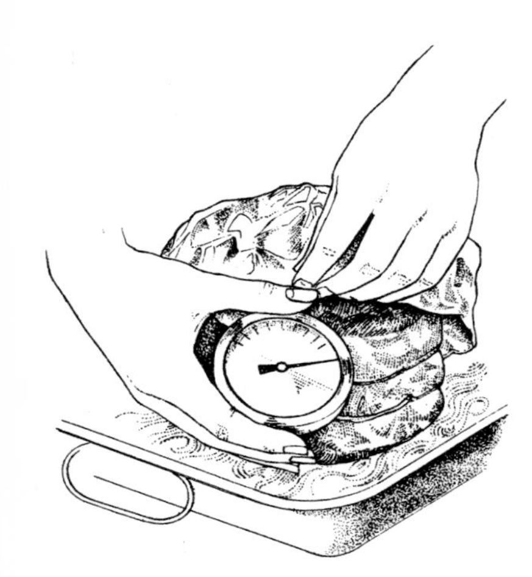

Beef

Prime roasting cuts Weigh joint, preheat oven to moderately hot (400°F, 200°C, Gas Mark 6), place joint in roasting tin in oven for 10 minutes to seal the meat. Cover with a dome of foil, reduce temperature to moderate, (350°F, 180°C, Gas Mark 4) and allow 1 hour per lb. ($\frac{1}{2}$ kg.) from this point. Towards end of cooking time, insert a meat thermometer in the centre of joint, check from then onwards every 10 minutes until temperature reaches 140°F (60°C) for rare, 155°F (68°C) for medium and 170°F (76°C) for well done.

Medium roasting cuts Weigh joint, preheat oven to moderate (350°F, 180°C, Gas Mark 4), place joint in dimpled roaster or roasting bag and roast for 1 hour per lb. ($\frac{1}{2}$ kg.). Towards end of cooking time insert meat thermometer to centre of joint, if necessary plunging through the bag above the level of the juices. Check from then onwards as above.

Lamb

Proceed as for beef prime cuts until thermometer registers 180°F (82°C). For those who like lamb slightly pink, 170°F (77°C).

Pork

Joints required with crackling Weigh joint, preheat oven to moderately hot (400°F, 200°C, Gas Mark 6), place joint in roasting tin in oven for 20 minutes. Cover with a dome of foil and proceed as for beef prime cuts until thermometer registers 190°F (88°C).

Joints from which skin has been removed Proceed as for beef prime cuts until thermometer registers 190°F (88°C).

Small joints

Those weighing less than 3 lb. ($1\frac{1}{2}$ kg.) are not very satisfactory, but do require cooking by the same calculation of 1 hour per lb. ($\frac{1}{2}$ kg.) as for large joints. The criterion should always be the temperature of the joint, which is why it is advisable to start checking the temperature 45 minutes to 1 hour before you expect the joint to be done.

How to deal with emergencies

The whole point of buying a freezer and filling it with a costly investment in frozen food is to save time, trouble and *money*. If your cooker goes wrong you may have to eat cold food, cook on a picnic stove, or invite yourself out to friends for several days running, but you are not actually losing any cash – just suffering inconvenience. If your freezer stock is at risk, you stand to lose not just its worth in money, but its worth in cooking, packing, sorting, labelling, all of which took up your valuable time. Resolve not to let this disaster overtake you by safeguarding against it in every possible way, and that means for a start, taking out a sensible insurance.

Before deciding whether or not to insure against emergencies, work out the total investment involved.

Typical yearly running cost

Depreciation of freezer (cost £90, has anticipated life of 15 years)	£6
Cost of electricity (based on $1\frac{1}{2}$p per cubic foot per week)	£9
Repairs and maintenance (averaged over the freezer's life)	£2
Insurance of freezer and contents (based on contents worth £100)	£3
Total	£20

If you place this yearly investment against your savings in bulk buying food, you should still make a profit. By the time the freezer is worn out, it is written off. The food inside also represents a considerable sum of money, and while a breakdown of the appliance may cost only a few pounds to put right, the loss of its contents might leave you £100 or more in the red. So insure, especially against this latter contingency.

Insuring freezer and contents

Your existing household policy may cover you for all the likely kinds of damage to the freezer itself, as it does with your cooker or washing machine, but not for the damage of food by deterioration if the freezer is not functioning. You need a special insurance to get this protection. An insurance broker can arrange for this to be added as an extension of your household insurance (the cheapest way) or as a separate operation. It ought not to cost more than £3 per annum for the freezer plus £100 worth of food. The cover is for accidents you can't avoid – mechanical failure, or cutting off of the power supply. Accidents such as temporarily unplugging and forgetting to plug the freezer in again are not covered, and loss of power due to strikes rather than an explosion at the local power station are probably not covered either. Read the small print on the policy to make sure. Some insurers operate block insurance schemes for a group of stores or other retailers. This means your supplier can offer you a package deal on the cost of freezer and insurance together. Or the service may be offered to private individuals, and you can participate. Do not be surprised if you are asked to state whether you have a maker's guarantee or warranty in force. There are now far more elderly freezers about, and these are the most likely sources of claims. Remember, there are two quite separate risks; the freezer itself and the food inside. A garage-sited freezer might get a serious knock from a carelessly driven car, or (as I know actually happened in one case) a heavy weight slung from the roof beams might fall on it. The freezer, if a total write-off, might cost £100 to replace, but so would the food stored in it if that also is a total loss.

Forewarned is forearmed

Let husbands and sarcastic friends laugh, the time to find out what will happen if disaster strikes is when you buy the freezer. Right there and then note the vital telephone numbers and addresses on a card, and tape the S.O.S. notice to the side of the freezer. The engineer, the dry ice supplier, the freezer centre or wholesale butcher where your food might find frozen sanctuary during the emergency, other freezer owners who might have room to spare and those true-blue friends who would come round and help you have an enormous cook-up with food which is eatable, but not re-freezable in its present state. A number of large insulated paper sacks will suffice to transport the food if it is not yet semi-defrosted, or if it is, stout cardboard cartons may be safer. Proper insulated bags are ideal, so make a list in the front of your freezer log of people you know who own them and would lend in an emergency.

Check up at the time of purchase what help you may expect from the suppliers of the freezer. If you buy through a source which also supplies frozen food, rather than the local Electricity Board showrooms or an electrical shop, some guarantee of help with emergency storage may be included in the deal, and if at the same time you arrange full insurance cover and maintenance, you can really feel secure.

Suppose the freezer stops working?

Do not immediately panic if the freezer mysteriously ceases to operate, and the top layer of food starts to defrost. Check the wiring, plug, socket and fuse before ringing the maintenance engineer. You may not need him. If you do, while you are waiting resist the impulse to keep opening the lid or door and peeping inside. Limit yourself to one quick glance round, move ice cream to the bottom or coldest part of the freezer, then keep it shut, shut, shut. The larger the volume of food stored and the more tightly packed together, the longer the whole lot stays frozen. Don't be tempted to get a friend to remove half your stock and leave a great big open space inside your freezer, or you risk having the other half defrost more speedily.

But what happens when the service engineer comes, shakes his head sadly and tells you a major repair is necessary? If it really is a long job you need a replacement freezer or the assurance that your food will be stored at the depot in one of the company's freezers.

I am personally reluctant to advise even a desperate freezer owner to attempt keeping a large stock of food frozen by packing it with dry ice. This comes in large pieces, almost too heavy to lift, which have to be hacked up. It is dangerous to handle with bare hands, let alone the danger of eating some if it were to come in direct contact with food. But if you choose to chop some up (wearing thick gloves), pack it into cardboard cartons and imbed these, so to speak, in the centre of the food stock, it would postpone that point of no return after which the only solution is to throw out all the food. Some can often be salvaged. Defrosted fruit might be cooked and refrozen as made-up dishes. But bearing in mind that enzyme and bacterial activity is so rapid in recently thawed food, take no risks.

Going on holiday

If you go away for more than a few days and cannot switch off at the main without disconnecting your freezer also, don't switch off, just remove all the other fuses. But leave tidily in a box together with a torch near the front door, so that you don't return at midnight after a fortnight's absence and have to grope around the fuse box. The ideal installation, of course, is one with the freezer on a completely separate circuit.

Moving house

Since I first wrote about this problem, the sight of a freezer on the tailboard of a moving van is as common as once was the cage with the old cock linnet; last out of the old house and first in the new one is the safest rule for freezers. Don't let the freezer come as a surprise to the removers, check the loaded weight and whether they can handle it. If not, order some dry ice for moving day. Make sure you have recently had a defrost and sort-out. At the last possible moment, pack food into insulated bags and then into tea chests lined with newspapers, each with its ration of well-protected dry ice, and into the van. Then on with the freezer and off to the new home. Check in advance that the existing plug will fit the new socket. If the move takes place in one day, nothing more need be done, but one member of the family or moving team ought to be responsible for seeing that the freezer is speedily connected, checking that it functions, and remembering to transfer the food back to safety before bedtime. If however the move extends over days, the only answer is to hire cold storage space, run down the stocks to almost nil, or try to sell the freezer with the house. You will probably find this an excuse to get a bigger model, and since the estimation of cubic capacity has risen to 2 cubic feet per person plus two for luck (and this is to my mind still too low an estimate) you may well need a larger freezer.

How long will food stay frozen?

Various estimates for a fully loaded freezer are between 12 and 24 hours. That is why a 24-hour answering service and prompt arrival of the engineer are so important. Possibly the discrepancy comes because of variable conditions. If it is summer, the ambient temperature will be high, and food will thaw relatively quickly to match it. In a cold ambient temperature, as the food approaches 20°F (−13°C) (at which it remains full of ice crystals, and therefore re-freezable) the difference between room and cabinet temperatures is no longer great, and thawing will slow down. Even then, packages at the bottom of a chest freezer will not be as liable to damage as those near the top. However, all these precautions sound daunting. So let us end on a reassuring note. Freezers rarely go wrong. Electricity cuts seldom last long enough to have any adverse effect on your frozen food. And the use of common sense and forethought will prevent damage through silly mistakes.

Part 2 *Catering better – spending less money, time and energy*

Bonus cooking-basics the chain way

We all know the theory that freezer owners tend to live better rather than saving money on buying food. This may be true; but while cooking in a king-size quantity saves time and effort rather than money, chain cooking ought to be economical. The whole idea is to start with an inexpensive basic ingredient such as minced beef. If this were presented to the family again and again as the eternal savoury mince, it would become very boring. But when a basic mixture is used to produce meatballs, a loaf mix, a spicy chilli mixture, or to make different meaty stuffings for vegetables like marrow and courgette or yet again to fill pasties and pies – there's nothing boring about minced beef. The chain method applies to many other basic foods besides meat.

Complicated sauces, for example, become quite easy to produce as developments of a basic white savoury or brown sauce.

Baking, which is undeniably a chore, gives so much more satisfaction when the end result is a selection of plain and fancy loaves and rolls or tea-rings, plaits, and fancy buns. The yield from each chain is described before the recipe, with suggestions on how the chain may be varied to suit your family's preferences.

White bread chain

Chain Makes 1 bumpy loaf and a selection of rolls.
Storage time Unrisen dough – 8 weeks; risen dough – 3 weeks; baked bread – 4 weeks; baked soft rolls – 6 weeks; crisp-crusted bread and rolls – 1 week before crusts begin to 'shell off'.

IMPERIAL	METRIC	AMERICAN
yeast liquid	*yeast liquid*	*yeast liquid*
½ oz. fresh yeast	15 g. fresh yeast	½ cake compressed yeast
¾ pint warm water	scant ½ litre warm water	scant 2 cups warm water
or	or	or
2 teaspoons dried yeast	2 teaspoons dried yeast	2 teaspoons dry yeast
1 teaspoon sugar	1 teaspoon sugar	1 teaspoon sugar
¾ pint warm water	scant ½ litre warm water	scant 2 cups warm water
dry mix	*dry mix*	*dry mix*
1½ lb. plain flour	700 g. plain flour	2 cups all-purpose flour
2 teaspoons salt	2 teaspoons salt	2 teaspoons salt
2 oz. lard or margarine	50 g. lard or margarine	¼ cup lard or margarine
egg wash	*egg wash*	*egg wash*
1 egg	1 egg	1 egg
1 teaspoon sugar	1 teaspoon sugar	1 teaspoon sugar
1 tablespoon water	1 tablespoon water	1 tablespoon water

For a slightly enriched bread use milk and water mixed instead of water. To prepare the yeast liquid if using fresh yeast, blend yeast into the water. If using dried yeast, dissolve the sugar in the water, sprinkle on the dried yeast and leave until frothy, about 10 minutes. Mix together the flour and salt and rub in the fat. Add the yeast liquid and work to a firm dough until sides of bowl are clean. Turn dough onto a lightly floured surface and knead thoroughly until firm and elastic and no longer sticky. It will take about 10 minutes.

To make the dough in an electric mixer, follow manufacturer's instructions for using the dough hook. Place yeast liquid in the mixer bowl, add dry ingredients, turn to lowest speed and mix for 1–2 minutes to form the dough. Increase speed slightly and mix for a further 2 minutes to knead the dough.

Shape dough into a ball and place in a large, lightly oiled polythene bag, tied loosely at the top. Leave to rise until dough is double in size and springs back when pressed gently with a floured finger.

Rising times vary with temperature

Quick rise	45–60 minutes in a warm place
Slower rise	2 hours at average room temperature
Cold rise	12–24 hours in a refrigerator

Return dough to room temperature before shaping. Turn risen dough onto a lightly-floured surface, flatten with the knuckles to knock out the air bubbles, and knead until dough is firm, about 2 minutes. Make the egg wash by blending together the egg, sugar and water.

Bumpy loaf

(Illustrated on page 89)

To freeze Wrap in freezer foil or polythene bag. Seal and label.
To serve Defrost, still wrapped, at room temperature for 6 hours. To make crisp place unwrapped in a moderately hot oven (400°F, 200°C, Gas Mark 6) for 5–10 minutes. If required quickly, place frozen loaf, wrapped in foil, in a moderately hot oven as above for 35 minutes. Remove foil and bake for further 10 minutes to crisp crust.

IMPERIAL	METRIC	AMERICAN
1½ lb. white bread dough	700 g. white bread dough	1½ lb. white bread dough
egg wash	egg wash	egg wash

Grease a 2-lb. (1-kg.) loaf tin. Divide the dough into four 6-oz. (175-g.) pieces. Shape each piece into an oblong the same width as the tin. Place in tin and brush top with egg wash.

To prove, place tin in a large, lightly oiled polythene bag and leave until dough is double in size or rises to top of tin.

It will take 30–40 minutes in a warm place
1–1½ hours at room temperature
12–24 hours in a refrigerator

Remove from polythene bag.

Bake in a hot oven (425°F, 220°C, Gas Mark 7) for 45–50 minutes. Turn out and cool on a wire tray.

Rolls

(Illustrated on page 89)

To freeze Wrap in freezer foil or polythene bags. Seal and label.
To serve Defrost, still wrapped, at room temperature for 1½ hours. To crisp crust, place unwrapped, in a moderately hot oven (400°F, 200°C, Gas Mark 6) for 5 minutes. If required quickly, place frozen rolls, wrapped in foil, in a hot oven (450°F, 230°C, Gas Mark 8) for 15 minutes.

IMPERIAL	METRIC	AMERICAN
2-oz. pieces white bread dough for each roll	50-g. pieces white bread dough for each roll	2-oz. pieces white bread dough for each roll

To shape rolls

Cloverleaf Divide each piece equally into three. Shape into three small balls. Place on a greased baking tray in the shape of a cloverleaf, pressing lightly together. Brush with egg wash and sprinkle with poppy seeds.

Plait Divide dough into three equal-sized pieces. Roll each piece into a 4-inch (10-cm.) strand. Pinch together one end of each strand, plait the strands. Pinch remaining ends together and tuck both ends underneath. Lift plait onto a greased baking tray, brush with egg wash and sprinkle with poppy seeds.

Cottage Cut one-third off each piece of dough. Shape largest piece into a ball and place on a greased baking tray. Shape smaller piece into a ball and place on top of larger ball. Dip a finger in flour and press down through the middle of roll to touch baking tray. Brush with egg wash.

Coburg Shape each piece of dough into a ball and place on a greased baking tray. Brush with egg wash. With a sharp knife cut a cross on top of roll.

To prove, cover rolls with lightly oiled polythene and leave until dough is double in size. Remove polythene and bake in a hot oven (425°F, 220°C, Gas Mark 7) for about 20 minutes. Cool on a wire tray.

Croissants and brioches made from any standard recipe can be packed in polythene bags, the air withdrawn and the bags sealed. If really fresh when frozen they will store well. Croissants – 8 weeks; brioches – 4 weeks.

Note Strong flour, if available, is better than ordinary plain flour as it has a higher gluten content.

Wheatmeal bread chain

Chain Makes two 7-inch (18-cm.) cob loaves and 8 rolls.
Storage time Unrisen dough – 8 weeks; risen dough – 3 weeks; baked bread – 4 weeks; baked soft rolls – 6 weeks; crisp-crusted bread and rolls – 1 week before crusts begin to 'shell off'.

IMPERIAL	METRIC	AMERICAN
1 lb. plain white flour	450 g. plain white flour	4 cups all-purpose flour
1 lb. wholemeal flour	450 g. wholemeal flour	4 cups wholewheat flour
4 teaspoons salt	4 teaspoons salt	4 teaspoons salt
1 oz. lard	25 g. lard	2 tablespoons lard
4 teaspoons sugar	4 teaspoons sugar	4 teaspoons sugar
1 oz. fresh yeast	25 g. fresh yeast	1 cake compressed yeast
1 pint warm water	6 dl. warm water	$2\frac{1}{2}$ cups warm water
or	or	or
4 teaspoons dried yeast	4 teaspoons dried yeast	4 teaspoons dried yeast
1 pint warm water	6 dl. warm water	$2\frac{1}{2}$ cups warm water

Mix flours and salt in a bowl and rub in the fat. To make the dough with fresh yeast, mix sugar into flour. Blend yeast into water and add to the flour, all at once. Mix to a soft, scone-like dough which leaves the bowl clean, adding a little more flour if necessary.

To make dough with dried yeast, mix half the sugar into the flour. Dissolve remaining sugar in the warm water and sprinkle the dried yeast on top. Leave until frothy, about 10 minutes. Add to flour and mix to a soft, scone-like dough.

Knead dough on a lightly floured surface until smooth, about 2 minutes.

Cob loaf

To freeze Wrap in freezer foil or polythene bag but make sure the crisp topping does not tear the material. Seal and label.
To serve Defrost, still wrapped, at room temperature for 6 hours. To crisp crust, place unwrapped, in a moderately hot oven (400°F, 200°C, Gas Mark 6) for 5–10 minutes. If required quickly, place frozen loaf, wrapped in foil, in a moderately hot oven as above for 35 minutes. Remove foil and bake for further 10 minutes to crisp crust.

IMPERIAL	METRIC	AMERICAN
$2\frac{1}{4}$ lb. wheatmeal bread dough	1 kg. 100 g. wheatmeal bread dough	$2\frac{1}{4}$ lb. wheatmeal bread dough
2 teaspoons cracked wheat	2 teaspoons cracked wheat	2 teaspoons cracked wheat

Divide the dough in half and shape each piece into a ball and place each in a greased, deep 7-inch (18-cm.) cake tin. Brush top with salt and water and sprinkle with cracked wheat. Place tins in a large, lightly oiled polythene bag and leave to rise until dough is double in size and springs back when pressed with a floured finger.

Rising times vary with temperature

Quick rise	30–45 minutes in a warm place
Slower rise	1–$1\frac{1}{2}$ hours at average room temperature
Cold rise	12–24 hours in a refrigerator

Remove from polythene bag and bake in a hot oven (450°F, 230°C, Gas Mark 8) for 30–40 minutes. Remove from tins and cool on a wire tray.

Wheatmeal rolls

Shape rolls following the instructions given for white bread chain; brush the tops with salt and water and sprinkle with cracked wheat instead of poppy seeds.

A selection of breads – bumpy loaf, rolls, wheatmeal loaf, croissants (see pages 87, 88).

1 Adding the measured corn oil and water to the egg yolks to make a light sponge cake mixture.

2 Turning out layers of cooked feather sponge to cool on a wire tray before packing.

3 Packing the unfilled layers with a foil divider, to freeze for making into a fancy gâteau.

1

2

3

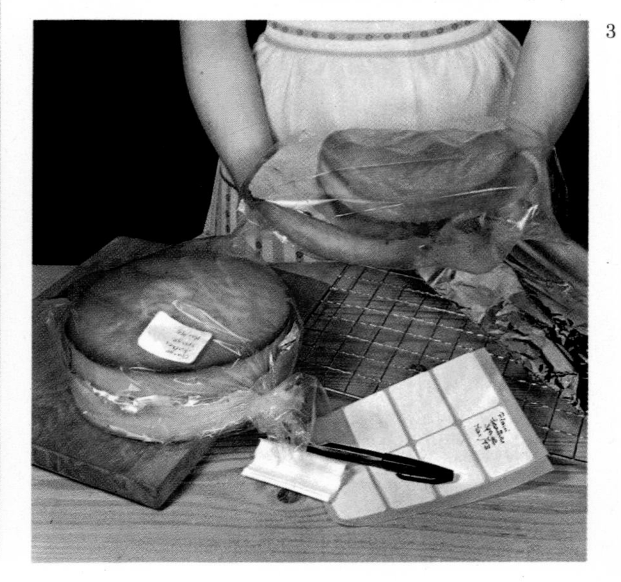

Fruit and cream sponge (see page 94).

Enriched bread chain

Chain Makes plaited sultana roll, orange and date twist and 12 Easter buns.
Storage time Unrisen dough – 5 weeks; risen dough – 3 weeks; baked teabreads and buns – 6 weeks.

IMPERIAL	METRIC	AMERICAN
yeast liquid	*yeast liquid*	*yeast liquid*
1½ oz. fresh yeast	40 g. fresh yeast	1½ cakes compressed yeast
18 fl. oz. warm milk	½ litre warm milk	1 pint plus 4 tablespoons warm milk
dough	*dough*	*dough*
3 lb. plain flour	1 kg. 400 g. plain flour	12 cups all-purpose flour
1 tablespoon salt	1 tablespoon salt	1 tablespoon salt
1 tablespoon castor sugar	1 tablespoon castor sugar	1 tablespoon granulated sugar
6 oz. margarine	175 g. margarine	¾ cup margarine
3 eggs, beaten	3 eggs, beaten	3 eggs, beaten

Blend yeast into milk. Mix together the flour, salt and sugar and rub in the margarine. Add the yeast liquid and beaten eggs all at once and mix to a soft dough. Turn out and knead or a lightly floured surface until firm and elastic and no longer sticky. This will take about 10 minutes.

To make the dough in an electric mixer, follow manufacturer's instructions for using the dough hook. Place yeast liquid and eggs in the mixer bowl, add remaining ingredients and turn machine to lowest speed for 1–2 minutes to form the dough. Increase speed slightly and mix for a further 2 minutes to knead the dough.

Shape dough into a ball and place in a large, lightly oiled polythene bag, tied loosely at the top. Leave to rise until double in size and springs back when pressed gently with a floured finger.

Note If using dried yeast use half the quantity given for fresh yeast and dissolve the sugar in the milk and sprinkle the dried yeast over. Leave for about 10 minutes until frothy. Mix together the remaining dry ingredients, rub in the margarine and add the yeast liquid and eggs. Form into a dough and continue as above.

Rising times vary with temperature

Quick rise	45–60 minutes in a warm place
Slower rise	1½–2½ hours at room temperature
Cold rise	12–24 hours in a refrigerator

Return dough to room temperature before shaping. Turn risen dough onto a lightly floured surface, flatten with the knuckles to knock out the air bubbles, and knead until dough is firm, about 2 minutes. Divide dough into three equal portions.

Plaited sultana roll

To freeze Wrap in freezer foil or polythene bag. Seal and label.
To serve Defrost, still wrapped, at room temperature for 3 hours. If required quickly, place frozen roll, wrapped in foil, in a moderate oven (350°F, 180°C, Gas Mark 4) for 15–20 minutes. Sprinkle roll with castor sugar and cut into slices.

IMPERIAL	METRIC	AMERICAN
1 portion enriched bread dough (one-third basic recipe)	1 portion enriched bread dough (one-third basic recipe)	1 portion enriched bread dough (one-third basic recipe)
8 oz. sultanas	225 g. sultanas	1⅓ cups seedless white raisins
2 tablespoons lemon juice	2 tablespoons lemon juice	3 tablespoons lemon juice
2 tablespoons marmalade	2 tablespoons marmalade	3 tablespoons marmalade
beaten egg to glaze	beaten egg to glaze	beaten egg to glaze

Roll out dough to an oblong about 12 inches by 8 inches (30 cm. by 20 cm.). Mix together the sultanas, lemon juice and marmalade. Spread the sultana mixture down the centre third of the dough leaving a panel of dough on either side approximately 4 inches (10 cm.) wide. Cut strips obliquely in these panels about 3 inches (7·5 cm.) in from the edge. Brush edges with beaten egg and plait over the filling, folding the ends under neatly. Place on a greased baking tray, cover with lightly oiled polythene and leave to rise until double in size. Remove polythene, brush with beaten egg and bake in a hot oven (425°F, 220°C, Gas Mark 7) for 15–20 minutes. Cool on a wire tray.

Orange and date twist

To freeze Open freeze until solid then wrap in polythene bag. Seal and label.
To serve Remove from bag while still frozen, place on serving dish and defrost at room temperature for 2 hours.

IMPERIAL	METRIC	AMERICAN
8 oz. dates, chopped	225 g. dates, chopped	1¼ cups pitted dates
finely grated zest and juice of 1 orange	finely grated zest and juice of 1 orange	finely grated zest and juice of 1 orange
1 portion enriched bread dough (one-third basic recipe)	1 portion enriched bread dough (one-third basic recipe)	1 portion enriched bread dough (one-third basic recipe)
4 oz. icing sugar	100 g. icing sugar	scant 1 cup sifted confectioners' sugar

Soak the chopped dates in 2 tablespoons orange juice for 15 minutes then drain. Mix the soaked dates and half the orange zest into the dough until evenly distributed. Divide dough in half and shape each piece into a roll about 15 inches (39 cm.) long. Twist the two pieces of dough lightly around each other two or three times, place on a greased baking tray and press the edges together neatly to form a circle. Cover with lightly oiled polythene and leave to rise until double in size. Remove polythene and bake in a hot oven (425°F, 220°C, Gas Mark 7) for 15–20 minutes. Cool on a wire tray. Mix together the sieved icing sugar and enough orange juice to give a thick coating consistency. Spoon the icing over the twist and sprinkle with the remaining grated orange zest.

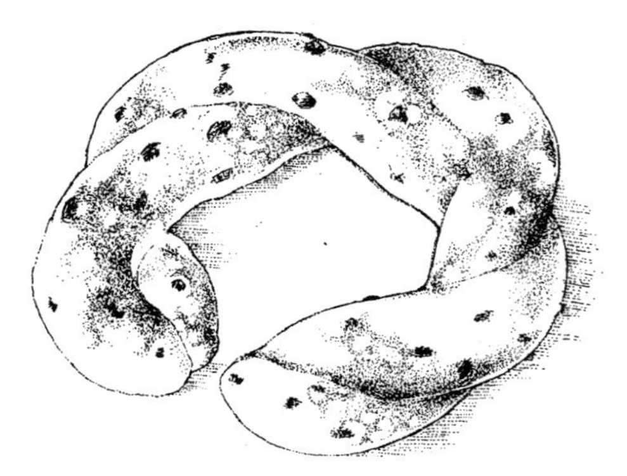

Easter buns

To freeze Open freeze then pack in layers in rigid-based polythene containers with foil dividers. Seal and label.
To serve Remove from container and defrost at room temperature for 1–1½ hours.

IMPERIAL	METRIC	AMERICAN
1 portion enriched bread dough (one-third basic recipe)	1 portion enriched bread dough (one-third basic recipe)	1 portion enriched bread dough (one-third basic recipe)
1 oz. butter, melted	25 g. butter, melted	2 tablespoons butter, melted
2 teaspoons ground allspice	2 teaspoons ground allspice	2 teaspoons ground allspice
2 oz. soft brown sugar	50 g. soft brown sugar	¼ cup brown sugar
2 oz. icing sugar	50 g. icing sugar	½ cup sifted confectioners' sugar
lemon juice	lemon juice	lemon juice

Roll out the dough to an oblong and brush with half the melted butter. Fold dough by bringing lower third over centre third and folding down top third. Give one quarter turn and roll out again. Brush with the remaining melted butter and sprinkle with the spice and soft brown sugar. Roll up like a Swiss roll and cut into 12 equal slices. Place each slice in a greased tartlet tin, cut surface upwards, and with a sharp knife cut a cross halfway through buns. Cover with lightly oiled polythene and leave to rise until double in size. Remove polythene and bake in a hot oven (425°F, 220°C, Gas Mark 7) for 10–12 minutes. Cool on a wire tray. Mix sieved icing sugar with enough lemon juice to give a coating consistency and trickle over the surface of each bun.

Six-in-one biscuit batch

Chain Makes approximately 24 small biscuits in each of six flavours.
To freeze Wrap each roll in freezer foil and label.
To serve Cut 24 thin slices from the frozen roll, place on a lightly greased baking tray and bake in a moderately hot oven (375°F, 190°C, Gas Mark 5) for 10–12 minutes, until light golden brown.
Storage time Uncooked biscuit mixture – 6 months.

IMPERIAL	METRIC	AMERICAN
1 lb. butter	450 g. butter	2 cups butter
8 oz. castor sugar	225 g. castor sugar	1 cup granulated sugar
8 oz. soft brown sugar	225 g. soft brown sugar	1 cup brown sugar
2 eggs, beaten	2 eggs, beaten	2 eggs, beaten
1 teaspoon vanilla essence	1 teaspoon vanilla essence	1 teaspoon vanilla extract
1 lb. self-raising flour	450 g. self-raising flour	4 cups all-purpose flour sifted with 4½ teaspoons baking powder

Cream the butter and sugars until light and fluffy. Gradually beat in the eggs and vanilla essence, then stir in the flour. Divide the mixture into six equal portions and add one of the following flavourings to each one, kneading until evenly combined.

IMPERIAL	METRIC	AMERICAN
2 oz. desiccated coconut	50 g. desiccated coconut	⅔ cup shredded coconut
½ teaspoon ground cinnamon and ½ teaspoon nutmeg	½ teaspoon ground cinnamon and ½ teaspoon nutmeg	½ teaspoon ground cinnamon and ½ teaspoon nutmeg
1 oz. chocolate, grated	25 g. chocolate, grated	scant ⅓ cup grated chocolate
1 oz. glacé cherries, chopped	25 g. glacé cherries, chopped	scant ¼ cup chopped candied cherries
4 butterscotch sweets, crushed	4 butterscotch sweets, crushed	4 butterscotch candies, crushed
2 oz. ground almonds and 1 drop almond essence	50 g. ground almonds and 1 drop almond essence	½ cup ground almonds and 1 drop almond extract

Shape each flavoured mixture into a roll 2 inches (5 cm.) in diameter.

Feather sponge cake chain

A large quantity of sponge mixture can be a problem to handle because it is difficult to fold in the stiffly beaten egg whites evenly. An easy amount to deal with and bake off at one time in the average size oven is based on 4 eggs. Prepare a second batch, possibly with a different basic flavour while the first four sponge layers are baking.

Variations are endless and you will probably be able to dream up a great many different cake chains for yourself. But here is a selection to start with, all produced from a simple basic mixture. There is also another recipe for a sponge on page 209 which could be doubled up and used in the same way but the richer mix using corn oil instead of fat gives a lovely light sponge which almost seems to improve with a stay in the freezer.

To cater for the inspiration of the moment, freeze a number of layers in packs with foil dividers and defrost two at a time to put together with fillings of your own devising.

Basic feather sponge mix

(Illustrated on page 90)

Chain Makes fruit and cream sponge and orange coconut cake, or 4 basic sponge layers.
Storage time Basic cake – 6–8 months.

IMPERIAL	METRIC	AMERICAN
10 oz. plain flour	275 g. plain flour	2½ cups all-purpose flour
2 oz. cornflour	50 g. cornflour	½ cup cornstarch
4 teaspoons baking powder	4 teaspoons baking powder	4 teaspoons baking powder
1 teaspoon salt	1 teaspoon salt	1 teaspoon salt
10 oz. castor sugar	275 g. castor sugar	1¼ cups granulated sugar
4 eggs	4 eggs	4 eggs
¼ pint plus 3 tablespoons corn oil	1½ dl. plus 3 tablespoons corn oil	scant 1 cup corn oil
¼ pint plus 3 tablespoons water	1½ dl. plus 3 tablespoons water	scant 1 cup water

Line the bottom of two 7-inch (18-cm.) sandwich tins with greaseproof paper and grease lightly. Grease and line a deep 7-inch (18-cm.) cake tin. (It is important to prepare the tins first.) Sieve the dry ingredients into a bowl. Separate the yolks from the whites of the eggs. Mix the egg yolks together lightly with a fork; add the corn oil and water. Stir this mixture into the dry ingredients and beat well to form a smooth, slack batter. Whisk the egg whites until stiff, fold lightly into the mixture.

Fruit and cream sponge

(Illustrated on page 90)

To freeze Open freeze and when quite firm pack in suitable polythene container, or in gussetted polythene bag, withdrawing surplus air. Seal and label.
To serve Remove from wrapping while still frozen to avoid damage. Place cake on a serving dish and defrost at room temperature for 2 hours.
Storage time 3–4 months.

IMPERIAL	METRIC	AMERICAN
half quantity basic feather sponge mix	half quantity basic feather sponge mix	half quantity basic feather sponge mix
icing sugar	icing sugar	confectioners' sugar
¼ pint double cream	1½ dl. double cream	⅔ cup whipping cream
4 oz. assorted freezable fresh fruit (seedless grapes, raspberries, cherries, etc.)	100 g. assorted freezable fresh fruit (seedless grapes, raspberries, cherries, etc.)	¼ lb. assorted freezable fresh fruit (seedless grapes, raspberries, cherries, etc.)

Divide the cake mixture between the two prepared sandwich tins and bake in a moderately hot oven (375°F, 190°C, Gas Mark 5) for 25–30 minutes. When cooked the surface of the sponge should be resilient when tested with the fingertip. Remove from tins, allow to cool on a wire tray. Sift the top layer with icing sugar and cut into six portions. Whip the cream, spread or pipe most of it over the bottom layer. Arrange five wedges only, spaced out and propped up at an angle on the filling. Decorate with fruit and finish with rosettes of whipped cream.

Orange coconut cake

To freeze Open freeze and when quite firm pack in suitable polythene container, or in gussetted polythene bag. Seal and label.
To serve Remove from wrapping while still frozen to avoid damage. Place cake on a serving dish and defrost at room temperature for 2 hours.
Storage time 4–6 months.

IMPERIAL	METRIC	AMERICAN
2 tablespoons desiccated coconut	2 tablespoons desiccated coconut	3 tablespoons shredded coconut
finely grated zest of 1 orange	finely grated zest of 1 orange	finely grated zest of 1 orange
half quantity basic feather sponge mix	half quantity basic feather sponge mix	half quantity basic feather sponge mix
orange butter icing	*orange butter icing*	*orange butter icing*
4 oz. butter	100 g. butter	½ cup butter
8 oz. icing sugar, sieved	225 g. icing sugar, sieved	1 cup sifted confectioners' sugar
2 tablespoons orange juice	1½ tablespoons orange juice	3 tablespoons orange juice

Quickly fold the coconut and orange zest into the sponge mixture and turn into the prepared cake tin. Bake in the centre of a moderately hot oven (375°F, 190°C, Gas Mark 5) for 40–45 minutes. Turn out and cool on a wire tray. To make the butter icing, cream butter, beat in icing sugar and orange juice. Split cake in half and sandwich together with some of the orange butter icing, reserving the rest for the top of the cake. If liked fork up the icing into peaks and sprinkle with a little extra coconut.

Basic chocolate feather sponge mix

Chain Makes mocha gâteau and choc-o-nut nibbles, or 4 basic chocolate sponge layers.
Storage time Basic cake – 6–8 months.

IMPERIAL	METRIC	AMERICAN
10 oz. plain flour	275 g. plain flour	2½ cups all-purpose flour
2 oz. cornflour	50 g. cornflour	½ cup cornstarch
1½ oz. cocoa powder	40 g. cocoa powder	generous ¼ cup unsweetened cocoa
4 teaspoons baking powder	4 teaspoons baking powder	4 teaspoons baking powder
1 teaspoon salt	1 teaspoon salt	1 teaspoon salt
10 oz. castor sugar	275 g. castor sugar	1¼ cups granulated sugar
4 eggs	4 eggs	4 eggs
¼ pint plus 3 tablespoons corn oil	1½ dl. plus 3 tablespoons corn oil	scant 1 cup corn oil
¼ pint plus 4 tablespoons water	1½ dl. plus 4 tablespoons water	scant 1 cup water

Line the bottom of two 7-inch (18-cm.) sandwich tins with greaseproof paper and grease lightly. Arrange 12 paper cake cases on a baking tray. Sieve the dry ingredients into a bowl. Separate the yolks from the whites of the eggs. Mix together the egg yolks, corn oil and water lightly with a fork. Stir this mixture into the dry ingredients and beat well to form a smooth, slack batter. Whisk the egg whites until stiff, fold lightly into the mixture.

Mocha gâteau

To freeze Open freeze and when quite firm pack in suitable polythene container, or in gussetted polythene bag. Seal and label.
To serve Remove from wrapping while still frozen to avoid damage. Place cake on a serving dish and defrost at room temperature for 2 hours.
Storage time 4–6 months.

IMPERIAL	METRIC	AMERICAN
half quantity basic chocolate feather sponge mix	half quantity basic chocolate feather sponge mix	half quantity basic chocolate feather sponge mix
coffee butter icing	*coffee butter icing*	*coffee butter icing*
4 oz. butter	100 g. butter	½ cup butter
8 oz. icing sugar, sieved	225 g. icing sugar, sieved	scant 2 cups sifted confectioners' sugar
2 tablespoons liquid coffee essence	1½ tablespoons liquid coffee essence	3 tablespoons strong black coffee
2-oz. block plain chocolate	50-g. block plain chocolate	2 squares semi-sweet chocolate

Divide the cake mixture between the two prepared sandwich tins, and bake in a moderately hot oven (375°F, 190°C, Gas Mark 5) for 25–30 minutes. Turn out and cool on a wire tray. To make the coffee butter icing, cream butter, beat in icing sugar and coffee essence. Spread bottom layer of cake with some of the coffee butter icing, reserving the rest for the top of the cake. Swirl the icing over the top of the cake and sprinkle with chocolate curls, made by scraping a potato peeler along the flat side of the block of chocolate.

Choc-o-nut nibbles

To freeze Pack in layers with foil dividers in a rigid-based polythene container and label.
To serve Allow to defrost at room temperature for about 40 minutes and if liked dredge lightly with icing sugar.
Storage time 6 months.

IMPERIAL	METRIC	AMERICAN
2 oz. chocolate chips	50 g. chocolate chips	⅓ cup chocolate chips
2 oz. walnuts, chopped	50 g. walnuts, chopped	½ cup chopped walnuts
half quantity basic chocolate feather sponge mix	half quantity basic chocolate feather sponge mix	half quantity basic chocolate feather sponge mix

Quickly fold the chocolate chips and nuts into the mixture and divide between the 12 paper cases placed on a baking tray. Bake in the top third of a moderately hot oven (375°F, 190°C, Gas Mark 5) for 15–20 minutes. Cool on a wire tray.

Golden guide number

to pies and fillings

Frozen pies and flans of all kinds can easily be defrosted and served up to taste just as good as those made the same day. There are several different ways to prepare and pack pastry to give you trouble-free, double-quick service from the freezer.

Freezing unbaked pastry

Shortcrust pastry is recommended for flans, tartlets, pasties and deep-dish fruit pies. Flaky pastry is recommended for deep-dish savoury pies, mince pies, jam puffs. Rough puff or puff pastry is recommended for sausage rolls and vol-au-vent.

1. Weigh pastry in 8-oz. (225-g.) and 1-lb. (450-g.) quantities, shape into a neat rectangle and pack closely as a parcel. Some cooks prefer to roll out the pastry thickly, and fold into three before packing. Defrost only enough to make rolling out possible, do not allow fat in pastry to become oily.
2. Roll out pastry to desired thickness, cut a sheet of cardboard in a round or oval shape, using the pie dish or foil pie plate as a pattern. Cut round a saucer for pasties, use biscuit cutters to stamp out tartlets. Use trimmings to make cheese straws for the freezer.

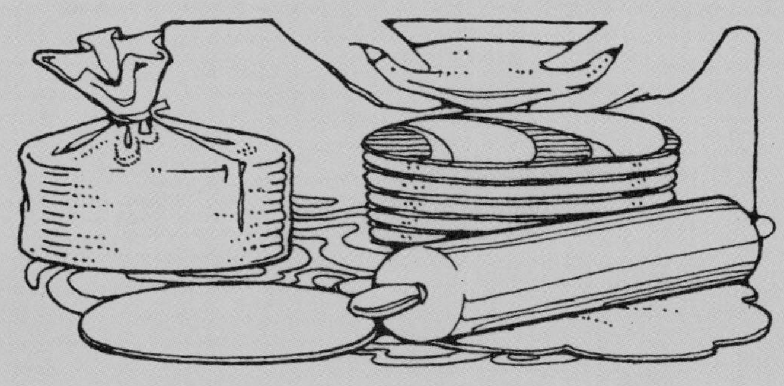

Large oval and round shapes

Put the first layer of pastry flat on the cardboard pattern, top with foil or freezer film dividers also cut from the pattern, alternating with pieces of pastry. Open freeze the stack, then slide it into a polythene bag, press out surplus air, seal with a twist tie. To use one slice, peel off and thaw at room temperature for 10 minutes before shaping. Place it on kitchen paper to absorb any moisture condensing on the underside of the pastry. Fill the pie dish, insert funnel, cover with pastry, making a steam vent for baking. Or use to line a pie plate.

Small round shapes

There is no need to use a cardboard base. Stack and pack with dividers as above. Tartlet cases should be open frozen in the patty tins and when hard, can be packed one inside the other. Alternatively, they can be filled with any cold filling which is sufficiently solid not to soak into the pastry and covered, raw, and packed in bags when frozen hard enough to prevent damage.

Open flans

Take the required number of foil pie plates from the pack, line all with pastry, pre-freeze and stack one inside the other, with dividers. Do not prick the bottoms before freezing, wrap closely, seal and freeze. Defrost in the refrigerator, or at room temperature.

Vol-au-vent and sausage rolls

Stamp out twice the number of circles required, damp half and arrange the others on top, with a smaller circle half-stamped into them to make lids when the cases are baked. Open freeze and pack in rigid-based containers ready for baking. Defrost for 20 minutes before baking in a hot oven (425°F, 220°C, Gas Mark 7). To make sausage rolls, cut strips of pastry 4 inches (10 cm.) wide, arrange raw sausages down the centre, fold over and seal by damping the edges with water. Cut up into short lengths, open freeze and pack in a rigid-based container in layers with dividers. Mark tops when ready for baking, treating as for vol-au-vent cases.

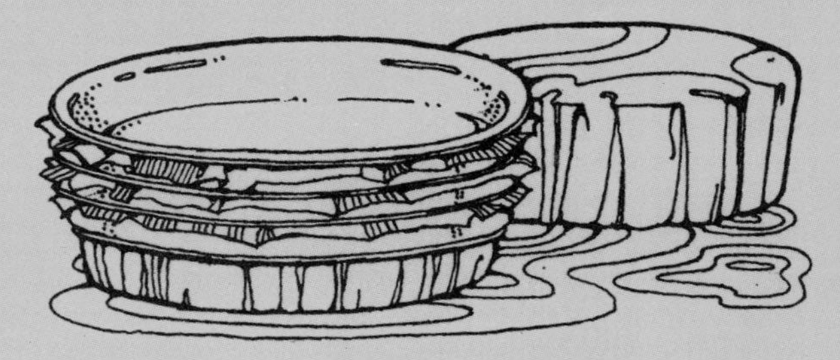

Filling unbaked pies for freezing

Line a number of foil pie plates with pastry and freeze solid, then remove. Line the plates with freezer film or foil, pour in the fillings, and open freeze. When solid, pack the cases and fillings separately, or assembled, with the freezer film or foil still in place. Peel the divider off the frozen filling and place in contact with the pastry, then put into the correct shaped pie plate and bake from the frozen state in a hot oven (425°F, 220°C, Gas Mark 7) for 30 minutes. If the pies are to be covered, rounds of pastry 1 inch (2·5 cm.) smaller than those used to line the pie plates can be cut out and frozen at the same time. This makes assembly of the pie of your choice very easy and quick.

Baking pies for freezing

Although the pre-frozen unbaked pie gives a fresher, crisper crust, many housewives prefer to finish the cooking, and have the choice of merely defrosting over a number of hours, or thawing and reheating in a hot oven. Flan cases, or small tart cases should be baked empty for about 10 minutes in a hot oven (425°F, 220°C, Gas Mark 7). Press down risen centres before the cases cool. Or fully cook, weighting with foil and rice to bake blind; empty and cool. Cooked pie cases need to be stored in rigid-based containers as they are more fragile than unbaked pastry. Empty cases take about 1 hour to defrost. Storage times: unbaked cases 2–3 months; baked cases 4–5 months.

Fully baked pies should be slightly underbaked for freezing, as they brown more if reheated. It is important to cool completely and freeze on a flat surface so the filling remains even. To serve, partially thaw at room temperature for about 30 minutes. Complete defrosting and reheating in a moderate oven (350°F, 180°C, Gas Mark 4) for 30 minutes.

Double-crust pies

Fillings in unbaked pies tend to sink into the pastry and make it soggy. Here are some suggestions for fillings to avoid this happening.

Cherry and rhubarb Use equal quantities of stoned cherries and chopped rhubarb. Make up a mixture of 1 teaspoon cornflour and 1 tablespoon sugar, in a polythene bag, add the fruit and toss until well coated, then turn into the pie case. If doubtful about the moisture content of the fruit, brush the inside of the case with lightly beaten egg white to seal it.

Strawberry and rhubarb Measure equal quantities of hulled strawberries and chopped rhubarb as above. This is an economical filling and it tastes mainly of strawberries.

Apple and cranberry Measure equal quantities of fresh or frozen cranberries and sliced cooking apples. (Steam the slices for 2 minutes, and cool or dip in an ascorbic acid solution to prevent discoloration.)

Apple and ginger Apple slices can make quite a variety of pies, if a little extra flavour is added. For example, a scattering of chopped preserved ginger, with a little ginger syrup spooned *over* the filling. The cornflour and sugar mixture in which the fruit has been turned will prevent it sinking to the bottom. Fresh dates, which can be bought frozen, blend well with apples too.

Creamy fillings Creamy mixtures including beaten egg whites, gelatine, cream and cornflour custards are suitable for freezing up to 3 months. The gelatine content causes them to taste rubbery after long storage. The following are two basic creamy fillings:

Lemon chiffon filling Measure everything in twos. 2 lemons, 2 eggs (separated), 2 tablespoons castor sugar, 2 teaspoons powdered gelatine dissolved in 2 tablespoons hot water. Grate zest thinly from one lemon and squeeze juice from both. Put all the ingredients except the gelatine and egg whites into a double boiler or basin over hot water; whisk until light and fluffy, remove from heat and whisk until cool. Stir in the gelatine mixture, whisk the egg whites stiffly, stir the lemon mixture until on the point of setting and fold in the egg whites. Pour into a baked flan case and allow to set. Chill, or partly freeze and wrap.

Orange velvet filling Measure everything by comparative weights. 2 eggs, their weight in butter, 2 oranges, their weight in sugar. Lightly beat the eggs, and put with the orange juice, grated zest, butter and sugar in the top of a double boiler or in a basin over hot water. Stir until sugar has dissolved, then continue cooking, stirring occasionally until the orange mixture thickens, about 20 minutes. Pour into a baked flan case and allow to set. Chill or partly freeze and wrap.

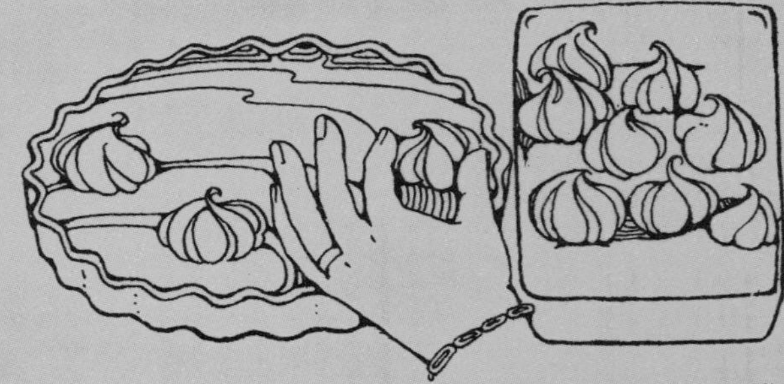

Storage and decoration It is better to top these flans with meringue when fully defrosted and put in a hot oven (425°F, 220°C, Gas Mark 7) for a few minutes to set and colour the points of the meringue. Alternatively transfer frozen whipped cream rosettes to the surface, and top with quartered glacé cherries. Green cherries look best with orange or lemon fillings.

Pies – savoury and sweet

Housewives tell me they hesitate to bake more at a time because they find it so tiring to handle large quantities of mixes. When you see my sweet and savoury pastry specials, you will probably feel it is worth while to have a big bake-up even if it is hard work. But a basic fork pastry mix is not at all tiring to make up in bulk, either for double-crust pies ready to bake, or circles of pastry ready to slip into a flan case or pie plate. It keeps your hands clean too!

Conventional shortcrust pastry, and your favourite flaky, rough puff or puff mixes, will all freeze successfully. I feel bound to add that bought frozen pastry is a boon to those who do not have a light touch with the rolling pin.

Basic fork-mix pastry

(Illustrated on page 54)

Chain Makes 2 savoury pies and several 10-inch (26-cm.) pastry circles or 2 savoury pies, one sweet pie and small batch of tartlets.
Storage time Unbaked pastry – 6 months.

IMPERIAL	METRIC	AMERICAN
1½ lb. plain flour	700 g. plain flour	6 cups all-purpose flour
1½ teaspoons salt	1½ teaspoons salt	1½ teaspoons salt
12 oz. white vegetable cooking fat	350 g. white vegetable cooking fat	1½ cups shortening
6 tablespoons cold water	6 tablespoons cold water	½ cup cold water

Put all the ingredients together in a bowl. Using a large fork, break down the fat into small pieces. Mix and stir with the fork until a firm ball of dough is formed. Put onto a well floured board and knead lightly.

Veal and pimento pie

(Illustrated on page 54)

To freeze Cover exposed pastry surfaces with foil and label.
To serve Remove wrapping, if liked cut steam vent and place, still frozen in a hot oven (425°F, 220°C, Gas Mark 7) for 30 minutes.
Storage time 3 months.

IMPERIAL	METRIC	AMERICAN
1 chicken stock cube	1 chicken stock cube	1 chicken boullion cube
½ pint boiling water	3 dl. boiling water	1¼ cups boiling water
12 oz. pie veal, diced	350 g. pie veal, diced	¾ lb. boneless veal, diced
2 oz. butter	50 g. butter	¼ cup butter
2 oz. flour	50 g. flour	½ cup all-purpose flour
½ pint milk	3 dl. milk	1¼ cups milk
1 5-oz. can red pimentos	1 150-g. can red pimentos	1 5-oz. can red pimientos
1¼ lb. fork-mix pastry	600 g. fork-mix pastry	1¼ lb. fork-mix pastry

Dissolve the stock cube in the boiling water and simmer the veal in this for about 30 minutes, or until tender. Melt the butter, stir in the flour and add the milk and sufficient of the stock from the veal to make a thick pouring sauce. Cook for 2 minutes then stir in the drained and chopped pimento and the veal. Cool. Roll out two-thirds of the pastry and use to line two 8-inch (20-cm.) shaped foil plates. Divide the filling between the two pastry cases and use remaining pastry to form two lids. Dampen edges and crimp together but do not cut steam vents.

Note For ready-to-use pastry circles, see **Golden Guide Number 10.**

Apricot and apple pie

To freeze Cover exposed pastry surfaces with foil and label.
To serve Remove wrapping, cut steam vent and place, still frozen, in a hot oven (425°F, 220°C, Gas Mark 7) for 30 minutes.
Storage time 4–6 months.
Note For alternative double-crust pie fillings, see **Golden Guide Number 10.**

IMPERIAL	METRIC	AMERICAN
1¼ lb. fork-mix pastry	600 g. fork-mix pastry	1¼ lb. fork-mix pastry
2 14-oz. cans apricot pie filling	2 400-g. cans apricot pie filling	2 14-oz. cans apricot pie filling
8 oz. frozen apple slices	225 g. frozen apple slices	½ lb. frozen apple slices
2 oz. demerara sugar	50 g. demerara sugar	¼ cup brown sugar
½ teaspoon cornflour	½ teaspoon cornflour	½ teaspoon cornstarch

Roll out two-thirds of the pastry and use to line two 8-inch (20-cm.) shaped foil plates. Divide the pie filling between the pastry cases. Cover with the frozen apple slices and then sprinkle the sugar and cornflour over the top. Roll out remaining pastry to make two lids, dampen edges and crimp together firmly but do not cut steam vents.

Golden guide number 

to sauces

One of the many advantages a chef has over the housewife in cooking is that he usually has at his disposal a basic white sauce (béchamel) and a basic brown sauce (Espagnole) from which he develops more elaborate ones, or transforms plain cooked meat and fish into exquisite dishes. With your freezer you can do even better, having pre-frozen quite a few varieties from a large quantity of the simple white and brown classic sauces. With a rich tomato sauce to make a well-rounded trio, a vast choice of dishes can be yours without more trouble than opening the freezer, taking out and carefully reheating the sauce of your choice.

Basic béchamel sauce chain

Chain Makes 4 pints ($2\frac{1}{4}$ litres) basic sauce.

To freeze Pour into $\frac{1}{2}$-pint (3-dl.) polythene containers leaving $\frac{1}{2}$-inch (1-cm.) headspace. Cover surface with a circle of freezer film or greaseproof paper. Seal and label.

To serve Partially defrost, place in a double boiler and stir frequently while re-heating.

Storage time 4–6 months.

Imperial	Metric	American
4 pints milk	$2\frac{1}{4}$ litres milk	10 cups milk
4 onions, sliced	4 onions, sliced	4 onions, sliced
8 bay leaves	8 bay leaves	8 bay leaves
24 peppercorns	24 peppercorns	24 peppercorns
4 oz. butter	100 g. butter	$\frac{1}{2}$ cup butter
2 oz. cornflour	50 g. cornflour	$\frac{1}{2}$ cup cornstarch
salt and pepper to taste	salt and pepper to taste	salt and pepper to taste

Place the milk in a saucepan and add the sliced onions, bay leaves and peppercorns. Bring to the boil and simmer for 15 minutes. In another saucepan melt the butter and stir in the cornflour. Cook for 1–2 minutes without browning. Gradually add the strained milk, stirring continuously. Bring to the boil and simmer for 2 minutes. Correct seasoning and cool.

Variations

The quantities given are sufficient to flavour $\frac{1}{2}$ pint (3 dl.) basic béchamel sauce. Beat in until well combined.

Imperial	Metric	American
Cheese sauce	**Cheese sauce**	**Cheese sauce**
2 oz. cheese, grated	50 g. cheese, grated	$\frac{1}{2}$ cup grated cheese
$\frac{1}{2}$ teaspoon made mustard	$\frac{1}{2}$ teaspoon made mustard	$\frac{1}{2}$ teaspoon made mustard
Sauce aurore	**Sauce aurore**	**Sauce aurore**
2 tablespoons tomato purée	2 tablespoons tomato purée	3 tablespoons tomato paste
pinch castor sugar	pinch castor sugar	pinch granulated sugar
Parsley sauce	**Parsley sauce**	**Parsley sauce**
2 tablespoons finely chopped parsley	2 tablespoons finely chopped parsley	3 tablespoons finely chopped parsley
pinch nutmeg	pinch nutmeg	pinch nutmeg
Shrimp or prawn sauce	**Shrimp or prawn sauce**	**Shrimp or prawn sauce**
2 oz. peeled shrimps or prawns	50 g. peeled shrimps or prawns	$\frac{1}{3}$ cup peeled shrimp or prawns
dash lemon juice	dash lemon juice	dash lemon juice
Sauce soubise	**Sauce soubise**	**Sauce soubise**
1 large onion, chopped, sautéed in butter until soft, then puréed	1 large onion, chopped, sautéed in butter until soft, then puréed	1 large onion, chopped, sautéed in butter until soft, then puréed
Caper sauce	**Caper sauce**	**Caper sauce**
1 tablespoon bottled capers, chopped	1 tablespoon bottled capers, chopped	1 tablespoon bottled capers, chopped
1 tablespoon lemon juice or vinegar	1 tablespoon lemon juice or vinegar	1 tablespoon lemon juice or vinegar
Mushroom sauce	**Mushroom sauce**	**Mushroom sauce**
3 oz. button mushrooms, sautéed in butter	75 g. button mushrooms, sautéed in butter	scant 1 cup button mushrooms, sautéed in butter

Imperial	Metric	American
Anchovy sauce	**Anchovy sauce**	**Anchovy sauce**
1–2 teaspoons anchovy essence	1–2 teaspoons anchovy essence	1–2 teaspoons anchovy extract
1 teaspoon lemon juice	1 teaspoon lemon juice	1 teaspoon lemon juice
few drops red food colouring	few drops red food colouring	few drops red food coloring
Sauce suprême	**Sauce suprême**	**Sauce suprême**
1 egg yolk	1 egg yolk	1 egg yolk
2 tablespoons single or double cream	2 tablespoons single or double cream	3 tablespoons coffee or whipping cream

Basic Espagnole sauce chain

Chain Makes 2 pints (generous 1 litre) basic sauce.

To freeze Pour into $\frac{1}{2}$-pint (3-dl.) polythene containers leaving $\frac{1}{2}$-inch (1-cm.) headspace. Cover surface with a circle of freezer film or greaseproof paper. Seal and label.

To serve Partially defrost, place in a double boiler and stir frequently while reheating.

Storage time 4–6 months.

Imperial	Metric	American
2 oz. dripping or lard	50 g. dripping or lard	$\frac{1}{4}$ cup drippings or lard
1 onion, sliced	1 onion, sliced	1 onion, sliced
1 carrot, sliced	1 carrot, sliced	1 carrot, sliced
2 sticks celery, sliced	2 sticks celery, sliced	2 stalks celery, sliced
2 rashers streaky bacon, chopped	2 rashers streaky bacon, chopped	2 bacon slices, chopped
2 pints brown stock	generous 1 litre brown stock	5 cups brown stock
1 tablespoon tomato purée	1 tablespoon tomato purée	1 tablespoon tomato paste
2 oz. cornflour	50 g. cornflour	$\frac{1}{2}$ cup cornstarch

Melt the dripping in a large pan and sauté the vegetables and bacon, drain well. Heat the stock to boiling point, add the vegetables and bacon and stir in the tomato purée. Moisten the cornflour with cold water, add a little stock and mix until smooth. Pour cornflour mixture into sauce and stir briskly until smooth and thick. Simmer for 30 minutes. Strain and cool.

The following quantities should be combined with $\frac{1}{2}$ pint (3 dl.) basic sauce.

Imperial	Metric	American
Reform sauce	**Reform sauce**	**Reform sauce**
4 tablespoons vinegar	4 tablespoons vinegar	$\frac{1}{3}$ cup vinegar
5 peppercorns	5 peppercorns	5 peppercorns
2 tablespoons port	2 tablespoons port	3 tablespoons port
1 tablespoon redcurrant jelly	1 tablespoon redcurrant jelly	1 tablespoon red currant jelly

Boil vinegar and peppercorns until liquid is reduced by half then strain. Add to the sauce with the port and redcurrant jelly.

Imperial	Metric	American
Sauce bigarade	**Sauce bigarade**	**Sauce bigarade**
3 tablespoons lemon juice	3 tablespoons lemon juice	scant $\frac{1}{4}$ cup lemon juice
4 tablespoons orange juice	4 tablespoons orange juice	$\frac{1}{3}$ cup orange juice
2 tablespoons port	2 tablespoons port	3 tablespoons port
pinch sugar	pinch sugar	pinch sugar

Add juices, port and sugar to the sauce, stir briskly.

Imperial	Metric	American
Sauce piquant	**Sauce piquant**	**Sauce piquant**
1 small onion, chopped	1 small onion, chopped	1 small onion, chopped
1 oz. butter	25 g. butter	2 tablespoons butter
$\frac{1}{4}$ pint wine vinegar	$1\frac{1}{2}$ dl. wine vinegar	$\frac{2}{3}$ cup wine vinegar
3 small gherkins, chopped	3 small gherkins, chopped	3 small sweet dill pickles, chopped

Sauté the onion in the butter until soft. Add vinegar and boil until reduced by half. Add to the sauce with the chopped gherkins.

Imperial	Metric	American
Sauce Robert	**Sauce Robert**	**Sauce Robert**
1 small onion, chopped	1 small onion, chopped	1 small onion, chopped
1 oz. butter	25 g. butter	2 tablespoons butter
$\frac{1}{4}$ pint dry white wine	$1\frac{1}{2}$ dl. dry white wine	$\frac{2}{3}$ cup dry white wine
pinch sugar	pinch sugar	pinch sugar

Sauté the onion in the butter until soft. Add wine and boil until reduced by half. Add to the sauce with the sugar.

Chicken chain cooking

When chicken is offered at an especially tempting price at your local supermarket, you may have room to spare for eight roasters in your freezer. But this could pose a problem if the family objects to a monotonous diet of roast chicken, so other ideas for serving chicken in appetising dishes are needed. Also, you will want to take advantage of the valuable bonus this purchase gives you, i.e. by-products in the form of giblets and carcases, all of which can be put to good account.

Here is an interesting exercise in chain cooking based on a purchase of eight 3-lb. ($1\frac{1}{2}$-kg.) chickens, which produces eight generous meals for a family of four, plus a delicious chicken liver pâté, and sufficient rich chicken stock to enhance many dishes to come. All the preparation and necessary cooking ahead can be done during one session in the kitchen.

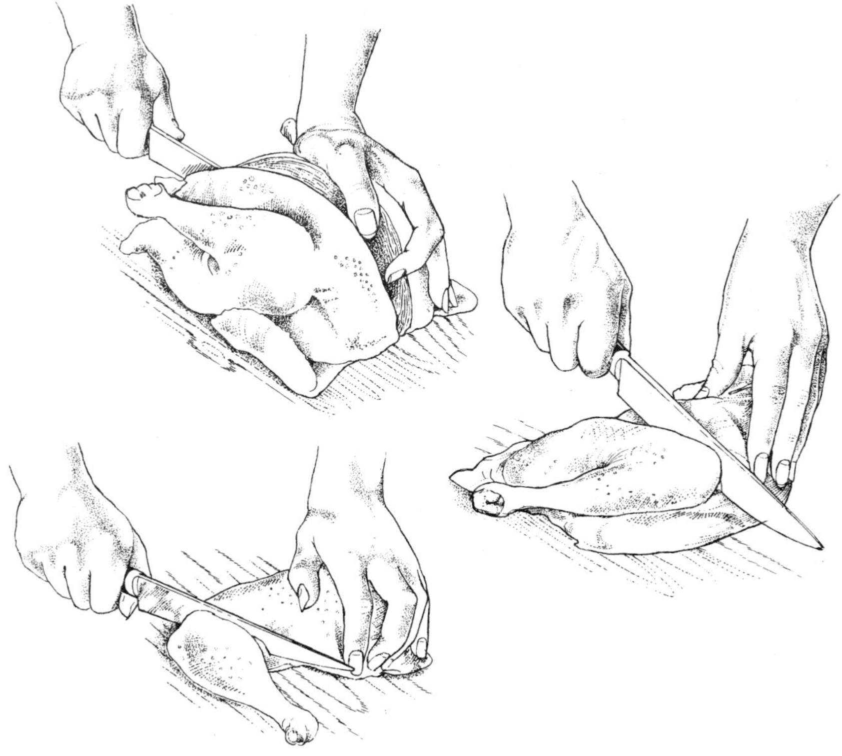

Eight-chicken chain

Chain From eight 3-lb. ($1\frac{1}{2}$-kg.) chickens – makes 4 portions chicken liver pâté, 8 portions fricassée, 8 fried chicken portions, 4 portions chicken Maryland, 8 portions chicken curry, 1 roasted chicken to serve immediately, 5 pints (3 litres) chicken stock.

Order of preparation Begin by dividing seven chickens into portions and then making the chicken stock. While this is cooking prepare the fried chicken joints. Then poach the chicken joints required for the fricassée in stock, which will enrich it. Finish making the fricassée and complete the preparation of the stock.

Chicken liver pâté

To freeze Pack into small rigid-based polythene serving containers, smooth the tops. Seal and label.

To serve Allow to defrost at room temperature for 2–3 hours and garnish with thin lemon slices. Serve with plain biscuits or fingers of toast, and raw vegetable snacks (such as spring onion curls, carrot sticks and sliced celery stalks).

Storage time 6 weeks.

IMPERIAL	METRIC	AMERICAN
3 oz. butter	75 g. butter	6 tablespoons butter
8 chicken livers	8 chicken livers	8 chicken livers
1 tablespoon corn oil	1 tablespoon corn oil	1 tablespoon corn oil
1 large onion, chopped	1 large onion, chopped	1 large onion, chopped
1 clove garlic, crushed	1 clove garlic, crushed	1 clove garlic, crushed
salt and pepper to taste	salt and pepper to taste	salt and pepper to taste
1 teaspoon chopped parsley	1 teaspoon chopped parsley	1 teaspoon chopped parsley
1 teaspoon thyme or mixed sweet herbs	1 teaspoon thyme or mixed sweet herbs	1 teaspoon thyme or mixed sweet herbs
2 tablespoons dry sherry	2 tablespoons dry sherry	3 tablespoons dry sherry

Melt half the butter and sauté the chicken livers for 4 minutes then remove from the pan. Add the oil and fry the onion and garlic until soft. Add the seasonings, remaining butter, softened, and herbs and mix well. Place the contents of the pan and the chicken livers in an electric blender and liquidise until smooth or pass through a sieve. When completely cold, fold in the sherry.

Chicken stock

To freeze Pour into suitable polythene containers, leaving 1-inch (2·5 cm.) headspace. Chill and skim off fat. Seal and label.
To serve Heat gently in a saucepan while still frozen or allow to defrost at room temperature for 4–6 hours, depending on quantity. Use in soups, sauces and gravies.
Storage time 3 months.

IMPERIAL	METRIC	AMERICAN
bones and remaining giblets from 8 chickens	bones and remaining giblets from 8 chickens	bones and remaining giblets from 8 chickens
5 pints water	3 litres water	12½ cups water
3 large carrots, sliced	3 large carrots, sliced	3 large carrots, sliced
3 large onions, quartered	3 large onions, quartered	3 large onions, quartered
3 bay leaves	3 bay leaves	3 bay leaves
bouquet garni	bouquet garni	bouquet garni
3 teaspoons salt	3 teaspoons salt	3 teaspoons salt
1 teaspoon pepper	1 teaspoon pepper	1 teaspoon pepper

Wash bones and giblets. Place in a large saucepan with the remaining ingredients. Bring slowly to the boil and skim. Cover and simmer gently for 3 hours, skimming from time to time and adding extra water as liquid evaporates. Strain through a fine sieve and add extra seasoning to taste.

Chicken fricassée

To freeze Pour into two 2-pint (generous 1-litre) polythene containers. Seal and label.
To serve Place the container under running water until the contents can be turned out into a saucepan. Heat carefully and add 20 canned drained button onions, 4 canned red pimentos cut into wedges and a 16-oz. (450-g.) can of drained and quartered asparagus tips. When very hot turn into a casserole dish and garnish with fried bread croûtons.
Storage time 4–6 months.

IMPERIAL	METRIC	AMERICAN
8 chicken portions, poached in chicken stock until tender	8 chicken portions, poached in chicken stock until tender	8 chicken portions, poached in chicken stock until tender
2 oz. butter	50 g. butter	¼ cup butter
2 oz. flour	50 g. flour	½ cup all-purpose flour
1 pint milk	generous ½ litre milk	2½ cups milk
salt and pepper to taste	salt and pepper to taste	salt and pepper to taste

Allow the chicken to cool a little then remove the flesh from the bones and cut up into small pieces. Place the butter, flour and milk in a saucepan and whisk over medium heat until sauce boils and is smooth and thick. Add the chicken and seasoning and cool thoroughly.

Fried chicken portions

To freeze Wrap each portion in freezer foil then pack together in a large rigid-based polythene container. This enables you to remove only the number of portions required.
To serve Defrost and deep fry in heated oil until golden brown, about 12 minutes. (Or shallow fry for 8 minutes on each side.) Drain well and arrange on a heated serving dish and garnish with parsley and lemon slices.
Storage time 4–6 months.

IMPERIAL	METRIC	AMERICAN
12 chicken portions	12 chicken portions	12 chicken portions
seasoned flour	seasoned flour	seasoned flour
3 eggs	3 eggs	3 eggs
6 tablespoons water	6 tablespoons water	½ cup water
toasted breadcrumbs	toasted breadcrumbs	dry bread crumbs

Coat the chicken portions in well-seasoned flour. Mix together the beaten eggs and the water. Dip the chicken portions in this and coat with toasted breadcrumbs.

Sweet chicken curry

To freeze Pour into two 2-pint (generous 1-litre) polythene containers. Seal and label.
To serve Defrost at room temperature for 6 hours and reheat gently in a saucepan to boiling point and simmer for 5 minutes. Serve with fluffy boiled rice and accompaniments – see recipes page 202
Storage time 4–6 months.

IMPERIAL	METRIC	AMERICAN
3 oz. lard	75 g. lard	6 tablespoons lard
8 chicken portions	8 chicken portions	8 chicken portions
2 large onions, chopped	2 large onions, chopped	2 large onions, chopped
2 oz. flour	50 g. flour	$\frac{1}{2}$ cup all-purpose flour
2 tablespoons curry powder	2 tablespoons curry powder	3 tablespoons curry powder
1 teaspoon curry paste	1 teaspoon curry paste	1 teaspoon curry paste
1 pint chicken stock	generous $\frac{1}{2}$ litre stock	$2\frac{1}{2}$ cups chicken stock
2 pineapple rings, chopped	2 pineapple rings, chopped	2 pineapple rings, chopped
2 tablespoons clear honey	2 tablespoons clear honey	3 tablespoons clear honey
2 tablespoons mango chutney	2 tablespoons mango chutney	3 tablespoons mango chutney
salt and pepper to taste	salt and pepper to taste	salt and pepper to taste
2 dessert apples	2 dessert apples	2 dessert apples
2 oz. sultanas	50 g. sultanas	$\frac{1}{3}$ cup seedless white raisins

Melt two-thirds of the lard in a large frying pan and cook the chicken portions on both sides until tender. Cool slightly then remove flesh from the bones. Add the remaining lard to the pan and in it fry the chopped onion until softened. Stir in the flour, curry powder and curry paste and cook over gentle heat for 2 minutes. Gradually blend in the stock and bring to boiling point, stirring constantly until sauce thickens. Stir in the pineapple, honey and chutney and season to taste. Replace the chicken in the sauce, bring to the boil, cover and simmer gently for about 30 minutes. Peel, core and chop the apples and add to the pan with the sultanas. Remove from the heat and cool.

Chicken Maryland

IMPERIAL	METRIC	AMERICAN
4 coated chicken portions	4 coated chicken portions	4 coated chicken portions
4 rashers streaky bacon	4 rashers streaky bacon	4 bacon slices
2 bananas	2 bananas	2 bananas
sweetcorn fritters	*sweetcorn fritters*	*corn fritters*
2 oz. self-raising flour	50 g. self-raising flour	$\frac{1}{2}$ cup all-purpose flour sifted with $\frac{1}{2}$ teaspoon baking powder
pinch salt	pinch salt	pinch salt
pinch cayenne pepper	pinch cayenne pepper	pinch cayenne pepper
1 egg	1 egg	1 egg
6 oz. drained sweetcorn	175 g. drained sweetcorn	about 1 cup kernel corn
little milk	little milk	little milk

Deep fry the chicken portions in heated oil until golden brown, about 12 minutes. Drain and keep hot. Mix together the ingredients for the fritters with enough milk to make a thick batter consistency. De-rind the bacon and cut each rasher in half. Roll the half slices, thread on a skewer and fry gently to release the fat. Remove and keep hot. Drop spoonfuls of the fritter batter into the hot bacon fat and cook until brown, turning to cook on other side. Drain on absorbent paper. Cut the bananas in half lengthwise and fry gently in the remaining bacon fat. Serve each portion of chicken with two bacon rolls, half a banana and two sweetcorn fritters.

Note This recipe uses frozen chicken portions prepared for frying.

Golden guide number

to chain cooking

Minced beef is a good example of how a cheap and simple ingredient can be prepared in bulk to give a variety of dishes. The price of beef makes all but the cheaper cuts a luxury. So minced beef is bound to appear more and more often on the menu, and in these many disguises very good it is too; so are similar dishes made with minced pork.

Minced beef chain

Chain Makes 3 portions meat and onion loaf, 3 portions Mexican beef and 3 portions sweet and sour meatballs.
Storage time Basic meat mixture – 4–6 months.

Note I usually combine the mixture with the breadcrumbs by hand in a large plastic bowl.

Imperial	Metric	American
$\frac{1}{4}$ pint oil	$1\frac{1}{2}$ dl. oil	$\frac{2}{3}$ cup oil
3 lb. onions, chopped	$1\frac{1}{2}$ kg. onions, chopped	3 lb. onions, chopped
6 lb. finely minced beef	3 kg. finely minced beef	6 lb. finely ground beef
$1\frac{1}{2}$ lb. soft white breadcrumbs	700 g. soft white breadcrumbs	12 cups fresh soft bread crumbs

Heat the oil in a large saucepan and sauté the onions until softened. Add the meat and cook, stirring, until thoroughly browned. Remove from the heat and stir in the breadcrumbs.

Meat and onion loaf

To freeze Seal and label.
To serve Remove cover. To defrost and cook, place in a moderately hot oven (375°F, 190°C, Gas Mark 5) for 1 hour.
Storage time 3 months.

Imperial	Metric	American
$4\frac{1}{2}$ lb. basic minced beef mixture	$2\frac{1}{4}$ kg. basic minced beef mixture	$4\frac{1}{2}$ lb. basic ground beef mixture
2 eggs	2 eggs	2 eggs
2 tablespoons chopped parsley	2 tablespoons chopped parsley	3 tablespoons chopped parsley
2 tablespoons horseradish sauce	2 tablespoons horseradish sauce	3 tablespoons horseradish sauce
1 packet French onion soup mix	1 packet French onion soup mix	1 package French onion soup mix

Combine all the ingredients and mix well. Divide the mixture between three 1-lb. ($\frac{1}{2}$-kg.) shaped foil oblong containers, pressing down well.

Mexican beef

To freeze Divide the mixture between three shaped foil dishes. Seal and label.

To serve Place container in a moderately hot oven (375°F, 190°C, Gas Mark 5) for 40 minutes, and add a little extra tomato juice or water to correct the consistency.

Storage time 1 month.

Note Garlic as an ingredient decreases length of storage. If garlic is omitted store for 3 months.

Imperial	Metric	American
3 lb. basic minced beef mixture	$1\frac{1}{2}$ kg. basic minced beef mixture	3 lb. basic ground beef mixture
2 16-oz. cans red kidney beans	2 450-g. cans red kidney beans	2 16-oz. cans red kidney beans
3 16-oz. cans tomatoes	3 450-g. cans tomatoes	3 16-oz. cans tomatoes
2 cloves garlic, crushed	2 cloves garlic, crushed	2 cloves garlic, crushed
1 tablespoon salt	1 tablespoon salt	1 tablespoon salt
2 teaspoons chilli powder	2 teaspoons chilli powder	2 teaspoons chili powder
$\frac{1}{2}$ tablespoon paprika pepper	$\frac{1}{2}$ tablespoon paprika pepper	$\frac{1}{2}$ tablespoon paprika pepper
3 teaspoons ground cummin seeds	3 teaspoons ground cummin seeds	3 teaspoons ground cummin seeds
$\frac{1}{2}$ teaspoon black pepper	$\frac{1}{2}$ teaspoon black pepper	$\frac{1}{2}$ teaspoon black pepper

Combine all the ingredients in a large saucepan. Bring to the boil and simmer gently for 20 minutes. Cool.

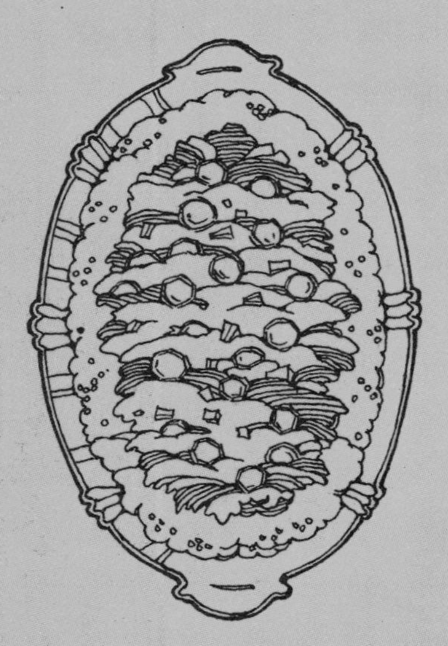

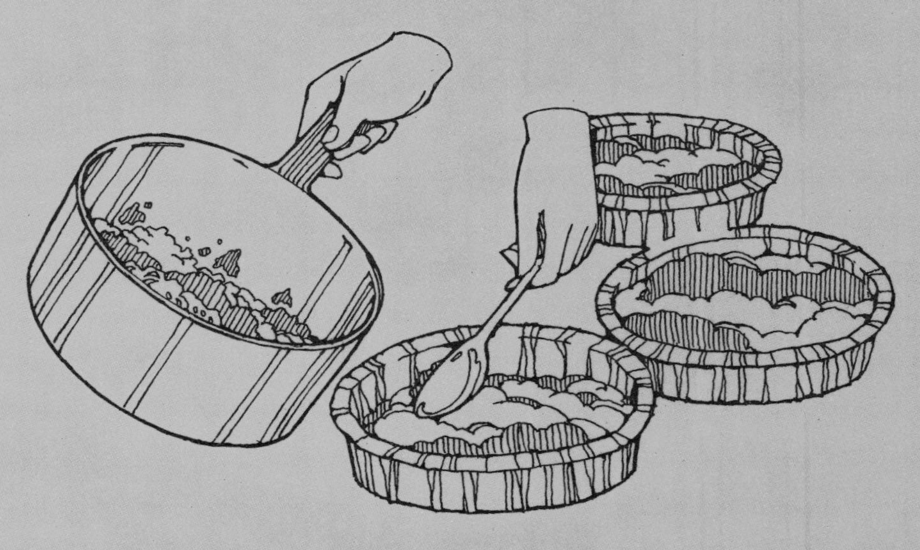

Sweet and sour meatballs

To freeze Pack meatballs and sauce separately into three rigid-based polythene containers and three pre-formed polythene bags. Seal and label.

To serve Fry meatballs from the frozen state in hot oil for 10 minutes. Drain. Reheat sauce gently in a saucepan. Serve meatballs on a bed of fluffy boiled rice and pour the sauce over.

Storage time 3 months.

Imperial	Metric	American
3 lb. basic minced beef mixture, cold	$1\frac{1}{2}$ kg. basic minced beef mixture, cold	3 lb. basic ground beef mixture, cold
4 tablespoons tomato purée	4 tablespoons tomato purée	$\frac{1}{3}$ cup tomato paste
2 tablespoons Worcestershire sauce	2 tablespoons Worcestershire sauce	3 tablespoons Worcestershire sauce
salt and pepper to taste	salt and pepper to taste	salt and pepper to taste
sauce	*sauce*	*sauce*
2 16-oz. cans pineapple pieces	2 450-g. cans pineapple pieces	2 16-oz. cans pineapple pieces
4-inch length cucumber	10-cm. length cucumber	4-inch length cucumber
4 carrots	4 carrots	4 carrots
4 tablespoons vinegar	4 tablespoons vinegar	$\frac{1}{3}$ cup vinegar
4 tablespoons Worcestershire sauce	4 tablespoons Worcestershire sauce	$\frac{1}{3}$ cup Worcestershire sauce
salt and pepper to taste	salt and pepper to taste	salt and pepper to taste
3 tablespoons cornflour	3 tablespoons cornflour	scant $\frac{1}{4}$ cup cornstarch
4 tablespoons water	4 tablespoons water	$\frac{1}{3}$ cup water

Combine all the ingredients for the meatballs and mix well. Form into small balls with floured hands, place on a baking tray and chill for 30 minutes to harden. Drain the pineapple and chop it. Cut the cucumber into small dice and place in a pan with the pineapple and grated carrot. Add the vinegar, the pineapple syrup made up to 2 pints (generous 1 litre) with water and the Worcestershire sauce. Bring to the boil and simmer gently for about 5 minutes then season to taste. Mix together the cornflour and water and blend into the sauce. Bring to the boil again, stirring constantly, until smooth and thick. Cool.

Poacher's roll, apricot-baked bacon (see pages 154, 195).

1 Preparing the best end of neck of lamb for noisettes – removing the bone, in one piece from the joint, removing as little of the meat as possible.

2 The joint tightly rolled up and secured with wooden cocktail sticks. The noisettes are made by cutting the meat, at 1-inch (2·5-cm.) intervals, between the cocktail sticks, so that each noisette is held in position with a cocktail stick.

3 Packing the noisettes, with foil dividers in between, for freezing.

1

2

3

Noisettes bordelaise (see page 231).

Opposite Piquant avocado dip, lemon avocado soup (see pages 181, 190).

1 Frozen spinach being turned into a saucepan to heat and thaw to make the roulade while the frozen haddock fillets defrost sufficiently to separate.

2 Egg yolks separated and ready to beat into the cooled spinach mixture while the mushrooms are sliced ready to make the filling.

3 Spinach soufflé mixture being turned into a Swiss roll tin lined with greaseproof paper, brushed with oil and mitred at the corners.

1

2

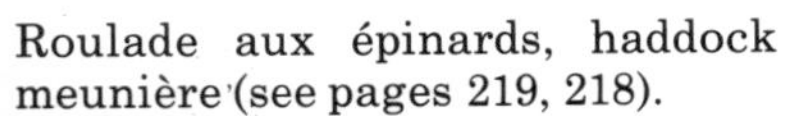

Roulade aux épinards, haddock meunière (see pages 219, 218).

3

Versatile meat mix

Chain Makes approximately 4 lb. (2 kg.) savoury minced beef mixture.
Storage time 3 months.

IMPERIAL	METRIC	AMERICAN
2 lb. minced beef	1 kg. minced beef	2 lb. ground beef
8 oz. pork sausagemeat	225 g. pork sausagemeat	1 cup sausagemeat
8 oz. soft white breadcrumbs	225 g. soft white breadcrumbs	4 cups fresh soft bread crumbs
4 oz. mushrooms, sliced	100 g. mushrooms, sliced	1 cup sliced mushrooms
bunch fresh parsley, chopped	bunch fresh parsley, chopped	bunch fresh parsley, chopped
1 teaspoon salt	1 teaspoon salt	1 teaspoon salt
½ teaspoon black pepper	½ teaspoon black pepper	½ teaspoon black pepper
1 tablespoon soy sauce	1 tablespoon soy sauce	1 tablespoon soy sauce
2 eggs, beaten	2 eggs, beaten	2 eggs, beaten
1 10-oz. can condensed cream of mushroom soup	1 275-g. can condensed cream of mushroom soup	1 10-oz. can condensed cream of mushroom soup

Place all the ingredients in a large mixing bowl and work firmly until evenly combined.

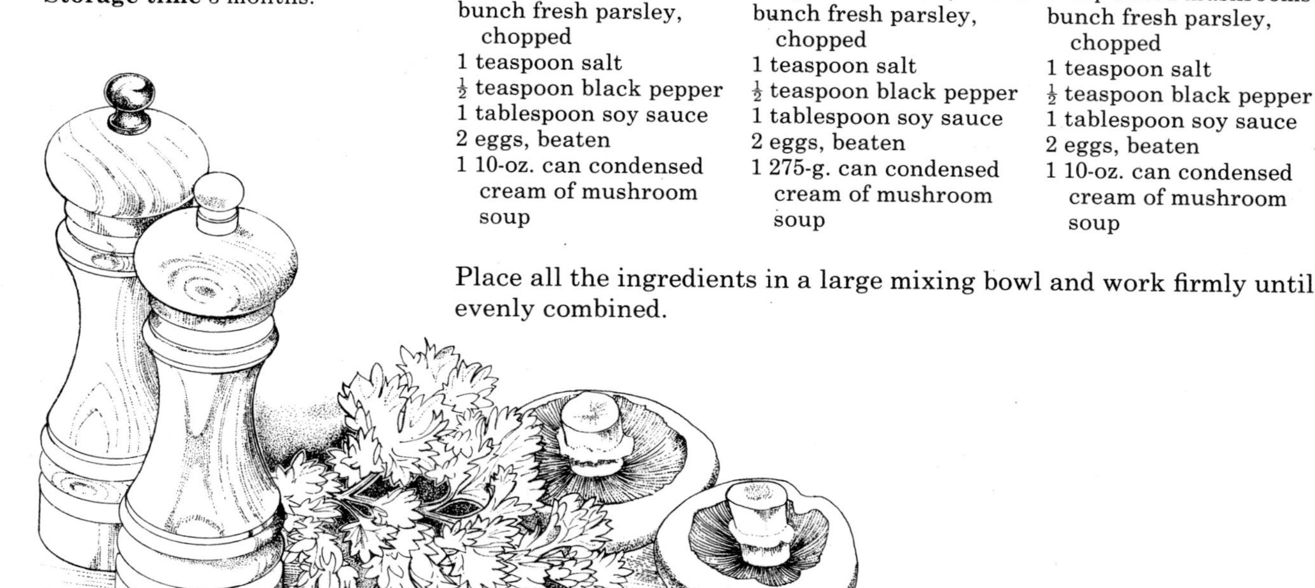

Savoury meat loaf

To freeze Seal and label.
To serve Uncover. To defrost and cook, place in a moderately hot oven (375°F, 190°C, Gas Mark 5) for 1 hour. Serve with Espagnole sauce (see page 101).
Storage time 3 months.

IMPERIAL	METRIC	AMERICAN
1½ lb. basic savoury minced beef mixture	700 g. basic savoury minced beef mixture	1½ lb. basic savoury minced beef mixture

Pack mixture into a 1-lb. (½-kg.) shaped foil oblong container.

Crispy meatballs

To freeze Pack in rigid-based polythene containers. Seal and label.
To serve Deep fry from the frozen state in hot oil for approximately 10 minutes, until crisp and brown. Drain and serve with favourite tomato sauce (see page 250).
Storage time 3 months.

IMPERIAL	METRIC	AMERICAN
1½ lb. basic savoury minced beef mixture	700 g. basic savoury minced beef mixture	1½ lb. basic savoury minced beef mixture
1 egg, beaten	1 egg, beaten	1 egg, beaten
breadcrumbs for coating	breadcrumbs for coating	bread crumbs for coating

With floured hands, form the mixture into small balls. Coat in egg and breadcrumbs and place on a baking tray. Chill for 30 minutes to harden. To make a change, use packet savoury stuffing instead of breadcrumbs.

Beefy stuffing for vegetables and pies

Pack the remaining basic mixture into rigid-based polythene containers or polythene bags, in reasonably small quantities. When required for use, partially defrost and use to stuff courgettes, aubergines, cabbage leaves, marrow, onions or use as a ready made pie or pasty filling.

Lamb chain using one whole lamb

Recipes marked * are given below.

Cuts of lamb	Recipes
2 best ends of neck	Guard of honour
1 boned shoulder	Stuffed shoulder of lamb with reform sauce*
1 boned and diced shoulder	2 minted lamb pies*
1 loin – chopped	Easy parcel chops*
1 loin – boned	Roast loin of lamb with fruity sauce*
Chump end – chopped	Chops for grilling or frying
1 leg	Roast with garlic and rosemary
½ leg (fillet), diced	Kebabs in marinade (see page 167)
½ leg (shank)	Small roast
Breasts, scrag end and middle neck	Basic lamb stew – Lamb hotpot,* Peasant's pottage,* Mediterranean lamb stew*
Bones	Scotch broth*

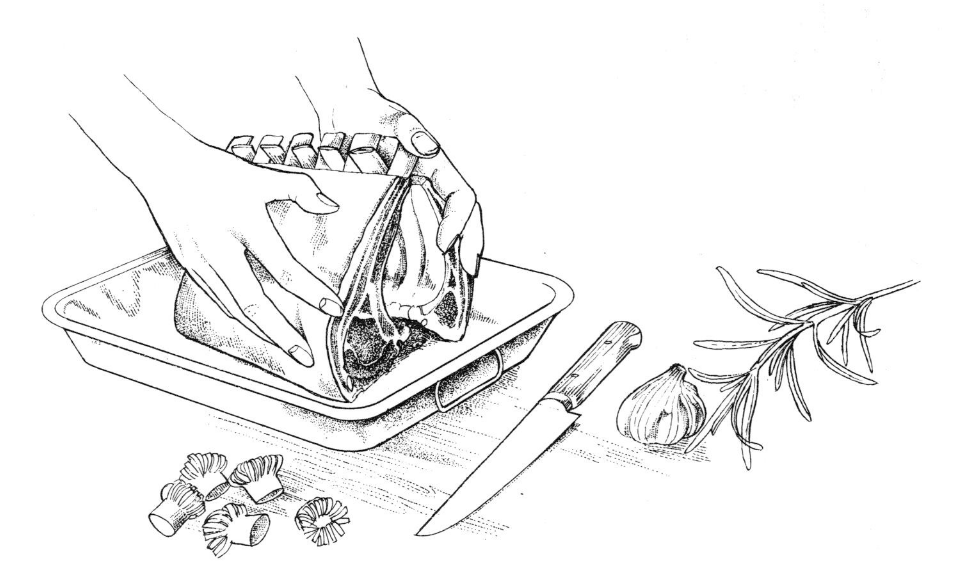

Stuffed shoulder of lamb

To freeze Wrap in freezer foil. Seal and label.
To serve Leave in wrapping until defrosted. Remove foil, weigh and place in a roasting tin. Cover lightly with foil and cook in a moderate oven (350°F, 180°C, Gas Mark 4) for 30 minutes per lb. (½ kg.). Remove thread or skewers, and serve with reform sauce, (see page 101).
Storage time Stuffed joint – 8 weeks.

IMPERIAL	METRIC	AMERICAN
1 boned shoulder of lamb	1 boned shoulder of lamb	1 boneless lamb shoulder
stuffing	*stuffing*	*stuffing*
1 lamb's kidney	1 lamb's kidney	1 lamb kidney
½ oz. butter	15 g. butter	1 tablespoon butter
1 teaspoon oil	1 teaspoon oil	1 teaspoon oil
2 oz. button mushrooms, chopped	50 g. button mushrooms, chopped	½ cup chopped button mushrooms
½ teaspoon dried oregano	½ teaspoon dried oregano	½ teaspoon dried oregano
2 tablespoons soft white breadcrumbs	2 tablespoons soft white breadcrumbs	3 tablespoons fresh soft bread crumbs
salt and pepper to taste	salt and pepper to taste	salt and pepper to taste
2 teaspoons grated lemon zest	2 teaspoons grated lemon zest	2 teaspoons grated lemon zest

Skin and core the kidney and chop finely. Melt the butter and oil in a frying pan and stir in the kidney, mushrooms and herbs. Cook gently for 5 minutes. Remove from the heat and stir in the breadcrumbs, seasoning and lemon zest. Cool. Spread open the joint and use the stuffing to fill the centre. Secure joint with fine string and trussing needle to make a neat shape. Skewers can be used but may pierce the foil when being wrapped for the freezer.

Minted lamb pie

To freeze Open freeze until solid then wrap in freezer foil or polythene bags.
To serve Brush with beaten egg and place in a moderately hot oven (400°F, 200°C, Gas Mark 6) for about 45 minutes until pastry is golden and filling heated through.
Storage time 4–6 months.

IMPERIAL	METRIC	AMERICAN
1 shoulder of lamb, boned	1 shoulder of lamb, boned	1 boneless lamb shoulder
2 oz. seasoned flour	50 g. seasoned flour	½ cup seasoned flour
2 tablespoons oil	2 tablespoons oil	3 tablespoons oil
1 large onion, chopped	1 large onion, chopped	1 large onion, chopped
1 lb. new potatoes	½ kg. new potatoes	1 lb. new potatoes
1 pint beef stock	generous ½ litre beef stock	2½ cups beef stock
4 tablespoons chopped mint	4 tablespoons chopped mint	5 tablespoons chopped mint
12 oz. new carrots	350 g. new carrots	¾ lb. new carrots
8 oz. frozen peas	225 g. frozen peas	½ lb. frozen peas
salt and pepper to taste	salt and pepper to taste	salt and pepper to taste
1 lb. frozen puff pastry	450 g. frozen puff pastry	1 lb. frozen puff paste

Cut the lamb into 1-inch (2·5-cm.) cubes and coat in seasoned flour. Heat the oil and fry the onion gently until softened. Add the lamb and cook for about 5 minutes, turning to brown the meat on all sides. Scrape the potatoes and cut them into chunks. Add to the pan with the stock and mint and bring to the boil. Cover and simmer for about 30 minutes. Add the scraped and sliced carrots and simmer for a further 10 minutes. Stir in the frozen peas, season and divide the mixture between two 2-pint (generous 1-litre) pie dishes. Cool. Roll out the pastry and cut two lids. Place pastry over filling and flute edges. Use the pastry trimmings to make leaves to decorate tops of pies.

Easy parcel chops

To freeze Wrap each chop parcel in freezer foil and seal. Pack together in a large rigid-based polythene container or bag. Seal and label.
To serve Place required number of still sealed frozen chops in a hot oven (425°F, 220°C, Gas Mark 7) for 40 minutes. Place the parcels in a grill pan, open the foil and grill until topping is golden brown.
Storage time 4–6 months.

IMPERIAL	METRIC	AMERICAN
8 frozen loin of lamb chops	8 frozen loin of lamb chops	8 frozen lamb loin chops
topping	*topping*	*topping*
2 canned red pimentos, chopped	2 canned red pimentos, chopped	2 canned red pimientos, chopped
16 black olives, stoned and chopped	16 black olives, stoned and chopped	16 ripe olives, pitted and chopped
1 large onion, chopped	1 large onion, chopped	1 large onion, chopped
3 oz. soft white breadcrumbs	75 g. soft white breadcrumbs	1½ cups fresh soft bread crumbs
1 teaspoon salt	1 teaspoon salt	1 teaspoon salt
¼ teaspoon pepper	¼ teaspoon pepper	¼ teaspoon pepper
2 tablespoons chopped parsley	2 tablespoons chopped parsley	3 tablespoons chopped parsley
2 oz. butter, melted	50 g. butter, melted	¼ cup butter, melted

Combine all the topping ingredients thoroughly. Place each chop on a large square of freezer foil. Divide topping between the parcels. Fold over foil to make airtight seal.

Roast loin of lamb with fruity sauce

IMPERIAL	METRIC	AMERICAN
1 frozen loin of lamb, boned	1 frozen loin of lamb, boned	1 frozen boneless lamb loin
½ teaspoon dried lemon thyme	½ teaspoon dried lemon thyme	½ teaspoon dried lemon thyme
½ teaspoon ground ginger	½ teaspoon ground ginger	½ teaspoon ground ginger
½ teaspoon ground black pepper	½ teaspoon ground black pepper	½ teaspoon ground black pepper
1½ teaspoons salt	1½ teaspoons salt	1½ teaspoons salt
sauce	*sauce*	*sauce*
1 13-oz. can crushed pineapple	1 375-g. can crushed pineapple	1 13-oz. can crushed pineapple
1 tablespoon mild Continental mustard	1 tablespoon mild Continental mustard	1 tablespoon mild Continental mustard
2 tablespoons soy sauce	2 tablespoons soy sauce	3 tablespoons soy sauce

Tie the loin into a neat shape for roasting. Mix together the thyme, ginger, pepper and salt and rub this mixture well into the surface of the joint. Place in a roasting tin and cook in a moderate oven (350°F, 180°C, Gas Mark 4) for 1½ hours. Pour the contents of the can of pineapple, the mustard and soy sauce into a saucepan. Stir and heat gently. About 30 minutes before end of cooking time for the joint, spoon a little of the sauce over to glaze it and return to the oven. Hand the remaining sauce separately.

Stewing lamb chain

The breast, scrag end and middle neck from a lamb weighing about 30 lb. (15 kg.) will produce sufficient basic stew to make three meals for six people. It is an economy of time and effort to make up a simple stew from this basic recipe and elaborate it with the addition of different vegetables and other ingredients when required to serve.

Basic lamb stew

Chain Makes 6 portions lamb hot-pot, 6 portions peasant's pottage, 6 portions Mediterranean lamb stew.
To freeze Pack the meat into three polythene containers, strain the stock over the meat, leaving 1-inch (2·5-cm.) headspace. Seal and label.
Storage time 6 months.

IMPERIAL	METRIC	AMERICAN
2 breasts, scrag end and middle neck from 1 whole lamb	2 breasts, scrag end and middle neck from 1 whole lamb	2 breasts and 1 shoulder from 1 whole lamb
2 tablespoons dried onion flakes	2 tablespoons dried onion flakes	3 tablespoons dried onion flakes
3 teaspoons salt	3 teaspoons salt	3 teaspoons salt
1 teaspoon pepper	1 teaspoon pepper	1 teaspoon pepper
juice of ½ lemon	juice of ½ lemon	juice of ½ lemon

Put all the ingredients into a large pan, or in two saucepans, add enough water to cover and bring to the boil slowly. Skim, cover and simmer for 1–1½ hours, or until meat can easily be removed from bones. Allow to cool until surplus fat can be skimmed from the surface, but has not completely set. Remove as many of the bones as possible, washing your hands carefully before touching the meat. Reduce the remaining stock by half.

Lamb hotpot

IMPERIAL	METRIC	AMERICAN
1 oz. butter	25 g. butter	2 tablespoons butter
8 oz. diced frozen carrots or mixed vegetables	225 g. diced frozen carrots or mixed vegetables	½ lb. diced frozen carrots or mixed vegetables
½ teaspoon dried mixed sweet herbs	½ teaspoon dried mixed sweet herbs	½ teaspoon dried mixed sweet herbs
1 tablespoon tomato purée	1 tablespoon tomato purée	1 tablespoon tomato paste
2 tablespoons hot water	2 tablespoons hot water	3 tablespoons hot water
8 oz. frozen peas	225 g. frozen peas	½ lb. frozen peas
1 container basic lamb stew, defrosted	1 container basic lamb stew, defrosted	1 container basic lamb stew, defrosted
3 large old potatoes, thinly sliced	3 large old potatoes, thinly sliced	3 large old potatoes, thinly sliced
oil	oil	oil
salt	salt	salt

Melt the butter, use to sauté the carrots lightly for 1–2 minutes. Add the herbs, tomato purée dissolved in the hot water, and the peas. Remove from the heat, stir in the stewed lamb and place this mixture in a pie dish. Cover with layers of overlapping potato slices, brush with oil and sprinkle with salt. Cook in a moderately hot oven (375°F, 190°C, Gas Mark 5) for 45 minutes.

Peasant's pottage

IMPERIAL	METRIC	AMERICAN
8 oz. haricot beans	225 g. haricot beans	generous 1 cup navy beans
1 oz. dripping	25 g. dripping	2 tablespoons drippings
1 large onion, chopped	1 large onion, chopped	1 large onion, chopped
1 container basic lamb stew, defrosted	1 container basic lamb stew, defrosted	1 container basic lamb stew, defrosted
2 tablespoons diced garlic sausage	2 tablespoons diced garlic sausage	3 tablespoons diced garlic sausage
2 tablespoons chopped parsley	2 tablespoons chopped parsley	3 tablespoons chopped parsley
salt and pepper to taste	salt and pepper to taste	salt and pepper to taste

Soak the beans overnight in sufficient water to cover. Next day, drain and put into a large saucepan with the dripping, onion and sufficient fresh water to cover. Simmer for about 1½ hours or until beans are almost cooked. Add the stewed lamb, garlic sausage and parsley and reheat until it comes to the boil; taste, adding more salt and pepper if necessary.

Mediterranean lamb stew

IMPERIAL	METRIC	AMERICAN
1 tablespoon corn oil	1 tablespoon corn oil	1 tablespoon corn oil
1 large aubergine, sliced	1 large aubergine, sliced	1 large eggplant, sliced
8 oz. courgettes, sliced	225 g. courgettes, sliced	½ lb. small zucchini, sliced
½ lemon	½ lemon	½ lemon
½ teaspoon dried mixed sweet herbs	½ teaspoon dried mixed sweet herbs	½ teaspoon dried mixed sweet herbs
1 small can tomato purée	1 small can tomato purée	¼ cup tomato paste
pinch freshly ground black pepper	pinch freshly ground black pepper	pinch freshly ground black pepper
1 container basic lamb stew, defrosted	1 container basic lamb stew, defrosted	1 container basic lamb stew, defrosted
salt	salt	salt

Heat the oil and use to sauté the aubergine and courgette slices until they begin to brown. Add the juice of the lemon and 1 teaspoon grated lemon zest, herbs, tomato purée and seasoning. Cover the pan and simmer for 10 minutes. Add the stewed lamb, reheat to boiling point, taste and add more salt and pepper if necessary.

Scotch broth

To freeze Pour into suitable polythene containers leaving 1-inch (2·5-cm.) headspace (or pre-formed polythene bags). Seal and label.
To serve Place, still frozen, in a saucepan with a little water and heat very gently. Bring to the boil and simmer for 10–15 minutes.
Storage time 3 months.

IMPERIAL	METRIC	AMERICAN
4 lb. meaty lamb bones (if bones are stripped add 8 oz. meat)	2 kg. meaty lamb bones (if bones are stripped add 225 g. meat)	4 lb. meaty lamb bones (if bones are stripped add ½ lb. meat)
4 pints water	2¼ litres water	10 cups water
salt and pepper to taste	salt and pepper to taste	salt and pepper to taste
2 onions, chopped	2 onions, chopped	2 onions, chopped
2 carrots, chopped	2 carrots, chopped	2 carrots, chopped
1 small turnip or parsnip, chopped	1 small turnip or parsnip, chopped	1 small turnip or parsnip, chopped
1 leek, sliced	1 leek, sliced	1 leek, sliced
2 sticks celery, sliced	2 sticks celery, sliced	2 stalks celery, sliced
2 oz. pearl barley	50 g. pearl barley	generous ¼ cup pearl barley

Remove meat from the bones and cut into chunks. Place in a saucepan with the bones and add the water and seasoning. Bring to the boil, cover and simmer for about 2 hours. Add the vegetables together with the barley and simmer for a further 1½ hours. Remove the bones from the soup, correct the seasoning, cool and skim off any fat.

King-size family recipes

King-size recipes to provide everyday meals for the family must be easy and quick to prepare. Saving of money is just as important as time saving in these days when the household budget always seems stretched to the limit. So all the recipes in this section are planned to make use of items which you can buy relatively cheaply in bulk from your frozen food centre.

To qualify them for inclusion in this section, I have applied the three-point test to each of these recipes.

1. Preparation is quick and easy, washing up is minimal.
2. Cooking method is the simplest possible and sure to succeed.
3. Main ingredient can be bulk-purchased economically.

In addition the total ingredients have been carefully costed to ensure that they are easy on your purse.

Unlike chain cooking which produces a number of different dishes from one basic ingredient, my king-size family recipes are intended to cut short daily kitchen chores. It is quite a feat to cope with frying, sauce-making and other activities which keep you tied to the vicinity of the cooker, just when you want to be free to lay the table, greet home-coming members of the family and hear their latest news. As a young housewife, I shed more than a few tears over chops which burned while my back was turned, or sauces which boiled down to the consistency of glue during the few minutes it took me to search out a fresh pack of paper napkins.

Many of these recipes are for dishes which you can put in the oven and forget and none of them will spoil if left for a little longer than the indicated cooking time.

My experience has always been that multiplying the basic amount to serve your family by three produces a quantity that is practical to cook, (unless you have a very large family). This gives you a meal to serve and two to store. Of course all three portions can be frozen if you prefer, and especially if you cook another king-size recipe at the same time. But the rule I first evolved for myself of cooking six meals, serving one at once and refrigerating a portion of the second dish for the next day seems to suit most housewives best. Once-a-week cooking keeps the freezer nicely stocked with a couple of fresh-cooked meals during the week into the bargain. And if you decide not to eat from the freezer so frequently – once-a-month cooking gives you a respite from kitchen chores on one or two nights each week.

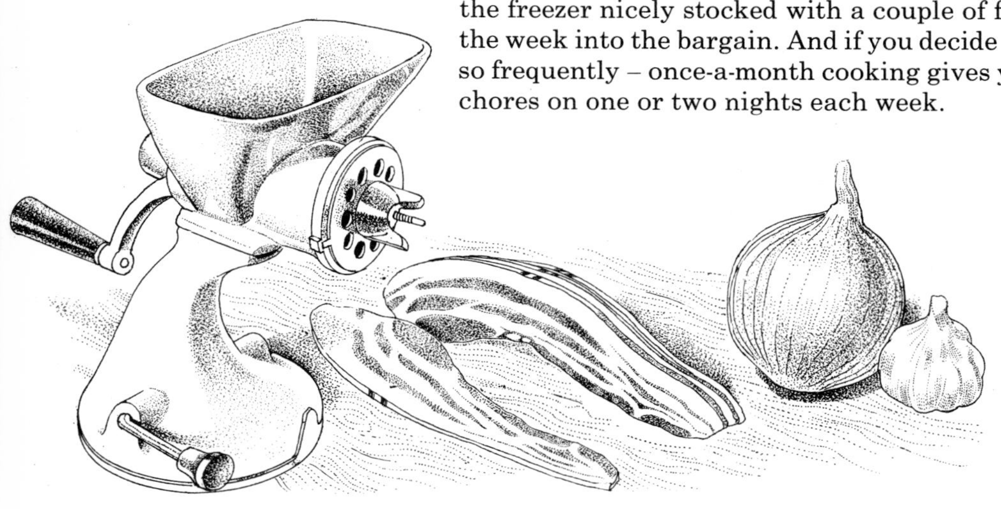

Danish liver pâté

King-size Makes 12 portions.
To freeze Cover with foil and crimp edges tightly together or pack in polythene bag. Seal and label.
To serve Unwrap and allow to defrost at room temperature for about 1–2 hours. Serve with crisp melba toast and a garnish of salad vegetables.
Storage time 6 weeks.

IMPERIAL	METRIC	AMERICAN
8 oz. streaky bacon	225 g. streaky bacon	$\frac{1}{2}$ lb. bacon slices
1 lb. pig's liver	450 g. pig's liver	1 lb. pork liver
1 large onion, chopped	1 large onion, chopped	1 large onion, chopped
1 clove garlic, crushed	1 clove garlic, crushed	1 clove garlic, crushed
1 oz. flour	25 g. flour	$\frac{1}{4}$ cup all-purpose flour
$\frac{1}{2}$ pint milk	3 dl. milk	$1\frac{1}{4}$ cups milk
2 eggs	2 eggs	2 eggs
1 teaspoon salt	1 teaspoon salt	1 teaspoon salt
$\frac{1}{2}$ teaspoon black pepper	$\frac{1}{2}$ teaspoon black pepper	$\frac{1}{2}$ teaspoon black pepper

De-rind the bacon and mince together with the liver, onion and garlic several times, using the finest cutting blade. Combine with the remaining ingredients and mix well. Divide the mixture between small greased baking dishes or shaped foil containers. Place in a moderate oven (325°F, 170°C, Gas Mark 3) for 1–$1\frac{1}{4}$ hours, depending on size. Cool.

Potage St. Germain

King-size Makes approximately 4 pints (2¼ litres) soup.
To freeze Pour into polythene containers, leaving 1-inch (2·5-cm.) headspace, or line saucepans with polythene bags, pour in soup and freeze until solid. Remove from saucepans. Seal and label.
To serve Hold container under cold water until contents can be turned out into a saucepan. Reheat very gently and stir frequently to prevent burning. Add about 2 tablespoons single cream to each pint of soup just before serving and garnish with croûtons.
Storage time 4 months.

Note Peas which have just passed the tender stage can very well be used for this soup.

IMPERIAL	METRIC	AMERICAN
1 oz. butter	25 g. butter	2 tablespoons butter
2 lb. peas, shelled	1 kg. peas, shelled	2 lb. peas, shelled
6 spring onions, sliced	6 spring onions, sliced	6 scallions, sliced
1 small lettuce, shredded	1 small lettuce, shredded	1 small lettuce, shredded
3 pints stock	1½ litres stock	7½ cups stock
salt and pepper to taste	salt and pepper to taste	salt and pepper to taste

Melt the butter in a saucepan and add the peas, spring onions and lettuce. Cover with a tightly fitting lid and cook over a very low heat for 10 minutes without removing the lid. Add the stock and seasoning, bring to the boil, cover and simmer for 1 hour. Liquidise or sieve and adjust seasoning. Cool.

Family mushroom soup

King size Makes 4 pints (2¾ litres).
To freeze Pour into polythene containers, leaving 1-inch (2·5-cm.) headspace, or line containers with polythene bags, pour in soup and freeze until solid. Remove bag from container. Seal and label.
To serve Turn still frozen into a saucepan, add ¼–½ pint (1½–3 dl.) milk and reheat gently to boiling point. Simmer for 2 minutes.
Storage time 4–6 months.

IMPERIAL	METRIC	AMERICAN
1 large onion	1 large onion	1 large onion
1½ lb. mushrooms	700 g. mushrooms	1½ lb. mushrooms
4 oz. butter	100 g. butter	½ cup butter
2 tablespoons chopped parsley	2 tablespoons chopped parsley	3 tablespoons chopped parsley
salt and pepper to taste	salt and pepper to taste	salt and pepper to taste
pinch ground nutmeg	pinch ground nutmeg	pinch ground nutmeg
2 oz. white bread	50 g. white bread	2 oz. white bread
3 pints hot chicken stock	1½ litres hot chicken stock	7½ cups hot chicken stock

Finely chop the onion and mushrooms. Melt the butter and use to sauté the vegetables, until the onions are softened but not browned. Stir in the parsley and seasonings. Soak the bread in a little of the hot stock, stir into the pan then gradually add the remaining stock and stir well. Bring to the boil, cover and simmer for 15 minutes. Liquidise in an electric blender, or sieve. Cool.

Farmhouse soup

King-size Makes approximately 5 pints (3 litres) soup.
To freeze Pour into polythene containers, leaving 1-inch (2·5-cm.) headspace, or line saucepans with polythene bags, pour in soup and freeze until solid. Remove from saucepans. Seal and label.
To serve Hold container under cold water until contents can be turned out into a saucepan. Add a little water or milk and water and reheat very gently, stirring frequently to prevent burning.
Storage time 4 months.

IMPERIAL	METRIC	AMERICAN
4 large leeks	4 large leeks	4 large leeks
1 large potato	1 large potato	1 large potato
2 large carrots	2 large carrots	2 large carrots
1 stick celery	1 stick celery	1 stalk celery
2 large onions	2 large onions	2 large onions
1 large parsnip	1 large parsnip	1 large parsnip
3 pints chicken stock	1½ litres chicken stock	7½ cups chicken stock
3 rashers bacon	3 rashers bacon	3 bacon slices
sprig parsley	sprig parsley	sprig parsley
2 bay leaves	2 bay leaves	2 bay leaves
½ teaspoon dried thyme	½ teaspoon dried thyme	½ teaspoon dried thyme
salt and pepper to taste	salt and pepper to taste	salt and pepper to taste

Wash all the vegetables, peel where necessary and chop them finely. Put into a large saucepan with all the other ingredients and bring to the boil. Cover and simmer gently for about 1 hour then remove the bacon and bay leaves. Allow to cool, liquidise or sieve and check seasoning.

Sailor's soup

King-size Makes approximately 5 pints (3 litres) soup.
To freeze Pour into polythene containers, leaving 1-inch (2·5-cm.) headspace, or line saucepans with polythene bags, pour in soup and freeze until solid. Remove from saucepans. Seal and label.
To serve Hold container under cold water until contents can be turned out into a saucepan. Add a little water or milk and water and reheat very gently to boiling point, stirring frequently to prevent burning. Sprinkle with chopped parsley or grated cheese before serving.
Storage time 4–6 months.

IMPERIAL	METRIC	AMERICAN
1½ lb. white fish fillets	700 g. white fish fillets	1½ lb. white fish fillets
3 pints water	1½ litres water	7½ cups water
1 small turnip	1 small turnip	1 small turnip
1 large carrot	1 large carrot	1 large carrot
3 sticks celery	3 sticks celery	3 stalks celery
1 medium onion, chopped	1 medium onion, chopped	1 medium onion, chopped
1 16-oz. can tomato juice	1 450-g. can tomato juice	1 16-oz. can tomato juice
bouquet garni	bouquet garni	bouquet garni
½ teaspoon dried dill weed	½ teaspoon dried dill weed	½ teaspoon dried dill weed
salt and pepper to taste	salt and pepper to taste	salt and pepper to taste

Skin the fish and put into a large saucepan with the water. Wash all the vegetables, peel where necessary and chop them finely. Bring the fish to the boil, add the vegetables, tomato juice, bouquet garni and dill. Bring back to the boil, cover and simmer for about 30 minutes. Remove the bouquet garni and allow to cool slightly. Liquidise until smooth, or sieve, and check seasoning. Cool.

Goulash sauce

King-size Makes approximately 3 pints (1½ litres) sauce.
To freeze Pack in ½-pint (3-dl.) polythene containers, leaving ½-inch (1-cm.) headspace. Seal and label.
To serve Defrost and reheat gently in a saucepan with stewed veal or beef and spoon over a little soured cream just before serving.
Storage time 3 months.

IMPERIAL	METRIC	AMERICAN
2 red or green peppers	2 red or green peppers	2 red or green sweet peppers
4 large tomatoes	4 large tomatoes	4 large tomatoes
4 oz. butter	100 g. butter	½ cup butter
4 large onions, sliced	4 large onions, sliced	4 large onions, sliced
4 tablespoons paprika pepper	4 tablespoons paprika pepper	5 tablespoons paprika pepper
4 tablespoons flour	4 tablespoons flour	5 tablespoons all-purpose flour
2 pints stock	generous 1 litre stock	5 cups stock
2 tablespoons tomato purée	2 tablespoons tomato purée	3 tablespoons tomato paste
1 clove garlic, crushed	1 clove garlic, crushed	1 clove garlic, crushed
2 bay leaves	2 bay leaves	2 bay leaves
sprig parsley	sprig parsley	sprig parsley
sprig thyme	sprig thyme	sprig thyme

Deseed and finely chop the peppers and peel and slice the tomatoes. Melt the butter and fry the onions until softened but not browned. Add the paprika pepper and cook gently for about a minute. Stir in the flour and gradually add the stock, tomato purée, chopped pepper, tomatoes, garlic, bay leaves, parsley and thyme. Bring to the boil, stirring constantly, cover and simmer for about 10 minutes. Remove herbs and cool.

Curry sauce

King-size Makes approximately 4 pints (2¼ litres) sauce.
To freeze Pack into ½-pint (3-dl.) polythene containers leaving ½-inch (1-cm.) headspace. Seal and label.
To serve Defrost and reheat gently in a saucepan with cooked meat.
Storage time 4–6 months.

IMPERIAL	METRIC	AMERICAN
6 oz. desiccated coconut	175 g. desiccated coconut	2 cups shredded coconut
1 pint boiling water	generous ½ litre boiling water	2½ cups boiling water
4 oz. butter	100 g. butter	½ cup butter
2 large onions, chopped	2 large onions, chopped	2 large onions, chopped
6 tablespoons curry powder	6 tablespoons curry powder	½ cup curry powder
3 oz. flour	75 g. flour	¾ cup all-purpose flour
2 pints stock	generous 1 litre stock	5 cups stock
2 large apples	2 large apples	2 large apples
1½ tablespoons lemon juice	1½ tablespoons lemon juice	2 tablespoons lemon juice
4 oz. sultanas	100 g. sultanas	generous ½ cup seedless white raisins
4 tablespoons mango chutney	4 tablespoons mango chutney	5 tablespoons mango chutney

Place the coconut in a basin and pour over the boiling water. Allow to stand for 24 hours if possible then strain and reserve the 'milk'. Melt the butter and fry the onion until softened but not browned. Stir in the curry powder and flour and cook gently for about 4 minutes. Gradually add the stock and bring to the boil, stirring constantly. Peel, core and chop the apple and add to the sauce with the lemon juice, coconut milk and sultanas. Bring back to the boil, cover and simmer for about 1 hour. Cut up any large pieces in the chutney and stir into the sauce. Cool.

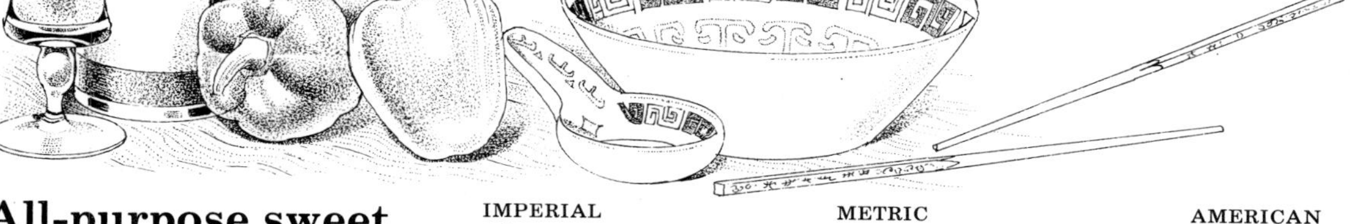

All-purpose sweet and sour sauce

King-size Makes approximately 2 pints (generous 1 litre) sauce.
To freeze Pack into small polythene containers. Seal and label.
To serve Hold container under cold water until contents can be turned out into a saucepan. Add a little water and reheat very gently, stirring frequently to prevent burning.
Storage time 4–6 months.

IMPERIAL	METRIC	AMERICAN
3 large green peppers	3 large green peppers	3 large green sweet peppers
1 15-oz. can pineapple pieces	1 425-g. can pineapple pieces	1 15-oz. can pineapple pieces
2 carrots, canned or cooked	2 carrots, canned or cooked	2 carrots, canned or cooked
6 tablespoons oil	6 tablespoons oil	½ cup oil
2 onions, chopped	2 onions, chopped	2 onions, chopped
2 tablespoons ginger wine or dry sherry	2 tablespoons ginger wine or dry sherry	3 tablespoons ginger wine or dry sherry
4 tablespoons sugar	4 tablespoons sugar	5 tablespoons sugar
2 tablespoons soy sauce	2 tablespoons soy sauce	3 tablespoons soy sauce
2 tablespoons vinegar	2 tablespoons vinegar	3 tablespoons vinegar
4 tablespoons tomato ketchup	4 tablespoons tomato ketchup	5 tablespoons tomato catsup
4 tablespoons cornflour	4 tablespoons cornflour	5 tablespoons cornstarch

Deseed and chop the peppers, drain and chop the pineapple pieces and chop the carrots. Heat the oil in a saucepan and use to fry the onion, pepper and carrot until the onion is softened but not browned. Add the wine or sherry, sugar, soy sauce, vinegar, pineapple and ketchup and stir well until sugar is dissolved. Make the pineapple syrup up to ½ pint (3 dl.) with water and blend with the cornflour. Add to the saucepan and bring to the boil, stirring constantly. Cover and simmer for about 5 minutes. Cool.

Braised beef in a bag with Andaluz sauce

King-size Makes 3 servings for 4 people.
To freeze Pack sliced meat into three shaped foil containers and spoon the sauce over. Seal and label.
To serve Remove lid and cover container with foil. Place in a moderately hot oven (400°F, 200°C, Gas Mark 6) for 30 minutes.
Storage time 4 months.

IMPERIAL	METRIC	AMERICAN
2 tablespoons dripping	2 tablespoons dripping	3 tablespoons drippings
4-lb. piece topside	2-kg. piece topside	4-lb. piece rolled rump
sauce	*sauce*	*sauce*
2 large green peppers	2 large green peppers	2 large green sweet peppers
1 large onion	1 large onion	1 large onion
2 tablespoons dripping	2 tablespoons dripping	3 tablespoons drippings
½ teaspoon garlic salt	½ teaspoon garlic salt	½ teaspoon garlic salt
1 tablespoon cornflour	1 tablespoon cornflour	1 tablespoon cornstarch
½ pint beef stock	3 dl. beef stock	1¼ cups beef stock
1 16-oz. can tomatoes	1 450-g. can tomatoes	1 16-oz. can tomatoes

Melt the dripping in a frying pan and use to brown all the surfaces of the beef. Remove from the pan and place in a transparent roasting bag. Seal and place the bag in a dry roasting tin. Cook in a moderate oven (350°F, 180°C, Gas Mark 4) for about 2 hours. Remove beef from the bag, cool and slice. Reserve the accumulated juices in the bag. To make the sauce, deseed and slice the peppers finely and slice the onion. Melt the dripping in a saucepan, or use the remaining dripping from browning the joint. Add the pepper and onion slices and the garlic salt. Cook gently until the vegetables are softened. Blend the cornflour into the stock, or juices from the beef, add to the pan and stir well. Bring to the boil and simmer for 1 minute. Sieve or liquidise the tomatoes and add to the sauce. Cook until reduced to make a coating sauce. Remove from the heat and cool.

Beef stew with cheese cobbler

King-size Makes 3 servings for 4 people.
To freeze Divide the stew between three polythene or shaped foil containers. Pack the cobblers in three polythene bags. Seal and label.
To serve Defrost stew and place in an ovenproof casserole, or remove lid from foil container. Place frozen cobblers, overlapping, on top of defrosted stew and reheat in a moderately hot oven (375°F, 190°C, Gas Mark 5) for 30 minutes.
Storage time 3 months.

Note To serve one portion of the stew immediately. Cook one-third of the stew in a smaller casserole. After 1½ hours, cover with an overlapping layer of uncooked cobblers and return to the oven for 20 minutes.

IMPFRIAL	METRIC	AMERICAN
3 tablespoons flour	3 tablespoons flour	4 tablespoons flour
salt and pepper to taste	salt and pepper to taste	salt and pepper to taste
4½ lb. chuck steak, cubed	2¼ kg. chuck steak, cubed	4½ lb. chuck steak, cubed
6 tablespoons corn oil	6 tablespoons corn oil	½ cup corn oil
3 lb. onions	1½ kg. onions	3 lb. onions
3 tablespoons pickled silverskin onions	3 tablespoons pickled silverskin onions	4 tablespoons small white onions
½ teaspoon ground cloves	½ teaspoon ground cloves	½ teaspoon ground cloves
½ teaspoon ground nutmeg	½ teaspoon ground nutmeg	½ teaspoon ground nutmeg
2 bay leaves	2 bay leaves	2 bay leaves
cobbler	*cobbler*	*cobbler*
1½ lb. self-raising flour	700 g. self-raising flour	6 cups all-purpose flour sifted with 7 teaspoons baking powder
1 teaspoon salt	1 teaspoon salt	1 teaspoon salt
½ teaspoon pepper	½ teaspoon pepper	½ teaspoon pepper
½ teaspoon dry mustard	½ teaspoon dry mustard	½ teaspoon dry mustard
6 oz. butter	175 g. butter	¾ cup butter
8 oz. strong Cheddar cheese, grated	225 g. strong Cheddar cheese, grated	2 cups grated strong Cheddar cheese
¾ pint milk or milk and water	scant ½ litre milk or milk and water	scant 2 cups milk or milk and water

Season the flour well with salt and pepper. Turn the cubed steak in the seasoned flour. Heat the oil and fry steak lightly until brown all over, transfer to a large ovenproof casserole. Quarter the onions and add to the casserole with the pickled onions, spices, bay leaves, and sprinkle over any remaining flour. Stir thoroughly. Add just enough water to cover, put on lid and cook in the centre of a moderate oven (350°F, 180°C, Gas Mark 4) for 2 hours. Cool.

To make the topping, sieve the flour with the salt, pepper and mustard into a bowl. Rub in the butter and stir in the cheese. Blend to a soft dough with milk and knead lightly. Roll out to ½ inch (1 cm.) thickness and cut out rounds with a biscuit cutter. Place on baking trays and brush with milk. Bake in the top of oven with the stew for 12 minutes only. Cool.

Variation The same stew tastes entirely different with lemon-flavoured, herby stuffing balls, instead of the cobbler topping.

Forcemeat balls

To freeze Open freeze on baking trays until solid. Pack in suitable quantities in polythene bags. Seal and label.
To serve Place frozen forcemeat balls on a baking tray and cook in the oven with the stew for 30 minutes. Add to the stew just before serving.
Storage time 6 weeks.

IMPERIAL	METRIC	AMERICAN
4 tablespoons soft white breadcrumbs	4 tablespoons soft white breadcrumbs	5 tablespoons fresh soft bread crumbs
1 tablespoon bacon dripping or ½ oz. shredded suet	1 tablespoon bacon dripping or 15 g. shredded suet	1 tablespoon bacon drippings or a little finely chopped suet
good pinch dried sweet herbs	good pinch dried sweet herbs	good pinch dried sweet herbs
½ teaspoon grated lemon zest	½ teaspoon grated lemon zest	½ teaspoon grated lemon zest
salt and pepper to taste	salt and pepper to taste	salt and pepper to taste
1 egg, lightly beaten	1 egg, lightly beaten	1 egg, lightly beaten

Mix together all the dry ingredients and bind together with the beaten egg. Divide the mixture into 12 portions and, with floured hands, roll into balls.

Breast of lamb with corn stuffing

King-size Makes 3 servings for 4 people.
To freeze Wrap each stuffed joint in freezer foil or polythene bag. Seal and label.
To serve Defrost overnight. Weigh and place in a roasting tin. Cook in a moderate oven (350°F, 180°C, Gas Mark 4) for 40 minutes per lb. (½ kg.).
Storage time 2 months.
Note The joints, when cooked, make a lot of fat which should be skimmed from the roasting tin and used for dripping. A sharp-flavoured sauce, or simply mint sauce, is the best accompaniment.

IMPERIAL	METRIC	AMERICAN
3 oz. butter	75 g. butter	6 tablespoons butter
1 large onion, chopped	1 large onion, chopped	1 large onion, chopped
3 12-oz. cans corn kernels, drained	3 350-g. cans corn kernels, drained	3 12-oz. cans kernel corn, drained
9 oz. soft white breadcrumbs	250 g. soft white breadcrumbs	4½ cups fresh soft bread crumbs
1 tablespoon dried marjoram or 2 tablespoons fresh	1 tablespoon dried marjoram or 2 tablespoons fresh	1 tablespoon dried marjoram or 3 tablespoons fresh
1 teaspoon lemon juice	1 teaspoon lemon juice	1 teaspoon lemon juice
salt and pepper to taste	salt and pepper to taste	salt and pepper to taste
3 eggs, lightly beaten	3 eggs, lightly beaten	3 eggs, lightly beaten
3 breasts of lamb, boned	3 breasts of lamb, boned	3 boneless lamb breasts

Melt the butter in a pan and fry the onion gently until soft but not browned. Remove from the heat, stir in the well-drained corn kernels, breadcrumbs, herbs, lemon juice and seasoning. Bind together with the egg and divide between the three boned breasts of lamb. Roll up and tie, sufficiently loosely to allow for expansion of the stuffing.

Celery-baked chicken

King-size 3 servings for 4 people.
To freeze Cool, transfer to three containers, each suitable in size and shape to hold four chicken portions plus one-third of the sauce. Cover, seal and label.
To serve Allow to defrost, covered, transfer to a covered casserole and reheat at 350°F, 180°C, Gas Mark 4 for 45 minutes.
Storage time 3–4 months.

IMPERIAL	METRIC	AMERICAN
12 chicken portions	12 chicken portions	12 chicken portions
1 large head celery, diced	1 large head celery, diced	1 large bunch celery, diced
6 parsley sprigs	6 parsley sprigs	6 parsley sprigs
1 tablespoon salt	1 tablespoon salt	1 tablespoon salt
1 teaspoon pepper	1 teaspoon pepper	1 teaspoon pepper
2 teaspoons dried basil	2 teaspoons dried basil	2 teaspoons dried basil
1 8-oz. can tomato juice	1 225-g. can tomato juice	1 8-oz. can tomato juice
1 tablespoon Worcestershire sauce	1 tablespoon Worcestershire sauce	1 tablespoon Worcestershire sauce
1 tablespoon soy sauce	1 tablespoon soy sauce	1 tablespoon soy sauce

Arrange chicken portions on a bed of diced celery in a large roasting tin. Mix together the other ingredients and pour over the chicken. Cover with foil and bake in a moderate oven (350°F, 180°C, Gas Mark 4) until chicken is tender, about 1½ hours.

Somerset chicken in cider

(Illustrated on page 127)

King-size 3 servings for 4 people.
To freeze Pack uncooked joints, plus marinade, in shaped foil containers. Seal and label.
To serve Defrost completely, remove the chicken from the marinade, drain well, toss in the flour. Melt the butter and fry the chicken until evenly browned. Put into a flameproof casserole with the rest of the marinade, orange juice, salt and pepper. Cover and simmer for 30 minutes. Remove chicken and place on a bed of peppered saffron rice. Boil the liquid until reduced by about one-third. Moisten the cornflour with a little cold water and use to thicken the sauce. Strain over the chicken to glaze and garnish with parsley.
Storage time 4–6 months.

IMPERIAL	METRIC	AMERICAN
1½ pints Taunton cider	scant 1 litre Taunton cider	3¾ cups cider
3 tablespoons corn oil	3 tablespoons corn oil	scant ¼ cup corn oil
12 chicken leg portions, halved	12 chicken leg portions, halved	12 chicken leg portions, halved
fresh rosemary sprigs or 1 teaspoon dried rosemary	fresh rosemary sprigs or 1 teaspoon dried rosemary	fresh rosemary sprigs or 1 teaspoon dried rosemary
3 tablespoons flour	3 tablespoons flour	4 tablespoons all-purpose flour
2 oz. butter	50 g. butter	¼ cup butter
juice of 3 oranges	juice of 3 oranges	juice of 3 oranges
salt and pepper to taste	salt and pepper to taste	salt and pepper to taste
1½ oz. cornflour	40 g. cornflour	generous ¼ cup cornstarch

Mix together the cider and oil. Place the chicken portions in a shallow dish and pour over the oil mixture. Sprinkle over the rosemary and allow to stand overnight, in a cool place.

Peppered saffron rice

(Illustrated on page 127)

Makes 12 portions with chicken.
To freeze Pack in polythene bags and partially freeze. Squeeze the bags between the hands to separate the grains. Return to the freezer.
To serve Defrost for 1 hour, spread out on a baking tray, dot with butter, cover with foil and reheat in a moderate oven (325°F, 170°C, Gas Mark 3) for 15 minutes.
Storage time 4–6 months.

IMPERIAL	METRIC	AMERICAN
1 frozen green pepper	1 frozen green pepper	1 frozen green sweet pepper
1 frozen red pepper	1 frozen red pepper	1 frozen red sweet pepper
1 teaspoon powdered saffron or saffron strands	1 teaspoon powdered saffron or saffron strands	1 teaspoon powdered saffron or saffron strands
1 lb. long-grain rice	450 g. long-grain rice	2¼ cups long-grain rice
2 tablespoons corn oil	2 tablespoons corn oil	3 tablespoons corn oil

Defrost the peppers and deseed if necessary. Chop very finely. Steep the saffron in a large pan of salted water, bring water to the boil and use to cook the rice in the usual way. Drain well and refresh. Shake dry. Meanwhile, heat the oil and use to fry the chopped pepper until just tender. Stir into the cooked rice. Cool.

Poulet grand'mère

King-size Makes 3 servings for 4 people.
To freeze Divide meat between three shaped foil containers. Spoon over the vegetables and stock. Seal and label.
To serve Place in a moderate oven (350°F, 180°C, Gas Mark 4) for 1 hour to defrost and reheat.
Storage time 4–6 months.

IMPERIAL	METRIC	AMERICAN
6 tablespoons oil	6 tablespoons oil	½ cup oil
1 large boiling fowl	1 large boiling fowl	1 large boiling fowl
3 large onions, quartered	3 large onions, quartered	3 large onions, quartered
6 carrots, diced	6 carrots, diced	6 carrots, diced
1 large turnip, diced	1 large turnip, diced	1 large turnip, diced
1 head celery, sliced	1 head celery, sliced	1 head celery, sliced
6 medium potatoes, diced	6 medium potatoes, diced	6 medium potatoes, diced
1 chicken stock cube	1 chicken stock cube	1 chicken bouillon cube
salt and pepper to taste	salt and pepper to taste	salt and pepper to taste
½ teaspoon ground bay leaves	½ teaspoon ground bay leaves	½ teaspoon ground bay leaves
½ lemon	½ lemon	½ lemon

Heat the oil in a large saucepan or flameproof casserole and use to brown the chicken on all sides. Remove chicken and add the prepared vegetables to the remaining oil in the pan. Fry gently for 3 minutes. Replace the chicken on the vegetables and add enough water to almost come to the top of the chicken. Add the crumbled stock cube, seasonings and the lemon. Bring to the boil, cover tightly and simmer for 2–2½ hours until the chicken is really tender. Take chicken from the liquid and remove flesh from bones. Boil the vegetables and stock rapidly until slightly reduced. Allow to cool then remove the lemon and excess fat.

Picnic pie

King-size Makes 3 servings for 4 people.
To freeze Place lid on container or cover with freezer foil and crimp edges together. Seal and label.
To serve Uncover and defrost for about 6 hours or overnight in the refrigerator. If required hot, place defrosted pie in a moderately hot oven (375°F, 190°C, Gas Mark 5) for 25–30 minutes.
Storage time 3 months.

IMPERIAL	METRIC	AMERICAN
6 oz. lard	175 g. lard	¾ cup lard
1½ lb. skinless sausages	700 g. skinless sausages	1½ lb. all-pork sausages
6 medium onions, sliced	6 medium onions, sliced	6 medium onions, sliced
12 oz. carrots, chopped	350 g. carrots, chopped	¾ lb. carrots, chopped
2½ lb. frozen shortcrust pastry	1¼ kg. frozen shortcrust pastry	2½ lb. frozen basic pie dough
3 eggs	3 eggs	3 eggs
salt and pepper to taste	salt and pepper to taste	salt and pepper to taste
¼ teaspoon cayenne pepper	¼ teaspoon cayenne pepper	½ teaspoon cayenne pepper
¾ pint warm milk	scant ½ litre warm milk	scant 2 cups warm milk
9 oz. Cheddar cheese, grated	250 g. Cheddar cheese, grated	2¼ cups grated Cheddar cheese
beaten egg to glaze	beaten egg to glaze	beaten egg to glaze

Melt the lard and use to fry the sausages gently for about 10 minutes. Drain sausages on absorbent kitchen paper and cool. Place the onion in the pan with the remaining lard and fry gently until just softened. Drain well. Place the carrots in a saucepan with a little salted water. Bring to the boil and cook for 5 minutes, then drain. Roll out two-thirds of the pastry and use to line three 1-lb. (½-kg.) oblong shaped foil containers, leaving the pastry ½ inch (1 cm.) above the top of the container all round. Place layers of sausages, onion and carrot in the pastry cases. Lightly beat the eggs with the seasonings and pour on the warmed milk. Stir in the cheese and divide the mixture between the three pies. Roll out the remaining pastry and use to make three lids. Cover filling with pastry, turn down edges of bottom pastry and crimp to seal well. Flute edges and decorate with leaves made from pastry trimmings. Brush with beaten egg and place in a moderately hot oven (375°F, 190°C, Gas Mark 5) for 1 hour until the pastry is golden brown. Cool.

Basic savoury pudding

King-size Makes 3 portions for 4 people. (Three puddings with choice of filling.)
To freeze When cold, cover with freezer foil. Seal and label.
To serve Boil or steam for 1 hour to defrost and reheat.
Storage time 4–6 months.

IMPERIAL	METRIC	AMERICAN
pastry	*pastry*	*dough*
12 oz. shredded suet	350 g. shredded suet	scant 2 cups finely chopped suet
1½ lb. self-raising flour	700 g. self-raising flour	6 cups all-purpose flour sifted with 7 teaspoons baking powder
¼ teaspoon salt	¼ teaspoon salt	¼ teaspoon salt
¼ teaspoon pepper	¼ teaspoon pepper	¼ teaspoon pepper
water to mix	water to mix	water to mix

To make the pastry, mix together the suet, flour and seasoning with sufficient water to make a firm dough. Roll out two-thirds of the pastry and use to line three 2-pint (generous 1-litre) pudding basins.

Filling ingredients given below are sufficient for one pudding.

Filling 1

IMPERIAL	METRIC	AMERICAN
12 oz. lean bladebone of pork, diced	350 g. lean bladebone of pork, diced	¾ lb. lean boneless pork, diced
3 leeks, sliced	3 leeks, sliced	3 leeks, sliced
2 tablespoons seasoned flour	2 tablespoons seasoned flour	3 tablespoons seasoned flour
1 teaspoon dried sage	1 teaspoon dried sage	1 teaspoon dried sage
chicken stock or water, or half stock and half dry white wine	chicken stock or water, or half stock and half dry white wine	chicken stock or water, or half stock and half dry white wine

Layer meat and leeks in pudding case, sprinkling each layer with seasoned flour and sage. Pour over sufficient stock to come three-quarters of the way up the pudding.

Filling 2

IMPERIAL	METRIC	AMERICAN
8 oz. lamb's liver, sliced	225 g. lamb's liver, sliced	½ lb. lamb liver, sliced
3 lamb's kidneys, sliced	3 lamb's kidneys, sliced	3 lamb kidneys, sliced
2 tablespoons seasoned flour	2 tablespoons seasoned flour	3 tablespoons seasoned flour
1½ teaspoons mixed dried herbs	1½ teaspoons mixed dried herbs	1½ teaspoons mixed dried herbs
1 8-oz. can tomatoes	1 225-g. can tomatoes	1 8-oz. can tomatoes

Toss liver and kidney in seasoned flour and herbs and add tomatoes. Transfer mixture to pudding case.

Filling 3

IMPERIAL	METRIC	AMERICAN
12 oz. stewing steak, diced	350 g. stewing steak, diced	¾ lb. beef stew meat, diced
4 oz. ox kidney, diced	100 g. ox kidney, diced	¼ lb. beef kidney, diced
2 tablespoons seasoned flour	2 tablespoons seasoned flour	3 tablespoons seasoned flour
4 oz. mushrooms, sliced	100 g. mushrooms, sliced	1 cup sliced mushrooms
beef stock or water, or half stock and half red wine	beef stock or water, or half stock and half red wine	beef stock or water, or half stock and half red wine

Toss steak and kidney in seasoned flour and add the mushrooms. Transfer mixture to pudding case and pour in sufficient stock to come three-quarters of the way up the pudding.

Roll out remaining pastry to form lids, moisten edges and seal well together. Cover securely with foil or greaseproof paper and steam or boil for 3–4 hours. Cool.

1 Marinating the chicken portions in the cider and oil mixture.

2 Packing the chicken joints, plus marinade, in shaped foil containers for freezing.

3 Placing the sealed shaped foil container in the freezer.

1

2

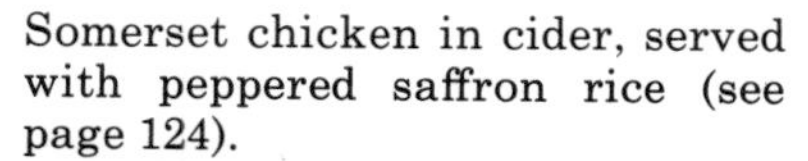

Somerset chicken in cider, served with peppered saffron rice (see page 124).

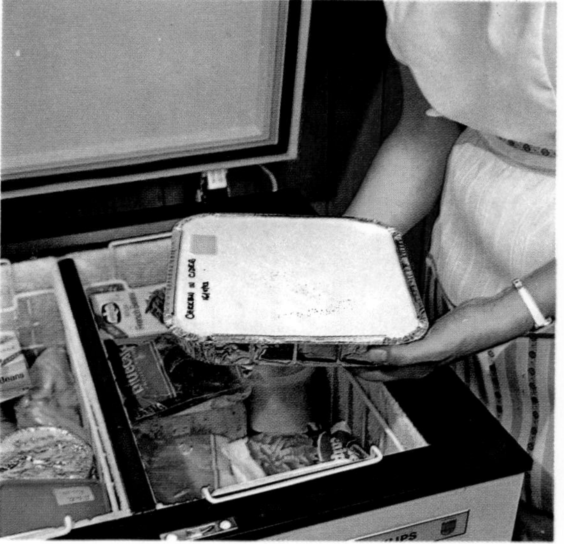

3

1 Red and green sweet pepper chopped ready to add to cooked rice for the topping for the pork chops.

2 Tearing lengths of foil from the roll, to be folded into squares to make pork chops with filling into parcels for freezing.

3 Chops placed on squares of double thickness foil, topped with rice mixture and being folded into well sealed parcels and labelled.

1

2

3

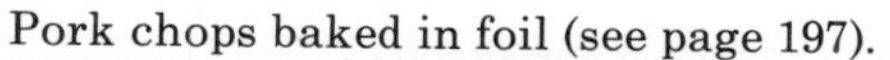

Pork chops baked in foil (see page 197).

Opposite Skipper pâté, brisling mousse (see page 194).

Golden guide number

to the two-way freeze method

Some housewives prefer to carry out all the messy preparation chores and freeze main dishes in the uncooked state. They contend that defrosting and reheating take nearly as long as defrosting and cooking. So they save oven heat, and also the trouble of cooling down the cooked dish rapidly and hygienically, ready for its dive into the freezer.

Others feel that more time is saved if the dish is fully cooked, giving an immediate hot fresh meal to serve, and leaving two or more ready-cooked family servings, one perhaps to be refrigerated and used up a few days later, and one only destined for long term freezer storage.

Stewing or braising, or possibly slow roasting, are the cooking methods most suitable. Marinating of the meat beforehand tenderises it and adds spicy flavour, especially to kebabs, if frozen this way.

The two-way method works like this. Prepare the dish completely for cooking and freeze at this stage. Or cook and freeze ready to defrost and reheat.

Method 1 Put the meat with the ingredients for the marinade or sauce in the covered casserole you will use for cooking, or in a suitable moisture-vapour-proof container. Place in the refrigerator for a few hours or overnight to allow the marinating process to begin, then freeze. Make sure that the casserole lid is sufficiently airtight. The tenderising process will continue and be intensified by freezing. When required let the casserole defrost at room temperature for at least 1 hour, unless made of a material which will stand transferring from the freezer to oven heat (ceramic, metal based, etc.) without risk of shattering. Put straight into the oven in the semi-frozen state, to thaw and cook in one operation. If you do not wish to put a useful casserole out of commission, a marinated joint can be frozen in a shaped foil container made by lining the casserole with freezer foil; leave a $\frac{1}{2}$-inch (1-cm.) headspace only above the joint. Smooth a cap of foil over the top of the joint so that the surface of meat exposed will not dehydrate while frozen. To serve, defrost, transfer to an open roasting tin or casserole, according to whether you prefer the braising or slow roasting method, and cook in the oven, at the same temperature for 45 minutes per lb. ($\frac{1}{2}$ kg.).

Method 2 Carry out the first stage as above until sufficiently marinated. Then cook according to the recipe, and serve one meal straight from the oven. In the case of joints, cool and slice the remainder of the meat rather thickly and divide between two suitable foil containers with the rest of the sauce. Freeze ready to defrost and reheat when required. For all other meat, serve 1 portion and freeze 2 portions.

One word of warning, do slice the braised meat and freeze meal-size portions of slices in the sauce, rather than freezing the joint whole. Braised slices defrost quickly and reheat perfectly in the sauce; the whole joint tends to be less tasty, perhaps not being fully impregnated with the sauce, and takes a long, long time to defrost even in a hot oven, right through to the centre. Foil trays of sliced meat and sauce can be placed still frozen in a moderately hot oven (400°F, 200°C, Gas Mark 6) for 30–45 minutes depending on the size of the container.

Barbecued beef

King-size Makes 3 servings for 4 people.
Storage time 3 months.

Imperial	Metric	American
$3\frac{1}{2}$–$4\frac{1}{2}$ lb. lean brisket of beef, boned and rolled	$1\frac{3}{4}$–$2\frac{1}{2}$ kg. lean brisket of beef, boned and rolled	$3\frac{1}{2}$–$4\frac{1}{2}$ lb. rolled rump
3 tablespoons tomato purée	3 tablespoons tomato purée	4 tablespoons tomato paste
3 tablespoons cider vinegar	3 tablespoons cider vinegar	scant $\frac{1}{4}$ cup cider vinegar
2 tablespoons light brown sugar	2 tablespoons light brown sugar	3 tablespoons light brown sugar
1 tablespoon creamed horseradish sauce	1 tablespoon creamed horseradish sauce	1 tablespoon creamed horseradish sauce
1 tablespoon mild Continental mustard	1 tablespoon mild Continental mustard	1 tablespoon mild Continental mustard
1 teaspoon salt	1 teaspoon salt	1 teaspoon salt
$\frac{1}{4}$ teaspoon pepper	$\frac{1}{4}$ teaspoon pepper	$\frac{1}{4}$ teaspoon pepper
2 large onions, chopped	2 large onions, chopped	2 large onions, chopped

Place joint in an ovenproof casserole. Mix together all the other ingredients. Pour over the meat and cover. Leave for 3–5 hours, turning occasionally. For method 1 freeze at this stage.

To cook, place in a moderate oven (350°F, 180°C, Gas Mark 4) allowing 45 minutes per lb. ($\frac{1}{2}$ kg.). Serve 1 portion. Slice meat and freeze in sauce as above.

Bacon grill fritters, hot dog rolls **(see page 196).**

Beef pomodoro

King-size Makes 3 servings for 4 people.
Storage time 3 months.

Imperial	Metric	American
3½–4½ lb. lean brisket of beef, boned and rolled	1¾–2¼ kg. lean brisket of beef, boned and rolled	3½–4½ lb. rolled rump
6 tablespoons oil	6 tablespoons oil	½ cup oil
¼ pint red wine	1½ dl. red wine	⅔ cup red wine
1 15-oz. can tomato juice	1 425-g. can tomato juice	1 15-oz. can tomato juice
1 teaspoon salt	1 teaspoon salt	1 teaspoon salt
¼ teaspoon pepper	¼ teaspoon pepper	¼ teaspoon pepper
4 oz. black olives, stoned	100 g. black olives, stoned	scant 1 cup pitted ripe olives

Place joint in an ovenproof casserole and follow instructions for previous recipe.

Pineapple pork chops

King-size Makes 3 servings for 4 people.
To freeze Pack into three shaped foil containers, each suitable in size and shape to hold four pork chops and one-third of the sauce.
To serve Uncooked chops – uncover and place in a moderately hot oven (375°F, 190°C, Gas Mark 5) for 1¼ hours. Cooked chops – uncover and place in oven for 40 minutes.
Storage time 3 months.

Imperial	Metric	American
12 lean pork chops	12 lean pork chops	12 lean pork chops
6 tablespoons corn oil	6 tablespoons corn oil	½ cup corn oil
juice and grated zest of 3 lemons	juice and grated zest of 3 lemons	juice and grated zest of 3 lemons
1 8-oz. can pineapple pieces	1 225-g. can pineapple pieces	1 8-oz. can pineapple pieces
3 tablespoons dark brown sugar	3 tablespoons dark brown sugar	4 tablespoons dark brown sugar
3 teaspoons ground ginger	3 teaspoons ground ginger	3 teaspoons ground ginger
2 teaspoons salt	2 teaspoons salt	2 teaspoons salt
¼ teaspoon pepper	¼ teaspoon pepper	¼ teaspoon pepper
1 medium onion, chopped	1 medium onion, chopped	1 medium onion, chopped
1 teaspoon dried oregano	1 teaspoon dried oregano	1 teaspoon dried oregano

Arrange pork chops close together in a large roasting tin. Mix together the remaining ingredients to make a marinade and pour over the meat. Leave for 2–3 hours, turning occasionally. For method 1 freeze dish at this stage.

To cook, place in a moderately hot oven (375°F, 190°C, Gas Mark 5) for 40 minutes, turning twice in the sauce during this time. Cool.

Pork in spiced cider

King-size Makes 3 servings for 4 people.
To freeze Pack into three polythene containers. Seal and label.
To serve Hold container under cold running water until contents can be turned out into a strong saucepan. Uncooked – heat gently to boiling point and simmer for 2 hours. Cooked – simmer for 5 minutes only.
Storage time 3 months.

Imperial	Metric	American
4 lb. bladebone of pork	2 kg. bladebone of pork	4 lb. lean boneless pork
4 tablespoons seasoned cornflour	4 tablespoons seasoned cornflour	5 tablespoons seasoned cornstarch
4 large carrots	4 large carrots	4 large carrots
4 dessert apples	4 dessert apples	4 dessert apples
1 pint dry cider	generous ½ litre dry cider	2½ cups cider
4 bay leaves	4 bay leaves	4 bay leaves
6 cloves	6 cloves	6 cloves
6 peppercorns	6 peppercorns	6 peppercorns
¼ teaspoon powdered mace or nutmeg	¼ teaspoon powdered mace or nutmeg	¼ teaspoon powdered mace or nutmeg
1 lb. small onions	½ kg. small onions	1 lb. small onions
1½ pints stock	scant 1 litre stock	3¾ cups stock

Cut the meat into 1-inch (2·5-cm) cubes and coat with the seasoned cornflour. Thickly slice the peeled carrots, core and quarter the apples. In a large saucepan heat together the cider, bay leaves, cloves, peppercorns and mace. Bring to boiling point, remove from the heat and allow to cool. Remove spices with a slotted draining spoon and add the meat, onions, carrots and apples to the cider. Pour over the stock and stir well. For method 1 freeze dish at this stage.

To cook, bring to boiling point, cover and simmer for 2 hours. Cool. If you prefer a thicker stew, moisten remaining cornflour with a little cold water and stir into the pan. Cook for a further 3 minutes.

Worcestershire beef

King-size Makes 3 servings for 4 people.
To freeze Pack into three polythene containers. Seal and label.
To serve Hold container under cold water until contents can be turned out into a strong saucepan. Add a little water. Uncooked – heat gently to boiling point and simmer for 1½ hours, or until meat is tender. Cooked – simmer for 10 minutes only.
Storage time 3 months.

Imperial	Metric	American
5 tablespoons corn oil	5 tablespoons corn oil	6 tablespoons corn oil
3 tablespoons vinegar	3 tablespoons vinegar	4 tablespoons vinegar
3 tablespoons Worcestershire sauce	3 tablespoons Worcestershire sauce	scant ¼ cup Worcestershire sauce
1 teaspoon salt	1 teaspoon salt	1 teaspoon salt
1 teaspoon pepper	1 teaspoon pepper	1 teaspoon pepper
1 teaspoon dried thyme	1 teaspoon dried thyme	1 teaspoon dried thyme
1 bay leaf	1 bay leaf	1 bay leaf
¾ pint beef stock	scant ½ litre beef stock	scant 2 cups beef stock
2 large onions, grated	2 large onions, grated	2 large onions, grated
1 tablespoon cornflour	1 tablespoon cornflour	1 tablespoon cornstarch
3 lb. stewing beef, diced	1½ kg. stewing beef, diced	3 lb. beef stew meat, diced

Beat together the oil, vinegar, Worcestershire sauce, seasonings and herbs. Stir into the beef stock, in a large saucepan and bring to boiling point. Add the onion and simmer together for 15 minutes. Moisten the cornflour with a little cold water, stir into the pan and remove from heat. When cool, add the beef to the marinade and stir. Cover and allow to stand for 3–12 hours. For method 1 freeze dish at this stage.

To cook, bring to boiling point, cover and simmer for 1½ hours, or until tender. Cool.

Variety meat loaf with orange sauce

King-size Makes 3 servings for 4 people.
To freeze Open freeze loaves until solid then wrap in freezer foil. Pack sauce in three small polythene containers. Seal and label.
To serve Defrost one loaf in foil wrapping and one container of sauce. Unwrap loaf and place on an ovenproof serving dish. Uncooked loaf – place in a moderately hot oven (375°F, 190°C, Gas Mark 5) for 30 minutes. Coat with the sauce and return to the oven for a further 30 minutes. Cooked loaf – coat with sauce and place in oven for 30 minutes.
Storage time 3 months.

Imperial	Metric	American
6 oz. soft white breadcrumbs	175 g. soft white breadcrumbs	3 cups fresh soft bread crumbs
scant ½ pint strong beef stock	2½ dl. strong beef stock	scant 1¼ cups strong beef stock
3 eggs	3 eggs	3 eggs
2 lb. pork, minced	1 kg. pork, minced	2 lb. pork, ground
1½ lb. pig's liver, minced	700 g. pig's liver, minced	1½ lb. pork liver, ground
4 tablespoons grated onion	4 tablespoons grated onion	5 tablespoons grated onion
½ teaspoon seasoned pepper	½ teaspoon seasoned pepper	½ teaspoon seasoned pepper
2 teaspoons celery salt	2 teaspoons celery salt	2 teaspoons celery salt
2 tablespoons chopped parsley	2 tablespoons chopped parsley	3 tablespoons chopped parsley
1 tablespoon flour	1 tablespoon flour	1 tablespoon all-purpose flour
6 rashers streaky bacon	6 rashers streaky bacon	6 bacon slices
sauce	*sauce*	*sauce*
2 tablespoons corn oil	2 tablespoons corn oil	3 tablespoons corn oil
1 large onion, grated	1 large onion, grated	1 large onion, grated
1 16-oz. can tomato juice	1 450-g. can tomato juice	1 16-oz. can tomato juice
½ 6-oz. can frozen orange juice concentrate	½ 175-g. can frozen orange juice concentrate	½ 6-oz. can frozen orange juice concentrate
1 tablespoon lemon juice	1 tablespoon lemon juice	1 tablespoon lemon juice
1 tablespoon Worcestershire sauce	1 tablespoon Worcestershire sauce	1 tablespoon Worcestershire sauce
1 tablespoon light brown sugar	1 tablespoon light brown sugar	1 tablespoon light brown sugar
2 teaspoons salt	2 teaspoons salt	2 teaspoons salt

Add breadcrumbs to hot beef stock and leave to stand for 3 minutes, then beat until smooth. Lightly beat the eggs. Place the minced pork, liver, onion and seasonings in a large bowl and sprinkle over the flour. Combine well together and stir in the stock and beaten eggs. Divide the mixture into three equal portions and shape each into a meat loaf with floured hands. Remove rind from bacon and top each loaf with two rashers placed diagonally from corner to corner. For method 1 freeze dish at this stage.

To cook, place all three loaves in a roasting tin and place in a moderate oven (350°F, 180°C, Gas Mark 4) for 1 hour, basting once with the pan juices. Cool.

To make up the sauce, heat the oil and fry the onion gently until soft but not browned. Add the other ingredients, bring to the boil, cover and simmer for 10 minutes. Cool.

Stuffed bacon rolls with cabbage

King-size Makes 3 servings for 4 people.

To freeze Wrap each roll in freezer film or foil and pack closely in three rigid-based polythene containers. Seal and label.

To serve Defrost for 2 hours, and place on a greased baking tray. Uncooked rolls – place in a moderately hot oven (375°F, 190°C, Gas Mark 5) for 40 minutes. Cooked rolls – place in oven for 30 minutes only. Meanwhile, finely shred 1 small white cabbage for each 4-portion serving. Boil quickly in salted water until just tender, about 10 minutes. Drain and stir in a knob of butter. Turn cabbage into a serving dish and top with the hot bacon rolls.

Storage time 4–6 months.

Imperial	Metric	American
6 small, tart apples	6 small, tart apples	6 small, tart apples
3 medium onions	3 medium onions	3 medium onions
$1\frac{1}{2}$ lb. sausagemeat	700 g. sausagemeat	$1\frac{1}{2}$ lb. sausagemeat
3 tablespoons chopped parsley	3 tablespoons chopped parsley	4 tablespoons chopped parsley
4 oz. soft white breadcrumbs	100 g. soft white breadcrumbs	2 cups fresh soft bread crumbs
3 eggs, beaten	3 eggs, beaten	3 eggs, beaten
salt and pepper to taste	salt and pepper to taste	salt and pepper to taste
24 small rashers streaky bacon	24 small rashers streaky bacon	24 small bacon slices

Peel, core and chop the apples and chop the onions finely. Mix together the sausagemeat, apples, parsley, onion, breadcrumbs and beaten egg. Season with salt and pepper to taste. Remove the rinds from the bacon rashers and flatten each rasher with a broad-bladed knife. Place 1 generous tablespoon of stuffing on the end of each rasher, roll up and secure with a wooden cocktail stick. For method 1 freeze at this stage.

To cook, place on greased baking trays and cook in a moderately hot oven (375°F, 190°C, Gas Mark 5) for 30 minutes. Cool and remove sticks.

Golden guide number 14

to freezing cooked food in portions

Many of the cooked dishes suitable for freezing come in the form of stews, casseroles and thick sauces such as bolognaise to serve with pasta. The quickest way to pack the food ready to tuck in the freezer is to pour it into a number of rigid-based polythene containers, each holding a family-sized portion. But you may not have sufficient of these for the purpose and since their purchase represents a considerable investment, here are four possible alternatives. In each case, it is advisable to remove any unnecessary bones as these tend to pierce the wrapping while in the frozen state. After all, you do not want to waste valuable freezer space on them anyway. And do make such dishes thick and concentrated, not only to save storage space, but because the addition of water or other liquid at the time of reheating facilitates the process. Preformed packs which readily fit into a container for reheating come first.

1. Stew in shapes Line the saucepan or casserole in which you will want to reheat the stew with freezer foil. Use a piece large enough to leave plenty to fold over and close the top to make airtight. Pour in the stew, allow to cool. Partially freeze, fold in and seal, remove from the saucepan, label and freeze. (If it is a king-size recipe, choose three containers of suitable size.) At serving time, hold the pack under running warm water until you can strip off the foil (which may be washed while still hot and used for other household purposes) and pop the stew 'shape' into the pan which it fits exactly. It can be thawed and reheated gradually with the addition of a little extra liquid and an occasional stir to keep it from sticking.

2. Stew in bricks The second method is to make your stew into 'bricks', for oven heating. Line a big roasting tin with the foil, pour in the stew, cool and open freeze until fairly well set but not completely frozen – about 2 hours. Using a large, sharp knife, cut into three portions, and wrap each in foil or freezer paper. There is little expense or waste here. These oblong 'bricks' fit nicely into an average small casserole. With the addition of a little extra liquid the contents of a closed casserole will not even need to be stirred more than once – about 10 minutes before serving time. Place still frozen in a moderately hot oven (375°F, 190°C, Gas Mark 5) for about 1½ hours.

3. Stew in foil containers The third method is to pour cooked stew straight into shaped foil containers with lids (if the kind you buy has no lid, cover the top and crimp under the edges with sheet foil). To reheat, put straight into the oven as above. This is the way, too, if you wish to remove the lid or cover, and replace it with a topping which turns a stew into more of a fancy dish. The containers, although more costly than sheet wrapping, do eliminate the trouble of transferring to another dish for reheating and if carefully treated can continue to be used several times for other kitchen purposes.

4. Stew in boiling bags There are various types of boiling bags available. Make sure which you are buying. Porosan is considered sufficiently strong to use as a container for cooked food for freezing and which can be transferred straight to a saucepan of boiling water to defrost and reheat. Another type is considered by the manufacturers to require the added protection, while frozen, of over-wrapping. Put six together in a Tupperware flavour-saver for instance. Very heavy gauge sleeve polythene can be transformed into completely watertight sealed bags by using a heat sealer and this comes nearest to the commercial pack.

Plate dinners

An idea which developed naturally from the cook-ahead freezing method is the frozen 'plate dinner'. When a joint has been braised and sliced, and a rich sauce is available, it is a good idea to cook more vegetables than usual – mashed potatoes and diced carrots or peas are particularly suitable. Fill a few 3-section foil plates with meat slices well masked in sauce, and the appropriate vegetables. I bury a tiny nut of butter in the mound of potato and among the carrots or peas. Cover closely with sheet foil and crimp over the edge of the plate, label with the selected contents and freeze. Any special fancy may be catered for – 'very lean meat' etc. marked on label. These plate dinners defrost and reheat, covered, in a moderately hot oven (400°F, 200°C, Gas Mark 6) for 30 minutes, and enable hungry non-cooks to serve themselves a substantial meal with no fuss or washing-up worries. By the way, a dollop of stew could take the place of braised meat, but remember that too much liquid will tend to boil out of the covered plates during the reheating process. On the other hand, food that is too dry will shrivel up and become rather tasteless.

Baby foods

Very many casserole dishes containing poultry and meat would be suitable for baby if not too highly seasoned. When making up such a recipe in quantity leave out strong herbs and spices and remove a portion for baby before adding them. If necessary, transfer to a smaller saucepan or casserole and finish cooking, adding the extra flavourings to the rest before completing the cooking time. Baby's portion, when cooked, can be liquidised, then packed with the greatest attention to hygiene. Tupperware 2-oz. (50-g.) tumblers hold a nice amount for one meal for young babies. Defrost and stand the sealed tumblers in a clean pan. Place a saucer on top with a weight, such as a jam jar, and pour really hot water round the tumblers. In 5 minutes the food will be just warm enough for baby, who can be fed from the container with a teaspoon.

Here is a typical meal for a baby of one to two years. One 2-oz. (50-g.) tumbler potage St. Germain (see page 119), one 2-oz. (50-g.) tumbler liquidised poulet grand'-mère (see page 125), one 2-oz. (50-g.) tumbler apple purée (see page 186) sweetened with added clear honey. As baby gets older larger containers can be used, the food need not be liquidised and crunchy foods can be added.

Pork with peaches

King-size Makes 3 servings for 4 people.
Storage time 4 months.

IMPERIAL	METRIC	AMERICAN
3 lb. stewing pork	$1\frac{1}{2}$ kg. stewing pork	3 lb. pork stew meat
6 oz. butter	175 g. butter	$\frac{3}{4}$ cup butter
12 oz. button onions	350 g. button onions	$\frac{3}{4}$ lb. small onions
3 red peppers	3 red peppers	3 red sweet peppers
3 carrots, thinly sliced	3 carrots, thinly sliced	3 carrots, thinly sliced
1 8-oz. can peach halves, sliced	1 225-g. can peach halves, sliced	1 8-oz. can peach halves, sliced
2 beef stock cubes	2 beef stock cubes	2 beef bouillon cubes
1 chicken stock cube	1 chicken stock cube	1 chicken bouillon cube
$2\frac{1}{2}$ pints boiling water	$1\frac{1}{4}$ litres boiling water	$6\frac{1}{4}$ cups boiling water
3 tablespoons lemon juice	3 tablespoons lemon juice	4 tablespoons lemon juice
3 tablespoons flour	3 tablespoons flour	4 tablespoons all-purpose flour

Cut the meat into small cubes. Melt the butter and use to fry the meat until lightly browned. Remove meat and add the onions to the pan. Cook gently for 2 minutes. Deseed and chop the red pepper and blanch in a little boiling water. Place all the vegetables and the meat in a large casserole, together with the peaches and 3 tablespoons of the syrup from the can. Dissolve the stock cubes in the boiling water, add the lemon juice and bring to the boil. Moisten the flour with a little cold water and work in some of the boiling stock. Return to the pan, bring to the boil and allow to thicken slightly. Pour the sauce over the meat and vegetables in the casserole, cover and cook in a moderately hot oven (375°F, 190°C, Gas Mark 5) for 1 hour. Cool.

Beef and prune casserole

King-size Makes 3 servings for 4 people.
Storage time 3 months.

IMPERIAL	METRIC	AMERICAN
8 oz. prunes	225 g. prunes	$1\frac{1}{3}$ cups prunes
$3\frac{1}{2}$ lb. stewing steak	$1\frac{3}{4}$ kg. stewing steak	$3\frac{1}{2}$ lb. beef stew meat
2 oz. seasoned cornflour	50 g. seasoned cornflour	$\frac{1}{2}$ cup seasoned cornstarch
3 tablespoons oil	3 tablespoons oil	scant $\frac{1}{4}$ cup oil
$1\frac{1}{2}$ pints water	scant 1 litre water	$3\frac{3}{4}$ cups water
4 tablespoons vinegar	4 tablespoons vinegar	$\frac{1}{3}$ cup vinegar
4 tablespoons clear honey	4 tablespoons clear honey	5 tablespoons clear honey
$1\frac{1}{2}$ tablespoons soy sauce	$1\frac{1}{2}$ tablespoons soy sauce	2 tablespoons soy sauce
1 green pepper	1 green pepper	1 green sweet pepper
8 oz. button mushrooms	225 g. button mushrooms	2 cups button mushrooms

Soak the prunes in cold water overnight then drain. Cut the steak into 1-inch (2·5 cm.) cubes and toss in the seasoned cornflour. Heat the oil in a large flameproof casserole and use to brown the meat on all sides. Blend any remaining cornflour into the water and add to the casserole with the vinegar, honey and soy sauce. Cover and place in a moderate oven (350°F, 180°C, Gas Mark 4) for $1\frac{1}{2}$ hours. Deseed and slice the pepper and add to the casserole with the drained prunes and mushrooms. Return to the oven for a further 30 minutes. Cool.

Apple and apricot lamb curry

King-size Makes 3 servings for 4 people.
To serve Surround hot curry with fluffy boiled rice and serve with accompaniments (see page 202).
Storage time 3 months.

IMPERIAL	METRIC	AMERICAN
2 oz. desiccated coconut	50 g. desiccated coconut	$\frac{2}{3}$ cup shredded coconut
$\frac{1}{2}$ pint boiling water	3 dl. boiling water	$1\frac{1}{4}$ cups boiling water
4 lb. lamb, boned	2 kg. lamb, boned	4 lb. boneless lamb
4 tablespoons corn oil	4 tablespoons corn oil	$\frac{1}{3}$ cup corn oil
12 oz. onions, chopped	350 g. onions, chopped	$\frac{3}{4}$ lb. onions, chopped
4 tablespoons flour	4 tablespoons flour	5 tablespoons all-purpose flour
2 teaspoons ground ginger	2 teaspoons ground ginger	2 teaspoons ground ginger
4 tablespoons curry powder	4 tablespoons curry powder	5 tablespoons curry powder
2 teaspoons crushed coriander	2 teaspoons crushed coriander	2 teaspoons crushed coriander
salt to taste	salt to taste	salt to taste
$1\frac{1}{2}$ pints chicken stock	scant 1 litre chicken stock	$3\frac{3}{4}$ cups chicken stock
juice of 1 lemon	juice of 1 lemon	juice of 1 lemon
4 oz. seedless raisins	100 g. seedless raisins	$\frac{3}{4}$ cup seedless raisins
8 oz. dried apricots	225 g. dried apricots	$1\frac{1}{2}$ cups dried apricots
4 cooking apples	4 cooking apples	4 baking apples

Soak the coconut in the boiling water for 30 minutes then strain and reserve the 'milk'. Cut the lamb into 1-inch (2·5-cm.) cubes. Heat the oil in a frying pan and use to brown the meat on all sides. Remove meat and sauté onions in the remaining oil until soft but not browned. Replace the meat and sprinkle in the flour, ginger, curry powder, coriander and salt. Stir well. Gradually add the stock and coconut milk. Bring to the boil and add the lemon juice, raisins and apricots. Cover and simmer gently for $1\frac{1}{2}$ hours. Peel, core and chop the apples and add to the curry 20 minutes before end of cooking time. Taste and adjust seasoning. Cool.

Anna's veal casserole

King-size Makes 3 servings for 4 people.
Storage time 3–4 months.

IMPERIAL	METRIC	AMERICAN
1 tablespoon salt	1 tablespoon salt	1 tablespoon salt
2 tablespoons paprika pepper	2 tablespoons paprika pepper	3 tablespoons paprika pepper
2 oz. flour	50 g. flour	$\frac{1}{2}$ cup all-purpose flour
4 lb. veal, diced	2 kg. veal, diced	4 lb. veal, diced
4 oz. butter	100 g. butter	$\frac{1}{2}$ cup butter
6 large onions, chopped	6 large onions, chopped	6 large onions, chopped
6 large carrots, sliced	6 large carrots, sliced	6 large carrots, sliced
8 tablespoons tomato purée	8 tablespoons tomato purée	$9\frac{1}{2}$ tablespoons tomato paste
3 chicken stock cubes	3 chicken stock cubes	3 chicken bouillon cubes
1 lb. dried apricot halves	450 g. dried apricot halves	1 lb. dried apricot halves
1 tablespoon whole pickling spices	1 tablespoon whole pickling spices	1 tablespoon whole pickling spices

Mix together the salt, paprika pepper and flour and use to coat the meat. Melt the butter in a large flameproof casserole and use to sauté the onions and carrots until lightly brown. Add the meat and cook until browned on all sides. Sprinkle in the remaining seasoned flour, stir and add the tomato purée, crumbled stock cubes, apricots, spices tied in muslin, and enough water to come barely level with the contents of the casserole. Bring to boiling point, cover and simmer for about 2 hours, until the meat is tender. Cool.

Cheese plaits

King-size Makes 2 loaves.
To freeze Wrap in freezer foil. Seal and label.
To serve Place frozen loaf, still wrapped, in a moderately hot oven (375°F, 190°C, Gas Mark 5) for 20 minutes, opening the foil for the last 5 minutes to crisp the crust.
Storage time 3 months.

IMPERIAL	METRIC	AMERICAN
1 pint warm water	generous ½ litre warm water	2½ cups warm water
1 oz. dried yeast	25 g. dried yeast	1 oz. dry yeast
1½ oz. instant milk powder	40 g. instant milk powder	1½ oz. dried milk solids
2 teaspoons salt	2 teaspoons salt	2 teaspoons salt
2 oz. sugar	50 g. sugar	¼ cup sugar
2 lb. plain flour	900 g. plain flour	8 cups all-purpose flour
2 oz. butter, softened	50 g. butter, softened	¼ cup softened butter
3 eggs, beaten	3 eggs, beaten	3 eggs, beaten
1½ teaspoons black pepper	1½ teaspoons black pepper	1½ teaspoons black pepper
8 oz. Cheddar cheese, grated	225 g. Cheddar cheese, grated	2 cups grated Cheddar cheese
beaten egg to glaze	beaten egg to glaze	beaten egg to glaze

Place one-quarter of the water in a large bowl. Sprinkle over the yeast and stir to dissolve it before adding the remaining water, the dried milk, salt, sugar and half the flour. Stir well until smooth. Add the remaining ingredients, form into a dough and knead well for about 5 minutes. Place in a large greased polythene bag and leave to rise in a warm place until double in size, about 1–2 hours. Knead lightly for 2 minutes.

To make two plaits, divide dough into six equal pieces and shape each one into a roll, fatter in the centre and as long as a baking tray. Take three rolls, and plait them, tucking under both ends to make a neat shape. Place on a greased baking tray and brush with a little oil. Cover with greased polythene and leave to rise until double in size and bake in a moderate oven (350°F, 180°C, Gas Mark 4) for 20 minutes then reduce heat to 325°F, 170°C, Gas Mark 3 for a further 25 minutes. Brush with beaten egg 5 minutes before end of baking time. Cool.

Orange bread

King-size Makes 2 loaves.
To freeze Wrap in freezer foil. Seal and label.
To serve Place frozen loaf, still wrapped, in a moderately hot oven (375°F, 190°C, Gas Mark 5) for 20 minutes, opening the foil for the last 5 minutes to crisp the crust.
Storage time 3 months.

IMPERIAL	METRIC	AMERICAN
¼ pint warm water	1½ dl. warm water	⅔ cup warm water
1 oz. dried yeast	25 g. dried yeast	1 oz. dry yeast
¾ pint orange juice, warmed	4½ dl. orange juice, warmed	scant 2 cups orange juice, warmed
1½ lb. plain flour	700 g. plain flour	6 cups all-purpose flour
4 oz. sugar	100 g. sugar	½ cup sugar
1 teaspoon salt	1 teaspoon salt	1 teaspoon salt
2 oz. butter, softened	50 g. butter, softened	¼ cup softened butter
grated zest of 1 orange	grated zest of 1 orange	grated zest of 1 orange

Place the water in a large bowl, sprinkle on the yeast and stir until dissolved. Add the orange juice and one-third of the flour and beat thoroughly. Add the remaining ingredients and form into a dough. Turn onto a floured board and knead for 5 minutes. Place in a greased polythene bag and leave to rise in a warm place until double in size. Knead lightly for 2 minutes.

Divide the dough in half, shape each piece to fit a greased 1-lb. (½-kg.) loaf tin and brush with a little oil. Cover with greased polythene and leave to rise in a warm place until double in size. Bake in a moderate oven (350°F, 180°C, Gas Mark 4) for 40–45 minutes. Cool.

Note Frozen orange juice concentrate, diluted to normal strength, is ideal for this recipe.

Apricot nut bread

King-size Makes 2 loaves.
To freeze Wrap in freezer foil or polythene bag. Seal and label.
To serve Defrost, still wrapped, for about 2 hours.
Storage time 3 months.

IMPERIAL	METRIC	AMERICAN
1½ lb. plain flour	700 g. plain flour	6 cups all-purpose flour
12 oz. sugar	350 g. sugar	1½ cups sugar
2 tablespoons baking powder	2 tablespoons baking powder	3 tablespoons baking powder
1 teaspoon bicarbonate of soda	1 teaspoon bicarbonate of soda	1 teaspoon baking soda
¼ teaspoon salt	¼ teaspoon salt	¼ teaspoon salt
8 oz. dried apricots	225 g. dried apricots	scant 1½ cups dried apricots
8 oz. nuts, chopped	225 g. nuts, chopped	2 cups chopped nuts
4 eggs	4 eggs	4 eggs
4 oz. butter, melted	100 g. butter, melted	½ cup butter, melted
1 pint buttermilk	generous ½ litre buttermilk	2½ cups buttermilk

Sieve the dry ingredients into a bowl. Cut the apricots into small pieces and add to the bowl with the chopped nuts. Mix together the eggs, melted butter and buttermilk. Add this to the dry ingredients and beat until smooth. Divide the mixture between two greased 1-lb. (½-kg.) loaf tins and bake in a moderate oven (350°F, 180°C, Gas Mark 4) for 1 hour. Cool.

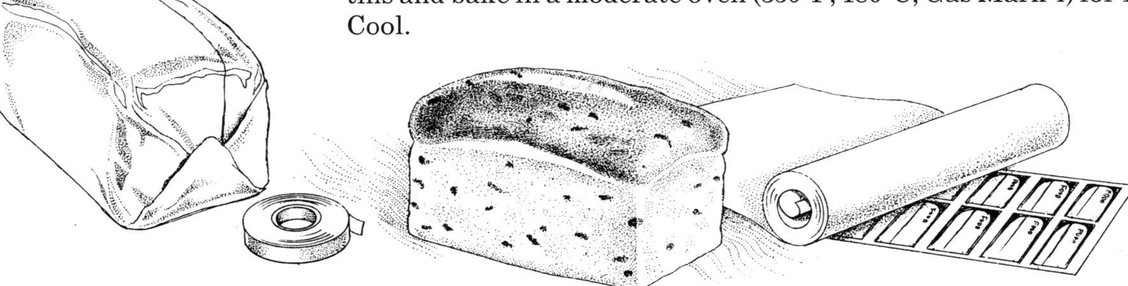

Australian boiled fruit cake

King-size Makes 2 cakes.
To freeze Wrap in freezer foil or polythene bag. Seal and label.
To serve Defrost, still wrapped, at room temperature for about 2 hours.
Storage time 4–6 months.

IMPERIAL	METRIC	AMERICAN
¾ pint water	4 dl. water	scant 2 cups water
6 oz. butter	175 g. butter	¾ cup butter
12 oz. sugar	350 g. sugar	1½ cups sugar
1 oz. cocoa powder	25 g. cocoa powder	¼ cup unsweetened cocoa
10 oz. raisins	275 g. raisins	2 cups raisins
1 teaspoon ground cinnamon	1 teaspoon ground cinnamon	1 teaspoon ground cinnamon
½ teaspoon ground cloves	½ teaspoon ground cloves	½ teaspoon ground cloves
½ teaspoon grated nutmeg	½ teaspoon grated nutmeg	½ teaspoon grated nutmeg
pinch salt	pinch salt	pinch salt
14 oz. plain flour	400 g. plain flour	3½ cups all-purpose flour
2 teaspoons baking powder	2 teaspoons baking powder	2 teaspoons baking powder
1 teaspoon bicarbonate of soda	1 teaspoon bicarbonate of soda	1 teaspoon baking soda
4 oz. walnuts, chopped	100 g. walnuts, chopped	1 cup chopped walnuts

Place the water, butter, sugar, cocoa powder, raisins and spices together in a saucepan. Bring to the boil and simmer for 4 minutes. Leave to cool to lukewarm. Sieve together the salt, flour, baking powder and bicarbonate of soda. Add to the raisin mixture with the walnuts and stir well until thoroughly blended. Divide the mixture between two 6-inch (15-cm.) square or 7-inch (18-cm.) round cake tins and bake in a moderate oven (350°F, 180°C, Gas Mark 4) for about 1½ hours, until well risen and golden brown. Allow to cool slightly before removing from tins. Cool.

Note Leave wrapped cake at room temperature for 1 week to mature before placing in freezer.

Spicy apple cake

King-size Makes 2 cakes.
To freeze Wrap in freezer foil. Seal and label.
To serve Place frozen cake, still wrapped, in a cool oven (300°F, 150°C, Gas Mark 2) for 30 minutes. Very good served warm with cream or ice cream.
Storage time 3 months.

IMPERIAL	METRIC	AMERICAN
4 oz. butter	100 g. butter	½ cup butter
4 oz. sugar	100 g. sugar	½ cup sugar
2 eggs	2 eggs	2 eggs
8 oz. plain flour	225 g. plain flour	2 cups all-purpose flour
2 teaspoons baking powder	2 teaspoons baking powder	2 teaspoons baking powder
¼ teaspoon salt	¼ teaspoon salt	¼ teaspoon salt
1 teaspoon ground cinnamon	1 teaspoon ground cinnamon	1 teaspoon ground cinnamon
1 teaspoon grated nutmeg	1 teaspoon grated nutmeg	1 teaspoon grated nutmeg
½ teaspoon vanilla essence	½ teaspoon vanilla essence	½ teaspoon vanilla extract
4 oz. walnuts, chopped	100 g. walnuts, chopped	1 cup chopped walnuts
4 cooking apples, grated	4 cooking apples, grated	4 baking apples, grated

Cream the butter and sugar until light and fluffy. Add the eggs and beat well. Sieve together the flour, baking powder, salt and spices and stir this into the creamed mixture. Add the vanilla essence, nuts and apples. Divide mixture between two greased and floured 8-inch (20-cm.) round or 7-inch (18-cm.) square cake tins and bake in a moderate oven (350°F, 180°C, Gas Mark 4) for 40–45 minutes. Cool on a wire tray.

Blackcurrant surprise

King-size Makes approximately 40 fingers.
To freeze Wrap in freezer foil. Seal and label.
To serve Unwrap and defrost at room temperature for 2–3 hours.
Storage time 4 months.

IMPERIAL	METRIC	AMERICAN
2 lb. blackcurrants	1 kg. blackcurrants	2 lb. black currants
1 lb. butter	450 g. butter	2 cups butter
1 lb. sugar	450 g. sugar	2 cups sugar
4 eggs	4 eggs	4 eggs
2 lb. plain flour	900 g. plain flour	8 cups all-purpose flour
8 teaspoons baking powder	8 teaspoons baking powder	8 teaspoons baking powder
1 teaspoon salt	1 teaspoon salt	1 teaspoon salt
1 pint milk	generous ½ litre milk	2½ cups milk
topping	*topping*	*topping*
8 oz. butter	225 g. butter	1 cup butter
1 lb. sugar	450 g. sugar	2 cups sugar
8 oz. plain flour	225 g. plain flour	2 cups all-purpose flour
2 teaspoons ground cinnamon	2 teaspoons ground cinnamon	2 teaspoons ground cinnamon

Top and tail the blackcurrants. Cream the butter and sugar until light and fluffy. Gradually beat in the eggs. Sieve together the flour, baking powder and salt and fold into the creamed mixture. Add sufficient milk to make a soft dropping consistency. Line two 7-inch (18-cm.) by 11-inch (28-cm.) baking tins with freezer foil and grease. Divide mixture between the two tins, smooth tops and sprinkle with the blackcurrants. To make the topping, cream butter and sugar as before and work in the the flour and cinnamon to give a crumbly consistency. Sprinkle the crumble topping over the fruit and bake in a moderate oven (350°F, 180°C, Gas Mark 4) for 1 hour. Cool in the tins. Cut into fingers.

Greengage pudding

King-size Makes 3 puddings.
To freeze Cover with foil. Seal and label.
To serve Steam or boil for 1 hour. Serve with cream.
Storage time 3 months.

IMPERIAL	METRIC	AMERICAN
10 oz. butter	275 g. butter	1¼ cups butter
10 oz. sugar	275 g. sugar	1¼ cups sugar
3 eggs, beaten	3 eggs, beaten	3 eggs, beaten
1¼ lb. self-raising flour	600 g. self-raising flour	5 cups all-purpose flour sifted with 5 teaspoons baking powder
1½ lb. greengages or plums, stoned	700 g. greengages or plums, stoned	1½ lb. greengages or plums, pitted

Cream the butter and sugar until light and fluffy. Gradually beat in the eggs and stir in the flour. Place a layer of mixture in the bottom of three foil pudding basins, cover with a layer of greengages. Repeat this process, ending with a layer of mixture. Cover with foil or greaseproof paper and steam for 1½ hours. Cool.

Baked stuffed apples

King-size Makes 3 servings for 4 people.
To freeze Place lid on container or cover with freezer foil and crimp edges together. Seal and label.
To serve Defrost at room temperature for 1–2 hours. If required hot, place in a moderate oven (350°F, 180°C, Gas Mark 4) for 20–25 minutes. Serve with ice cream, cream or custard sauce.
Storage time 4–6 months.

IMPERIAL	METRIC	AMERICAN
12 cooking apples	12 cooking apples	12 baking apples
6 tablespoons dark soft brown sugar	6 tablespoons dark soft brown sugar	7 tablespoons dark brown sugar
2 teaspoons ground cinnamon	2 teaspoons ground cinnamon	2 teaspoons ground cinnamon
4 oz. mixed dried fruit	100 g. mixed dried fruit	generous 1 cup mixed dried fruit
scant ½ pint water	2½ dl. water	1 cup water

Core the apples and score a horizontal line round the centre of each one. Divide the apples between three shaped foil dishes. Mix together the sugar and cinnamon and stuff the apples alternately with the dried fruit and sugar mixture, ending with sugar. Sprinkle the remaining sugar round the apples in the dishes, together with the water, to form a caramel sauce. Bake in a moderately hot oven (375°F, 190°C, Gas Mark 5) for 30–35 minutes. Cool.

Brown betty with fruit

King-size Makes 3 servings for 4 people.
To freeze Cover with freezer foil and crimp edges together or place dish in a polythene bag. Seal and label.
To serve Remove cover and place in a moderate oven (350°F, 180°C, Gas Mark 4) for 45 minutes.
Storage time 4–6 months.

Note Vary fruit according to the season:
June–July – Apricots, rhubarb, gooseberries, blackcurrants.
August–September – Apples, blackberries, damsons, plums.

IMPERIAL	METRIC	AMERICAN
6 oz. butter	175 g. butter	$\frac{3}{4}$ cup butter
12 oz. soft white breadcrumbs	350 g. soft white breadcrumbs	6 cups fresh soft bread crumbs
6 oz. soft brown sugar	175 g. soft brown sugar	$\frac{3}{4}$ cup brown sugar
3 lb. apricots, stoned (see note)	$1\frac{1}{2}$ kg. apricots, stoned (see note)	3 lb. apricots, pitted (see note)
6 tablespoons golden syrup	6 tablespoons golden syrup	7 tablespoons maple syrup
generous $\frac{1}{4}$ pint water	2 dl. water	$\frac{2}{3}$ cup water

Cut the butter into small pieces and rub into the breadcrumbs. Stir in the sugar. Divide one-third of this mixture between three greased ovenproof or shaped foil dishes. Cover with half the apricots and continue in layers, ending with a layer of crumbs. Place the syrup in a small saucepan with the water and stir until dissolved. Pour carefully over the puddings and bake them in a moderate oven (350°F, 180°C, Gas Mark 4) for 35–40 minutes, until golden brown. Cool.

Steamed apple crumb pudding

King-size Makes 3 servings for 4 people.
To freeze Cover with a sheet of freezer foil and crimp edges together. Label.
To serve Steam for 1 hour and serve with custard or syrup sauce.
Storage time 4–6 months.

IMPERIAL	METRIC	AMERICAN
$2\frac{1}{4}$ lb. cooking apples	generous 1 kg. cooking apples	$2\frac{1}{4}$ lb. baking apples
juice and grated zest of $1\frac{1}{2}$ lemons	juice and grated zest of $1\frac{1}{2}$ lemons	juice and grated zest of $1\frac{1}{2}$ lemons
12 oz. currants	350 g. currants	2 cups currants
$1\frac{1}{2}$ lb. soft white breadcrumbs	700 g. soft white breadcrumbs	20 cups fresh soft bread crumbs
12 oz. soft brown sugar	350 g. soft brown sugar	$1\frac{1}{2}$ cups brown sugar
9 eggs, beaten	9 eggs, beaten	9 eggs, beaten
$1\frac{1}{2}$ teaspoons grated nutmeg	$1\frac{1}{2}$ teaspoons grated nutmeg	$1\frac{1}{2}$ teaspoons grated nutmeg

Peel and grate the apples and mix immediately with the lemon juice and zest. Add the remaining ingredients and stir thoroughly. Divide mixture between three shaped foil pudding basins and steam for 2 hours. Cool.

Quick service dishes

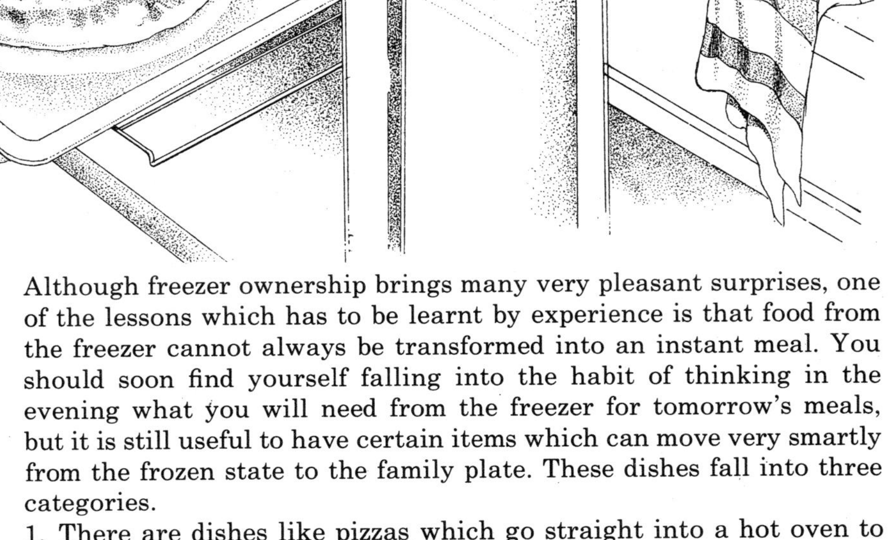

Although freezer ownership brings many very pleasant surprises, one of the lessons which has to be learnt by experience is that food from the freezer cannot always be transformed into an instant meal. You should soon find yourself falling into the habit of thinking in the evening what you will need from the freezer for tomorrow's meals, but it is still useful to have certain items which can move very smartly from the frozen state to the family plate. These dishes fall into three categories.

1. There are dishes like pizzas which go straight into a hot oven to defrost and reheat, (or defrost and cook) without requiring any further preparation.

2. There are small items which are fully cooked and defrost well within an hour of emerging from storage.
3. There are recipes such as apple and cranberry coupes, making use of home-frozen foods which can be quickly defrosted and turned into a mouthwatering finished dish.

The real criterion of quick-service is – can you suddenly decide to serve the food for a meal and have it ready within an hour or so? Try to keep one basket, or one shelf free for these useful quickies. This will prevent your stock from dwindling without being replenished. In my log book I put the initial Q in red against any entry which qualifies for inclusion in this category. Like many other busy housewives, my catering expertise is often put to the test in this way, and in fact I sometimes feel more inclined to label them L for lifesavers.

Family pâté

Makes 8 servings.
To freeze Cover with foil and smooth down edges. Label.
To serve Defrost at room temperature for 2 hours.
Storage time 1–2 months.

IMPERIAL	METRIC	AMERICAN
4 oz. lean streaky bacon	100 g. lean streaky bacon	$\frac{1}{4}$ lb. bacon slices
8 oz. pig's or ox liver	225 g. pig's or ox liver	$\frac{1}{2}$ lb. pork or beef liver
$1\frac{1}{4}$ lb. lean belly pork	600 g. lean belly pork	$1\frac{1}{4}$ lb. fresh picnic shoulder
1 medium onion, chopped	1 medium onion, chopped	1 medium onion, chopped
1 clove garlic, crushed	1 clove garlic, crushed	1 clove garlic, crushed
1 teaspoon salt	1 teaspoon salt	1 teaspoon salt
freshly ground black pepper to taste	freshly ground black pepper to taste	freshly ground black pepper to taste

De-rind the bacon and remove skin and any bones from the pork. Cut the bacon, pork and liver into pieces and mince three times with the onion and garlic. (If preferred, liquidise in an electric blender.) Mix in the seasoning thoroughly and divide mixture between eight small greased ramekin dishes. Cover each with foil and stand them in a shallow roasting tin containing a little cold water. Place in a cool oven (300°F, 150°C, Gas Mark 2) for about 1 hour. Remove the foil, press down the top of each dish of pâté and cool.

Pizza with traditional topping

Makes 3 pizzas with choice of topping.
To freeze Wrap in freezer foil. Seal and label.
To serve Unwrap, brush with oil and place in a moderately hot oven (400°F, 200°C, Gas Mark 6) for 35–40 minutes.
Storage time 3 months.

IMPERIAL	METRIC	AMERICAN
yeast liquid	*yeast liquid*	*yeast liquid*
1 teaspoon castor sugar	1 teaspoon castor sugar	1 teaspoon granulated sugar
½ pint warm water	3 dl. warm water	1¼ cups warm water
2 teaspoons dried yeast	2 teaspoons dried yeast	2 teaspoons active dry yeast
dough	*dough*	*dough*
1 lb. plain flour	900 g. plain flour	4 cups all-purpose flour
2 teaspoons salt	2 teaspoons salt	2 teaspoons salt
4 teaspoons olive oil	4 teaspoons olive oil	4 teaspoons olive oil
topping	*topping*	*topping*
2 green peppers	2 green peppers	2 green sweet peppers
1½ lb. tomatoes	700 g. tomatoes	1½ lb. tomatoes
6 tablespoons oil	6 tablespoons oil	½ cup olive oil
2 large onions, chopped	2 large onions, chopped	2 large onions, chopped
4 cloves garlic, crushed	4 cloves garlic, crushed	4 cloves garlic, crushed
2 tablespoons tomato purée	2 tablespoons tomato purée	3 tablespoons tomato paste
½ teaspoon dried thyme	½ teaspoon dried thyme	½ teaspoon dried thyme
2 bay leaves	2 bay leaves	2 bay leaves
4 teaspoons sugar	4 teaspoons sugar	4 teaspoons sugar
salt and black pepper to taste	salt and black pepper to taste	salt and black pepper to taste
3 slices lean ham	3 slices lean ham	3 slices lean ham
6 oz. Mozzarella cheese	175 g. Mozzarella cheese	6 oz. Mozzarella cheese
2 small cans anchovy fillets, drained	2 small cans anchovy fillets, drained	2 small cans anchovy fillets, drained
18 black olives, halved	18 black olives, halved	18 ripe olives, halved

To make the yeast liquid, dissolve the sugar in the water and sprinkle over the dried yeast. Allow to stand until frothy, about 10 minutes. Sieve the flour and salt into a bowl, pour in the yeast liquid and mix to a soft dough. Turn out onto a floured surface and knead until smooth and elastic, about 10 minutes. Shape into a ball, place in a greased polythene bag and leave to rise until double in size.

Deseed and chop the peppers and skin and chop the tomatoes. Heat the oil and fry the onions until softened, add the chopped pepper and cook gently for 5 minutes. Add the chopped tomatoes, garlic, tomato purée, thyme, bay leaves, sugar and seasoning to taste. Bring to the boil, stirring constantly, and simmer for about 30 minutes until very thick. Remove bay leaves and cool.

Knead the dough lightly for 2 minutes, divide into three equal portions and pat each one into a greased 7-inch (18-cm.) sandwich tin, or flan ring placed on a baking tray. Brush dough with olive oil. Place a slice of ham on each piece of dough. Cover with thin slices of cheese and spoon the tomato mixture over. Make a criss-cross pattern on each pizza with the anchovy fillets and garnish with the halved olives. Bake in a hot oven (450°F, 230°C, Gas Mark 8) for 30 minutes. Cool.

Two alternative pizza toppings

Smoked haddock topping Substitute 8 oz. (225 g.) cooked flaked smoked haddock for the ham.

Aubergine topping Slice 1 small aubergine, fry lightly on both sides in a little oil and substitute for the ham. Instead of anchovies, use 2 oz. (50 g.) garlic sausage, diced. Mix with the olives and scatter over the top of the pizza.

Golden guide number 

to pancakes

When eggs are down to their lowest in price, you can afford to be lavish with them. Fry up a large quantity of pancakes, which have long been a favourite sweet dish but are becoming increasingly popular with savoury fillings. Use them in four ways.

1. Basic pancakes Pack in layers and serve as a sweet with castor sugar and lemon juice.

2. Rolled pancakes Pack rolled around a savoury filling in shallow foil containers ready for oven baking.

3. Layered pancakes Make an appetising filling and layer the reheated pancakes with this, cutting in wedges to serve.

4. Folded pancakes Spread pancakes with a fruit and sugar-flavoured butter, fold in four and freeze. Oven bake to defrost and reheat, serve masked with a fruity sauce.

Examples of all these types of recipes are given here.

Basic pancakes

Makes Approximately 32 small pancakes.

To freeze Stack pancakes in layers with foil or freezer film dividers. Pack in a polythene bag. Seal and label.

To serve Unpack and spread out to defrost at room temperature for 30 minutes. If required hot, spread out on baking trays, cover with foil and place in a moderately hot oven (400°F, 200°C, Gas Mark 6) for 10 minutes.

Storage time 4–6 months.

Imperial	Metric	American
1 lb. plain flour	450 g. plain flour	4 cups all-purpose flour
1 teaspoon salt	1 teaspoon salt	1 teaspoon salt
6 eggs	6 eggs	6 eggs
2 pints milk	generous 1 litre milk	5 cups milk
4 tablespoons oil	4 tablespoons oil	$\frac{1}{3}$ cup oil

Sieve the flour and salt into a bowl and beat in the eggs and a little of the milk. Beat until smooth and gradually add the remaining milk; lastly fold in the oil. Allow batter to stand for about 30 minutes. Fry thin pancakes in a little oil until golden brown on both sides. Cool.

Italian stuffed pancakes

Makes 4 servings.

To freeze Pack filled pancakes in shaped foil dishes, cover with lid or sheet of freezer foil and crimp edges together. Label.

To serve Uncover, sprinkle with grated cheese and place in a moderately hot oven (400°F, 200°C, Gas Mark 6) for 30 minutes.

Storage time 4–6 months.

Imperial	Metric	American
8 small thin pancakes	8 small thin pancakes	8 small thin pancakes
filling	*filling*	*filling*
1 oz. butter	25 g. butter	2 tablespoons butter
1 small onion, chopped	1 small onion, chopped	1 small onion, chopped
1 oz. flour	25 g. flour	$\frac{1}{4}$ cup all-purpose flour
1 tablespoon tomato purée	1 tablespoon tomato purée	1 tablespoon tomato paste
1 teaspoon sugar	1 teaspoon sugar	1 teaspoon sugar
salt and pepper to taste	salt and pepper to taste	salt and pepper to taste
dash Worcestershire sauce	dash Worcestershire sauce	dash Worcestershire sauce
12 oz. cooked minced beef	350 g. cooked minced beef	$1\frac{1}{2}$ cups cooked ground beef

Melt the butter and use to cook the onion until softened but not coloured. Stir in the flour, tomato purée, sugar and seasonings. Cook for 1 minute, add the minced beef and 2 tablespoons boiling water. Stir over gentle heat for a further 2 minutes, adding a little more water if necessary but keeping the mixture very firm. Divide the filling between the pancakes and roll up. Cool.

Country house chicken (see page 196)

1 Dipping white fish fillets in iced water, after open freezing, to glaze them. This process is carried out to build up the glaze.

2 Heat sealing flat packs of fish fillets sufficient for a family meal. Fish must be fully defrosted before making into turbans.

3 Skinned fillets, divided lengthways, rolled round and secured with wooden cocktail sticks. The stuffing being piped into the centre of each fish roll.

1

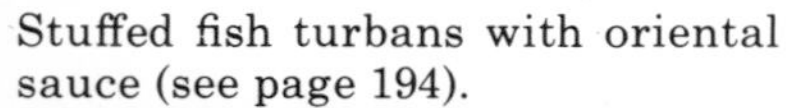

Stuffed fish turbans with oriental sauce (see page 194).

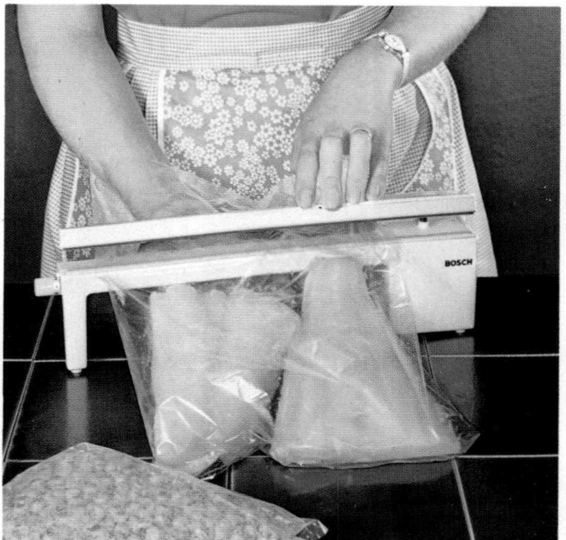

2

3

Layered shellfish pancakes

Makes 4 servings.

Imperial	Metric	American
8 small thin pancakes	8 small thin pancakes	8 small thin pancakes
filling	*filling*	*filling*
1 8-oz. pack peas and sweetcorn	1 225-g. pack peas and sweetcorn	1 8-oz. package peas and kernel corn
½ pint béchamel sauce (see page 100)	3 dl. béchamel sauce (see page 100)	1¼ cups béchamel sauce (see page 100)
8 oz. frozen prawns	225 g. frozen prawns	½ lb. frozen prawns or shrimp

Cook peas and sweetcorn according to directions on the pack. Turn the béchamel sauce into a saucepan and reheat very gently, stirring to prevent burning. When very hot, add most of the prawns, reserving a few for garnish. Return to the heat and bring slowly to boiling point. Place one pancake on a hot serving dish, spread with sweetcorn mixture, cover with another pancake and spread with prawn sauce. Continue in this way using all the pancakes and fillings. Garnish with the reserved prawns. Serve cut into wedges.

Orange butter pancakes

Makes 4 servings.

To freeze Pack filled pancakes slightly overlapping in a shallow shaped foil tray. Cover with lid or sheet of freezer foil and crimp edges together. Pack sauce in a small polythene container. Seal and label.

To serve Uncover and place frozen pancakes in a moderately hot oven (400°F, 200°C, Gas Mark 6) for 10 minutes. Reheat sauce in a small saucepan, pour over pancakes and return to oven for a further 10 minutes.

Storage time 4–6 months.

Imperial	Metric	American
8 small thin pancakes	8 small thin pancakes	8 small thin pancakes
filling	*filling*	*filling*
1 tablespoon coarse-cut orange marmalade	1 tablespoon coarse-cut orange marmalade	1 tablespoon coarse-cut orange marmalade
2 oz. unsalted butter	50 g. unsalted butter	¼ cup sweet butter
3 oz. icing sugar, sieved	75 g. icing sugar, sieved	¾ cup sifted confectioners sugar
2 tablespoons orange juice concentrate	2 tablespoons orange juice concentrate	3 tablespoons orange juice concentrate
sauce	*sauce*	*sauce*
1 oz. unsalted butter	25 g. unsalted butter	2 tablespoons sweet butter
1½ oz. castor sugar	40 g. castor sugar	3 tablespoons granulated sugar
2 tablespoons orange juice concentrate	2 tablespoons orange juice concentrate	3 tablespoons orange juice concentrate

Chop the marmalade finely and beat into the butter and icing sugar in a small bowl. Gradually beat in the orange juice. Spread each pancake very thinly with this filling and fold into four. To make the sauce, melt the butter in a small saucepan, add the sugar and orange juice and stir until dissolved. Cool.

Gouda supper snacks

Makes 12.
To freeze Pack in freezer foil or in rigid-based polythene containers with dividers of foil or freezer film. Seal and label.
To serve Separate the slices on a plate and allow to defrost for about 45 minutes. Place under a hot grill for 4–5 minutes or until the bacon is crisp and the surface of the cheese mixture golden brown.
Storage time 4–6 months.
Variation Instead of the bacon, make a criss-cross pattern of anchovy fillets. Reduce the butter for the cheese sauce to ½ oz. (15 g.) using the oil from the anchovy cans instead of the remaining butter. If liked the snacks can be garnished with tiny sprigs of parsley.

IMPERIAL	METRIC	AMERICAN
1 oz. butter	25 g. butter	2 tablespoons butter
1 large onion, grated	1 large onion, grated	1 large onion, grated
2 tablespoons flour	2 tablespoons flour	3 tablespoons all-purpose flour
1 teaspoon mild Continental mustard	1 teaspoon mild Continental mustard	1 teaspoon mild Continental mustard
¼ pint milk	1½ dl. water	⅔ cup milk
12 oz. Gouda cheese, grated	350 g. Gouda cheese, grated	3 cups grated Gouda cheese
1 large sliced white loaf	1 large sliced white loaf	1 large white sliced loaf
12 rashers streaky bacon	12 rashers streaky bacon	12 bacon slices

Melt the butter in a small saucepan and use to cook the onion very gently until transparent. Stir in the flour until fully absorbed, then the mustard and the milk, to form a thick sauce. Work in the grated cheese until it resembles a ball of soft dough. Toast the bread on one side only, removing the crusts if preferred, and spread the uncooked side with the cheese mixture. De-rind the bacon and divide in half. Cover each cheese slice with two pieces of bacon placed crosswise from corner to corner.

Beefy supper snacks

Makes 8 snacks.
To freeze Pack in freezer foil or in rigid-based polythene containers with dividers of foil or freezer film. Seal and label.
To serve Melt a little fat or oil in a frying pan and fry snacks for 10 minutes, turning once and frying the meat side first. Garnish with watercress and serve hot.
Storage time 4–6 months.

IMPERIAL	METRIC	AMERICAN
1 egg	1 egg	1 egg
12 oz. cooked minced beef	350 g. cooked minced beef	1½ cups cooked ground beef
salt and pepper to taste	salt and pepper to taste	salt and pepper to taste
1 tablespoon grated onion	1 tablespoon grated onion	1 tablespoon grated onion
1 tablespoon finely chopped pickled beetroot	1 tablespoon finely chopped pickled beetroot	1 tablespoon finely chopped pickled beets
1 tablespoon chopped capers	1 tablespoon chopped capers	1 tablespoon chopped capers
4 oz. mushrooms, chopped	100 g. mushrooms, chopped	1 cup chopped mushrooms
8 large slices white bread	8 large slices white bread	8 large slices white bread
butter for spreading	butter for spreading	butter for spreading

Beat the egg in a basin and stir in the meat. Season with salt and pepper and moisten with a little water. Add the onion, beetroot, capers and mushrooms. Butter one side of the bread slices, and spread each slice with meat mixture.

Cauliflower quickie

(Illustrated on page 72)
Makes 4 servings.

IMPERIAL	METRIC	AMERICAN
1 small cauliflower	1 small cauliflower	1 small cauliflower
½ pint cheese sauce, using Gouda cheese (see page 100)	3 dl. cheese sauce, using Gouda cheese (see page 100)	1¼ cups cheese sauce, using Gouda cheese (see page 100)
4 oz. frozen prawns	100 g. frozen prawns	¾ cup frozen prawns or shrimp
2 tomatoes, sliced	2 tomatoes, sliced	2 tomatoes, sliced

Slash base of cauliflower deeply, wash in salted water and cook whole, in lightly salted water until tender. Drain and place in an ovenproof dish. Turn frozen sauce into a saucepan and reheat gently to boiling point, stirring constantly to prevent burning. Stir in the prawns and pour sauce over the hot cauliflower. Arrange halved tomato slices around the sides of the dish.

Fish shells

Makes 8 servings.
To freeze Cover each shell with foil and crimp under the edges. Label.
To serve Uncover, sprinkle with toasted breadcrumbs and place in a hot oven (425°F, 220°C, Gas Mark 7) for 25 minutes. Garnish with tomato slices.
Storage time 4–6 months.

IMPERIAL	METRIC	AMERICAN
1 lb. cod fillet	½ kg. cod fillet	1 lb. cod fillet
1 pint milk	generous ½ litre milk	2½ cups milk
4 oz. bacon, chopped	100 g. bacon, chopped	¼ lb. bacon slices, chopped
4 oz. butter	100 g. butter	½ cup butter
2 medium onions, chopped	2 medium onions, chopped	2 medium onions, chopped
2 oz. flour	50 g. flour	½ cup all-purpose flour
1 tablespoon chopped parsley	1 tablespoon chopped parsley	1 tablespoon chopped parsley
salt and pepper to taste	salt and pepper to taste	salt and pepper to taste

Gently poach the cod fillet in the milk until cooked, about 15 minutes. De-rind the bacon and chop. Melt half the butter, add the onion and bacon and cook gently until onion is softened but not browned. Place the strained milk from cooking the fish, the remaining butter and the flour in a saucepan and whisk over medium heat until smooth and thick. Remove from the heat, stir in the parsley and seasoning and beat well. Divide the cooked fish between eight greased scallop shells or small shaped foil dishes, spoon over the bacon and onion mixture and the sauce. Cool.

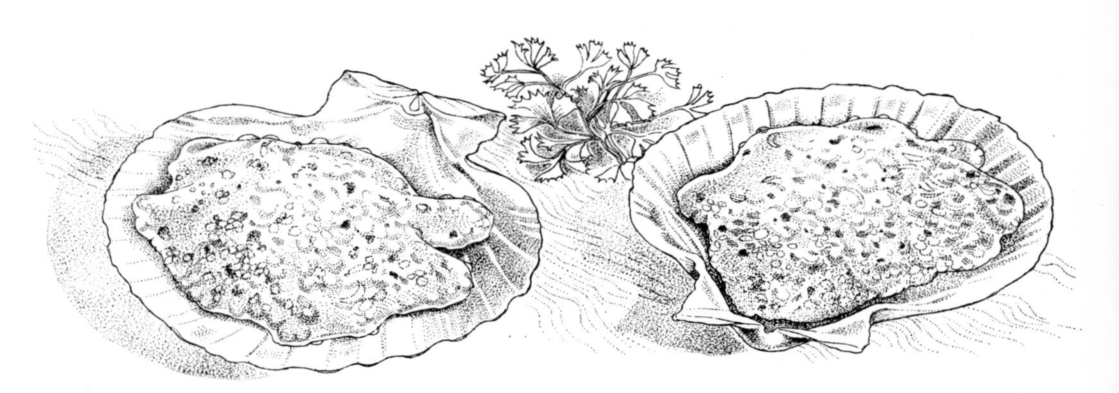

French onion cheese tart

Makes 2 tarts, 8 servings.
To freeze Open freeze until solid then wrap in freezer foil or polythene bags. Seal and label.
To serve Place frozen tart on a baking tray in a moderately hot oven (350°F, 180°C, Gas Mark 4) for 30 minutes.
Storage time 4–6 months.

IMPERIAL	METRIC	AMERICAN
pastry	*pastry*	*pastry*
12 oz. plain flour	350 g. plain flour	3 cups all-purpose flour
6 oz. butter	175 g. butter	¾ cup butter
pinch salt	pinch salt	pinch salt
pinch cayenne pepper	pinch cayenne pepper	pinch cayenne pepper
water to mix	water to mix	water to mix
filling	*filling*	*filling*
3 medium tomatoes	3 medium tomatoes	3 medium tomatoes
3 large onions, chopped	3 large onions, chopped	3 large onions, chopped
4 eggs	4 eggs	4 eggs
¼ pint milk	1½ dl. milk	⅔ cup milk
salt and pepper to taste	salt and pepper to taste	salt and pepper to taste
8 oz. Gruyère cheese, grated	225 g. Gruyère cheese, grated	2 cups grated Gruyère cheese

Make up the pastry and use to line two 7-inch (18-cm.) flan rings. Prick bottom of each flan case with a fork. To make the filling, skin and chop the tomatoes and mix with the onion. Beat the eggs into the milk and stir into the onion mixture. Season to taste and divide between the flan cases. Sprinkle with the grated cheese, bake on the top shelf of a moderately hot oven (400°F, 200°C, Gas Mark 6) for about 30 minutes. Cool.

Puff pastry pizza

Makes 2 pizzas, 8 servings.
To freeze Open freeze until solid then pack in polythene bags.
To serve Place pizza on a dampened baking tray, still frozen, and cook in a hot oven (425°F, 220°C, Gas Mark 7) for about 30 minutes until the pastry is well risen and golden brown.
Storage time 4–6 months.

IMPERIAL	METRIC	AMERICAN
1 lb. frozen puff pastry, defrosted	450 g. frozen puff pastry, defrosted	1 lb. frozen puff paste, defrosted
4 oz. butter	100 g. butter	½ cup butter
2 tablespoons oil	2 tablespoons oil	3 tablespoons oil
2 large onions, chopped	2 large onions, chopped	2 large onions, chopped
1 15-oz. can tomatoes	1 425-g. can tomatoes	1 15-oz. can tomatoes
1 teaspoon dried oregano	1 teaspoon dried oregano	1 teaspoon dried oregano
salt and pepper to taste	salt and pepper to taste	salt and pepper to taste
12 oz. Gouda cheese, grated	350 g. Gouda cheese, grated	3 cups grated Gouda cheese
2 small cans anchovies	2 small cans anchovies	2 small cans anchovies
black olives to garnish	black olives to garnish	ripe olives to garnish

Divide the pastry in half and roll out each piece to an oblong approximately 10 inches (26 cm.) by 7 inches (18 cm.). Heat the butter and oil in a small pan and gently fry the onion until softened. Add the tomatoes, oregano and seasoning and simmer together for 5–10 minutes. Cool. Spread the tomato mixture on the pastry bases to within 1 inch (2·5 cm.) of the edge. Sprinkle each with grated cheese, make a lattice work on top with the drained anchovies and garnish with black olives.

Cheese and meat strudel

Makes 2 strudels, 8 servings.
To freeze Wrap in freezer foil. Seal and label.
To serve Place, still frozen and wrapped in foil, in a moderately hot oven (375°F, 190°C, Gas Mark 5) for 30 minutes, unwrap and replace in oven for a further 10 minutes to crisp pastry.
Storage time 4–6 months.

IMPERIAL	METRIC	AMERICAN
1 lb. frozen flaky pastry, defrosted	450 g. frozen flaky pastry, defrosted	1 lb. frozen puff paste, defrosted
filling	*filling*	*filling*
1 large green pepper	1 large green pepper	1 large green sweet pepper
1 large onion	1 large onion	1 large onion
2 oz. butter	50 g. butter	¼ cup butter
8 oz. minced meat	225 g. minced meat	1 cup ground meat
2 tablespoons tomato purée	2 tablespoons tomato purée	3 tablespoons tomato paste
salt and pepper to taste	salt and pepper to taste	salt and pepper to taste
pinch nutmeg	pinch nutmeg	pinch nutmeg
¼ teaspoon Worcestershire sauce	¼ teaspoon Worcestershire sauce	¼ teaspoon Worcestershire sauce
6 oz. Gouda cheese, grated	175 g. Gouda cheese, grated	1½ cups grated Gouda cheese
1 teaspoon chopped parsley	1 teaspoon chopped parsley	1 teaspoon chopped parsley
beaten egg to glaze	beaten egg to glaze	beaten egg to glaze

Deseed and finely chop the pepper and finely chop the onion. Melt butter in a saucepan and gently cook the pepper, onion, meat, tomato purée and seasonings for about 10 minutes. Cool. Divide the pastry into two portions, place half on a tea towel and roll out very thinly to form a rectangle, approximately 9½ inches (24 cm.) wide. (The pattern on the tea towel should show through the pastry.) Spread half the cooled mixture over the pastry to within 1 inch (2·5 cm.) of the edges, sprinkle with half the grated cheese and parsley and dampen the edges. Lift up one end of the tea towel and roll up, as for a Swiss roll. Repeat with remaining ingredients. Place on a dampened baking tray, brush with beaten egg and bake in a moderately hot oven (375°F, 190°C, Gas Mark 5) for about 30 minutes until the pastry is well risen and golden brown. Cool.

Mock moussaka

Makes 6 servings.
To freeze Cover, seal and label.
To serve Remove covers and place in a moderately hot oven (400°F, 200°C, Gas Mark 6) for 25 minutes. Meanwhile, turn ½ pint (3 dl.) frozen cheese sauce (see page 100) into a saucepan and reheat gently. Remove from heat, beat in 2 egg yolks. Beat 2 egg whites until stiff, fold into the sauce and divide between the foil containers. Replace in oven for a further 20 minutes.
Storage time 4–6 months.

IMPERIAL	METRIC	AMERICAN
1 large potato	1 large potato	1 large potato
4 tablespoons corn oil	4 tablespoons corn oil	⅓ cup corn oil
1 lb. lamb, minced	450 g. lamb, minced	1 lb. lamb, ground
1 large onion, minced or grated	1 large onion, minced or grated	1 large onion, minced or grated
1 2¾-oz. can tomato purée	1 70-g. can tomato purée	¼ cup tomato paste
1 8-oz. can tomatoes	1 225-g. can tomatoes	1 8-oz. can tomatoes
1 teaspoon sugar	1 teaspoon sugar	1 teaspoon sugar
salt and pepper to taste	salt and pepper to taste	salt and pepper to taste

Slice the potato thinly. Heat the oil and use to fry the potato until softened but not browned. Mix together the remaining ingredients and spoon into small deep shaped foil containers in layers with the fried potato, ending with a layer of meat. Leave 1-inch (2·5-cm.) headspace to allow for addition of cheese sauce. Cool.

Double-crust chicken and mushroom pie

Makes 1 pie, 4 servings.
To freeze Place in polythene bag. Seal and label.
To serve Unwrap, cut steam vent, brush with beaten egg and place in a hot oven (425°F, 220°C, Gas Mark 7) for 30 minutes.
Storage time 4–6 months.

IMPERIAL	METRIC	AMERICAN
10 oz. frozen shortcrust pastry, defrosted	275 g. frozen shortcrust pastry, defrosted	10 oz. frozen basic pie dough, defrosted
1 chicken stock cube	1 chicken stock cube	1 chicken bouillon cube
¼ pint boiling water	1½ dl. boiling water	⅔ cup boiling water
¼ pint milk	1½ dl. milk	⅔ cup milk
1 oz. butter	25 g. butter	2 tablespoons butter
1 oz. flour	25 g. flour	¼ cup all-purpose flour
6 oz. cooked chicken, diced	175 g. cooked chicken, diced	¾ cup diced cooked chicken
4 oz. canned button mushrooms	100 g. canned button mushrooms	1 cup canned button mushrooms

Roll out two-thirds of the pastry and use to line a 7-inch (18-cm.) flan case. Dissolve the stock cube in the boiling water and place in a saucepan. Add the milk, butter and flour and whisk over medium heat until smooth and thick. Stir in the chicken and mushrooms. Cool and pour into the pastry case. Roll out remaining pastry to form a lid, dampen the edges and seal well together but do not cut a steam vent.

Veal escalopes with apple rings

Makes 6 servings.
To freeze Pack individually in freezer foil or together in shallow foil container. Seal and label.
To serve Place, still wrapped and frozen, in a moderate oven (350°F, 180°C, Gas Mark 4) as follows. Individually wrapped – 15 minutes. In foil container – 30 minutes. Meanwhile heat a little oil or fat in a small pan and fry fresh apple rings until golden both sides. Serve each hot escalope with an apple ring and fill centre of ring with a stuffed olive.
Storage time 4–6 months.

IMPERIAL	METRIC	AMERICAN
6 slices frying veal	6 slices frying veal	6 veal scallops
4 oz. soft white breadcrumbs	100 g. soft white breadcrumbs	2 cups fresh soft bread crumbs
salt and pepper to taste	salt and pepper to taste	salt and pepper to taste
2 eggs, beaten	2 eggs, beaten	2 eggs, beaten
2 oz. butter	50 g. butter	¼ cup butter
1 tablespoon oil	1 tablespoon oil	1 tablespoon oil

Trim the veal slices and season the breadcrumbs with salt and pepper. Dip the veal slices in beaten egg and coat with the breadcrumbs. Melt butter and oil in a frying pan and fry the escalopes over medium heat for about 5 minutes each side. Drain and cool.

Ham and apple pasties

Makes 6 pasties.
To freeze Pack in freezer foil, rigid-based polythene container or polythene bag. Seal and label.
To serve Place on a baking tray, brush with beaten egg and cook in a moderately hot oven (375°F, 190°C, Gas Mark 5) for 45 minutes.
Storage time 4–6 months.

IMPERIAL	METRIC	AMERICAN
1¼ lb. frozen shortcrust pastry, defrosted	600 g. frozen shortcrust pastry, defrosted	1¼ lb. frozen basic pie dough, defrosted
4 tomatoes	4 tomatoes	4 tomatoes
4 dessert apples	4 dessert apples	4 dessert apples
2 tablespoons oil	2 tablespoons oil	3 tablespoons oil
1 large onion, sliced	1 large onion, sliced	1 large onion, sliced
12 oz. ham, diced	350 g. ham, diced	¾ lb. cooked ham, diced
salt and pepper to taste	salt and pepper to taste	salt and pepper to taste

Roll out the pastry and use to cut out six 6-inch (15-cm.) circles. To make the filling, peel and slice the tomatoes and peel, core and slice the apples. Heat the oil and use to fry the onion until softened but not browned. Drain and cool. Mix together the onion, tomato, ham and apple and season well. Divide the filling between the pastry circles, dampen edges and bring together over filling to form triangular pasties. Press edges well together.

Devonshire pasties

Makes 6 pasties.
To freeze Pack in freezer foil, rigid-based polythene containers or polythene bag. Seal and label.
To serve Place on a baking tray and brush with beaten egg. Put into a moderately hot oven (400°F, 200°C, Gas Mark 6) for 10 minutes then reduce oven temperature to 350°F, 180°C, Gas Mark 4 for a further hour. If necessary protect pasties with a sheet of foil to prevent over-browning.
Storage time 4–6 months.

IMPERIAL	METRIC	AMERICAN
1¼ lb. frozen shortcrust pastry, defrosted	600 g. frozen shortcrust pastry, defrosted	1¼ lb. frozen basic pie dough, defrosted
1 lb. lamb, diced	450 g. lamb, diced	1 lb. lamb, diced
2 large potatoes, diced	2 large potatoes, diced	2 large potatoes, diced
2 medium onions, diced	2 medium onions, diced	2 medium onions, diced
1 small swede, diced	1 small swede, diced	1 small swede, diced
pinch mixed dried herbs	pinch mixed dried herbs	pinch mixed dried herbs
salt and pepper to taste	salt and pepper to taste	salt and pepper to taste
about 5 tablespoons stock or water	about 5 tablespoons stock or water	about 6 tablespoons stock or water

Roll out the pastry and use to cut out six 6-inch (15-cm.) circles. To make the filling, mix together the lamb, potato, onion, swede, herbs and seasoning. Moisten with a little stock or water and divide filling between the pastry circles. Dampen edges and bring together over filling. Press edges well together and flute.

Poacher's roll

(Illustrated on page 107)

Makes 1 roll, 4 servings.
To freeze Wrap in freezer foil. Seal and label.
To serve Place, still wrapped and frozen, in a moderate oven (350°F, 180°C, Gas Mark 4) for 30 minutes. Unwrap and return to oven for 5 minutes to crisp pastry. Serve on a bed of salad ingredients.
Storage time 4–6 months.

IMPERIAL	METRIC	AMERICAN
12 oz. frozen puff pastry, defrosted	350 g. frozen puff pastry, defrosted	¾ lb. frozen puff paste, defrosted
6 oz. streaky bacon	175 g. streaky bacon	6 oz. bacon slices
1 lb. pork sausagemeat	450 g. pork sausagemeat	1 lb. pork sausagemeat
2 oz. button mushrooms, chopped	50 g. button mushrooms, chopped	½ cup chopped button mushrooms
1 small onion, chopped	1 small onion, chopped	1 small onion, chopped
½ teaspoon dried sage	½ teaspoon dried sage	½ teaspoon dried sage
salt and pepper to taste	salt and pepper to taste	salt and pepper to taste
beaten egg to glaze	beaten egg to glaze	beaten egg to glaze

Roll out the pastry to an oblong 13 inches (28 cm.) by 10 inches (26 cm.). De-rind and chop the bacon and mix together with the sausagemeat, mushrooms, onion, sage and seasonings and mould into a sausage shape. Place along centre of pastry, dampen edges and roll up. Seal pastry at both ends to enclose filling completely. Turn roll over so that join is underneath and place on a dampened baking tray. Decorate roll with pastry trimmings and make three cuts across the top. Brush with beaten egg and bake in a hot oven (425°F, 220°C, Gas Mark 7) for 20 minutes. Reduce oven temperature to 375°F, 190°C, Gas Mark 5, for a further 35–40 minutes. If necessary protect roll with a sheet of foil to prevent over-browning. Cool.

Swiss cherry pie

Makes 1 pie, 4–6 portions.
To freeze Place in polythene bag, or wrap in foil. Seal and label.
To serve Unwrap, cut steam vent, brush with beaten egg and place in a hot oven (425°F, 220°C, Gas Mark 7) for 30 minutes.
Storage time 4–6 months.

IMPERIAL	METRIC	AMERICAN
10 oz. frozen shortcrust pastry, defrosted	275 g. frozen shortcrust pastry, defrosted	10 oz. frozen basic pie dough, defrosted
1 14-oz. can cherry pie filling	1 400-g. can cherry pie filling	1 14-oz. can cherry pie filling
4 oz. frozen pear slices	100 g. frozen pear slices	¼ lb. frozen pear slices
1 oz. demerara sugar	25 g. demerara sugar	2 tablespoons brown sugar
¼ teaspoon cornflour	¼ teaspoon cornflour	¼ teaspoon cornstarch

Roll out two-thirds of the pastry and use to line a 7-inch (18-cm.) flan case. Turn the cherry pie filling into it and cover with a layer of pear slices. (Pears quartered and frozen in sugar syrup will require to be partially defrosted and sliced.) Mix together the sugar and cornflour and sprinkle over the top. Roll out remaining pastry to form a lid, dampen the edges and seal well together but do not cut a steam vent.

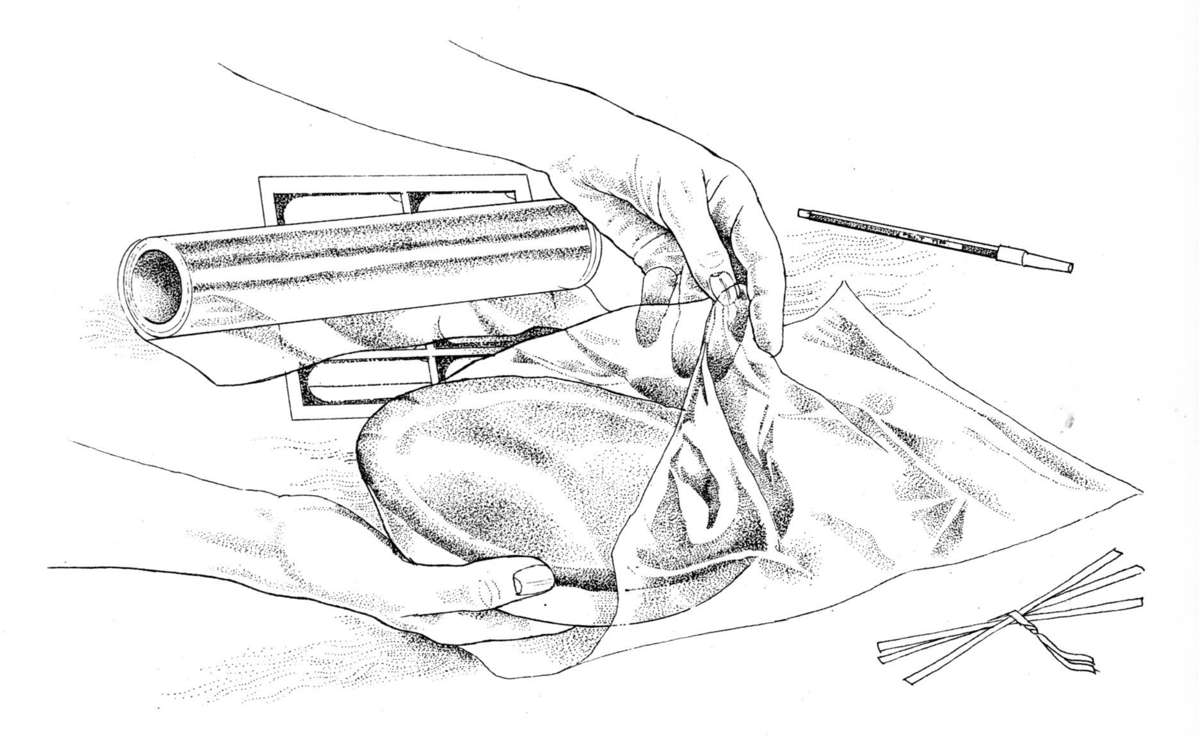

Chocolate fruit jiffies

(Illustrated on page 183)

Makes 12.
To freeze Pack in rigid-based polythene container with foil dividers. Seal and label.
To serve Place spread out on a dish and allow to defrost for 30 minutes.
Storage time 4–6 months.

Variation The basic quantities can be doubled, with the exception of the dried apricots. Make up half the quantity with the apricots, and the other half with chopped dried figs.

IMPERIAL	METRIC	AMERICAN
8 oz. plain chocolate	225 g. plain chocolate	8 squares semi-sweet chocolate pieces
2 oz. All-bran	50 g. All-bran	2 oz. All-bran
2 oz. seedless raisins	50 g. seedless raisins	6 tablespoons seedless raisins
2 oz. dried apricots, chopped	50 g. dried apricots, chopped	about ½ cup chopped dried apricots

Melt the chocolate in a basin over a pan of hot water. Stir in the bran, raisins and apricots and mix well. Divide the mixture between 12 paper bun cases and leave until set.

Spice cakes

(Illustrated on page 183)

Makes 18.
To freeze Pack in rigid-based polythene container with foil dividers. Seal and label.
To serve Place spread out on a dish and allow to defrost for 30 minutes.
Storage time 4–6 months.

IMPERIAL	METRIC	AMERICAN
8 oz. self-raising flour	225 g. self-raising flour	2 cups all-purpose flour sifted with 2 teaspoons baking powder
½ teaspoon salt	½ teaspoon salt	½ teaspoon salt
½ teaspoon ground cloves	½ teaspoon ground cloves	½ teaspoon ground cloves
1 teaspoon ground cinnamon	1 teaspoon ground cinnamon	1 teaspoon ground cinnamon
3 oz. butter or margarine	75 g. butter or margarine	6 tablespoons butter or margarine
6 tablespoons clear honey	6 tablespoons clear honey	7 tablespoons clear honey
1 egg	1 egg	1 egg
1½ oz. All-bran	40 g. All-bran	1½ oz. All-bran
4 oz. seedless raisins	100 g. seedless raisins	generous 1 cup seedless raisins
2 oz. icing sugar, sieved	50 g. icing sugar, sieved	½ cup sifted confectioners' sugar
18 glacé cherries	18 glacé cherries	18 candied cherries

Sieve together the flour, salt, cloves and cinnamon. Cream the butter and beat in the honey. Gradually beat in the egg then fold in the flour mixture and the bran and raisins. Place 18 paper bun cases on a baking tray and divide the mixture between them. Bake in a moderate oven (350°F, 180°C, Gas Mark 4) for 20–25 minutes. Cool. Mix a little cold water into the icing sugar to give a smooth coating consistency. Decorate each cake with a little icing, top with a glacé cherry, and leave to allow icing to set.

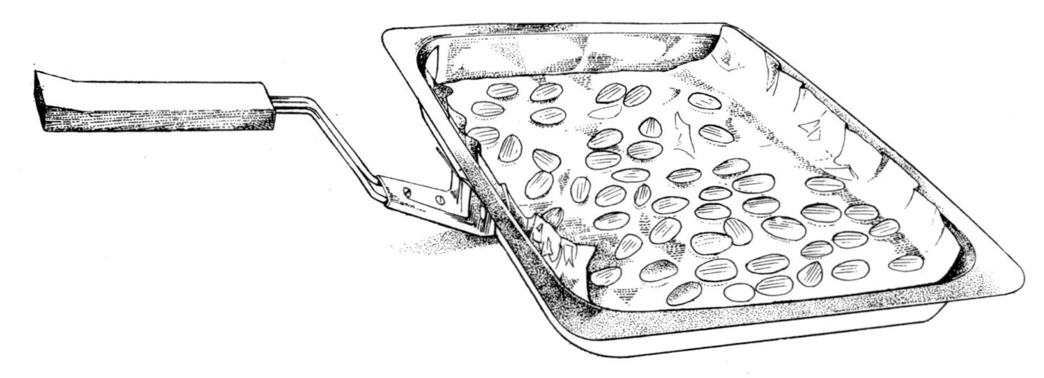

Tipsy roll

Makes 1 roll, 8–10 servings.
To freeze Re-assemble slices in roll shape with foil dividers and wrap in polythene bag.
To serve Unwrap, spread slices out on a dish and defrost for about 45 minutes. Serve with chocolate ripple ice cream.
Storage time 4–6 months.

IMPERIAL	METRIC	AMERICAN
12 oz. digestive biscuits	350 g. digestive biscuits	¾ lb. graham crackers
3 oz. toasted almonds	75 g. toasted almonds	½ cup toasted almonds
8 oz. butter	225 g. butter	1 cup butter
8 oz. castor sugar	225 g. castor sugar	1 cup granulated sugar
2 eggs	2 eggs	2 eggs
1½ tablespoons brandy	1½ tablespoons brandy	2 tablespoons brandy
3 tablespoons cocoa powder	3 tablespoons cocoa powder	4 tablespoons unsweetened cocoa powder

Place the biscuits in a polythene bag and crush them with a rolling pin. Finely chop the almonds. Cream together the butter and sugar until light and fluffy. Gradually beat in the eggs and brandy. Add the almonds, cocoa powder and biscuits and mix thoroughly. Shape the mixture into a roll, wrap tightly in foil and chill. Unwrap and slice roll carefully.

Cranberry coupes

Makes 4 servings.

IMPERIAL	METRIC	AMERICAN
1 16-oz. container frozen cranberries	1 450-g. container frozen cranberries	1 16-oz. container frozen cranberries
4 oz. castor sugar	100 g. castor sugar	½ cup granulated sugar
2 teaspoons cold water	2 teaspoons cold water	2 teaspoons cold water
1 16-oz. container frozen apple slices in sugar syrup	1 450-g. container frozen apple slices in sugar syrup	1 16-oz. container frozen apple slices in sugar syrup
whipped cream	whipped cream	whipped cream

Place half the cranberries in a saucepan with the sugar and water and heat gently until defrosted. Bring to boiling point and simmer for 5 minutes, until soft. Cool. Turn the apple slices and syrup into a saucepan and reheat gently until defrosted. Remove apple slices and cool. Boil the apple syrup until reduced to about 4 tablespoons, then cool. Place cooked cranberries in bases of four glasses, arrange cooked apple slices over, top with uncooked cranberries and glaze with syrup. Serve decorated with whipped cream.

Peach and raspberry sundaes

Makes 6 servings.

IMPERIAL	METRIC	AMERICAN
8 oz. frozen raspberries	225 g. frozen raspberries	½ lb. frozen raspberries
2 tablespoons sieved icing sugar	2 tablespoons sieved icing sugar	3 tablespoons sifted confectioners' sugar
2 tablespoons boiling water	2 tablespoons boiling water	3 tablespoons boiling water
1 16-oz. container frozen peach slices	1 450-g. container frozen peach slices	1 16-oz. container frozen peach slices
1 family brick raspberry ripple ice cream	1 family brick raspberry ripple ice cream	1 pint raspberry ripple ice cream

Hold pack of raspberries under cold running water to soften slightly, then sieve and beat in the sugar and water. If preferred, place raspberries, sugar and water in a blender goblet and liquidise, then sieve. Hold pack of peach slices under cold running water until slices can be separated. Place slices of ice cream in six sundae glasses, top each with a few peach slices and pour over the raspberry sauce.

Lemon flummery

Makes 6 servings.
To freeze Cover with lid and seal or use a sheet of freezer foil and crimp edges together. Label.
To serve Remove cover and allow to defrost at room temperature for about 1 hour and decorate with whipped cream.
Storage time 3 months.

IMPERIAL	METRIC	AMERICAN
2 teaspoons gelatine	2 teaspoons gelatine	2 teaspoons gelatin
3 tablespoons water	3 tablespoons water	4 tablespoons water
3 eggs	3 eggs	3 eggs
5 oz. castor sugar	150 g. castor sugar	⅔ cup granulated sugar
6 tablespoons lemon juice	6 tablespoons lemon juice	½ cup lemon juice
grated zest of 2 lemons	grated zest of 2 lemons	grated zest of 2 lemons

Place the gelatine and water in a small basin over a pan of hot water and leave until dissolved. Separate the eggs and whisk the yolks and sugar until thick and pale. Gradually whisk in the lemon juice then the gelatine and lemon zest. Leave until almost at setting point then beat the egg whites until stiff and fold into the lemon mixture evenly. Divide mixture between six individual dishes or polythene moulds and allow to set.

Raspberry cloud

Makes 4 servings.

IMPERIAL	METRIC	AMERICAN
1 lb. frozen raspberries	450 g. frozen raspberries	1 lb. frozen raspberries
2 egg whites	2 egg whites	2 egg whites
½ pint double cream	3 dl. double cream	1¼ cups heavy cream
2 oz. castor sugar or to taste	50 g. castor sugar or to taste	4 tablespoons sugar or to taste

Hold the pack of frozen raspberries under cold running water to soften slightly. Reserve a few raspberries and either sieve the rest or liquidise first and then sieve. Sweeten purée to taste. Whip the egg whites until stiff, add the cream and continue whipping until foamy. Stir in the raspberry purée and mix well. Pour into one large or individual glass dishes and decorate with the reserved raspberries.

Gingered pear coupes

(Illustrated on page 36)

Makes 4 servings.

IMPERIAL	METRIC	AMERICAN
8 tablespoons buttered crumbs (see below)	8 tablespoons buttered crumbs (see below)	9 tablespoons buttered crumbs (see below)
1 16-oz. container frozen sliced pears in sugar syrup	1 450-g. container frozen sliced pears in sugar syrup	1 16-oz. container frozen sliced pears in sugar syrup
2 tablespoons ginger syrup	2 tablespoons ginger syrup	3 tablespoons ginger syrup
¼ pint double cream	1½ dl. double cream	⅔ cup heavy cream
sugar to taste	sugar to taste	sugar to taste
1 egg white	1 egg white	1 egg white
4 small pieces stem ginger	4 small pieces stem ginger	4 small pieces candied ginger

Defrost containers of buttered crumbs and pears in warm water. Divide half the crumbs between four coupe glasses. Arrange pear slices upright over the crumb base, and pour over the ginger syrup and a little syrup from the frozen pears. Fill up coupes to the level of the top of the pear slices with more buttered crumbs. Whip the cream and sweeten to taste. Fold in the stiffly beaten egg white and spoon or pipe cream over the crumbs and fruit. Decorate with the stem ginger.

Sweet buttered crumbs can be used in a number of recipes with fruit and fruit purées. Make up a basic quantity as follows and freeze in four small polythene containers.

IMPERIAL	METRIC	AMERICAN
4 oz. unsalted butter	100 g. unsalted butter	½ cup sweet butter
12 oz. dried white crumbs	350 g. dried white breadcrumbs	3 cups fine dry bread crumbs
4 oz. demerara sugar	100 g. demerara sugar	½ cup brown sugar

Melt the butter, stir in the breadcrumbs and fry until beginning to turn brown. Add the sugar and continue frying until the mixture is golden brown, stirring constantly. Cool.

Slimming dishes planned for your diet

Almost every woman who needs to lose weight consoles herself by blaming those surplus pounds on something other than self-indulgence. Whether you blame your over-active glands or your extra-heavy bones the only way to get rid of a spare tyre and a double chin is to eat less of the fattening foods.

Get the freezer on your side for a start. Pack it with a tempting selection of low-calorie dishes especially for yourself. Choose something different to eat every day because boredom quickly banishes a dieter's good intentions. After all, few of us can afford to dine very often off a delicious fillet steak, even if we found this far from boring. Another enemy of success is ridicule. Family mealtimes can be an ordeal when everyone else has been served and they laugh at the wide open spaces on your pitiful plate where the potatoes ought to be. If, with the minimum of trouble, you can produce your own special appetising dish, remarks are likely to be more envious than scornful.

A fat friend of mine who found it all but impossible to cook her favourite dishes for the family without sharing them, kept up her morale by setting aside a small portion for herself each time in a special basket of goodies to eat after she had achieved her ideal weight. It might be a bit of a temptation though, to less strong-minded ladies!

THE FARMHOUSE CHEESE AND APPLE DIET

Readers often complain to me that because they now have such an appetising variety of calorie-loaded food in their freezers they tend to gain weight.

The freezer should be a friend to every woman with a weight problem because it enables her to keep a supply of 'permitted' dishes intended for her consumption only, while the rest of the family are eating their usual meals.

To exploit an entirely different aspect of freezing, I planned this very special 14-day diet for a magazine based on eating a nourishing and appetising fresh cheese-and-apple snack meal once a day. The rest of the meals come mainly from the freezer. So many readers have written to tell me how easy the diet is to follow, and how successful, that I include it here.

Before you consult the diet chart and glance through the selection of recipes, remember that it is based on a consumption of 1300 calories a day. If this diet is too generous for you to lose weight, sacrifice the daily crispbread and marmalade for breakfast. If it is too strict, give yourself a small scraping of butter on your breakfast crispbread.

It is understood that eggs are scrambled without added fat in a non-stick pan, chicken is grilled or oven-roasted without added fat and liver is steam-baked in a covered dish with a little of the clear soup from your lunchtime allowance. There is no need to cut down stringently on liquid intake. So long as you do not exceed the daily milk allowance you can drink tea, coffee, lemon juice, Slimline drinks and even low calorie fruit squashes as required. A dietician has advised me that a person in good health would not suffer any deficiency on this diet.

The housewife who kindly acted as a 'guinea pig' for this diet not only lost weight but very much enjoyed the rich subtle flavour of farmhouse cheeses – Cheddar, Cheshire and Lancashire.

Diet chart for 14 days

Daily basics: 3 oz. (75 g.) farmhouse Cheddar/Cheshire/Lancashire cheese
2 crisp dessert apples
½ pint (3 dl.) milk (preferably skimmed)

Breakfast
1 small glass frozen orange juice, or ½ grapefruit

1 egg (boiled or poached or scrambled in non-stick pan)

2 starch-reduced crispbreads

1 teaspoon low-sugar marmalade

Coffee or tea with milk from allowance

Three times a week substitute one of the following for the egg:
2 lamb's kidneys, grilled
2 oz. (50 g.) smoked haddock
1 rasher lean bacon, grilled
1 beef frankfurter

Elevenses
Coffee or tea with milk from allowance

Light meal
3 oz. (75 g.) farmhouse Cheddar/Cheshire/Lancashire cheese

2 crisp dessert apples

½ pint (3 dl.) clear soup or bouillon made with chicken or beef stock cube

Afternoon break
Coffee or tea with milk from allowance

Main meal
One protein portion (choose from freezer recipes)

One vegetable dish (choose from freezer recipes)

One of the following sweets:
fruit-flavoured yogurt
natural yogurt with 2 teaspoons clear honey added
4 oz. (100 g.) diced melon
4 oz. (100 g.) fruit purée made with frozen raspberries, strawberries, gooseberries or rhubarb and liquid sweetener to taste

Nightcap
Any preferred drink using up remaining milk allowance, or artificially-sweetened lemon juice and hot water

Your main meal choices

1 portion chicken in a parcel* with courgettes carnival*

4 oz. (100 g.) grilled rump steak with savoury tomato bake*

4 oz. (100 g.) grilled chicken with spicy cauliflower salad*

4 oz. (100 g.) lean ham with pineapple salad*

6 oz. (175 g.) steamed halibut steak with Provençale topping* and braised celery*

4 oz. (100 g.) baked liver with macedoine of vegetables

4 oz. (100 g.) minced beef à la grecque* with peas

6 oz. (175 g.) grilled cod steak with spinach mousse*

1 small roast chicken portion with sliced green beans

1 portion Chinese chicken* with broccoli spears

4 oz. (100 g.) lamb kebabs* with baby carrots

1 medium grilled mackerel with French bean salad*

4 oz. (100 g.) veal escalope with cauliflower and black olives*

6 oz. (175 g.) tuna steak with Spanish salad*

Recipes marked * are given in the book.

Chicken in a parcel

IMPERIAL	METRIC	AMERICAN
1 onion, chopped	1 onion, chopped	1 onion, chopped
4 oz. mushrooms, sliced	100 g. mushrooms, sliced	1 cup sliced mushrooms
1 stick celery, chopped	1 stick celery, chopped	1 stalk celery, chopped
salt and pepper to taste	salt and pepper to taste	salt and pepper to taste
2 small chicken portions	2 small chicken portions	2 small chicken joints

Divide the vegetables between two large squares of freezer foil. Season well and place a chicken portion on each. Seal the foil completely to make two airtight parcels. Bake one parcel in a moderately hot oven (375°F, 190°C, Gas Mark 5) for about 1 hour, until cooked. Freeze remaining parcel. Defrost and cook as above.

Courgettes carnival

IMPERIAL	METRIC	AMERICAN
½ pint tomato juice	3 dl. tomato juice	1¼ cups tomato juice
1 chicken stock cube	1 chicken stock cube	1 chicken bouillon cube
2 tablespoons lemon juice	2 tablespoons lemon juice	3 tablespoons lemon juice
1 stick celery, chopped	1 stick celery, chopped	1 stalk celery, chopped
1 small onion, chopped	1 small onion, chopped	1 small onion, chopped
12 oz. courgettes	350 g. courgettes	¾ lb. small zucchini

Place the tomato juice, stock cube and lemon juice in a saucepan and heat until the cube is dissolved. Add the celery and onion, bring to the boil and simmer gently for about 10 minutes. Meanwhile, top and tail the courgettes and cut into ½-inch (1-cm.) slices. Add to the pan, bring back to the boil, cover and simmer until vegetables are tender. A little water can be added if necessary. Serve half – cool remainder; place in plastic container and freeze. Defrost and reheat gently in saucepan.

Savoury tomato bake

IMPERIAL	METRIC	AMERICAN
1 oz. onion, finely chopped	25 g. onion, finely chopped	¼ cup finely chopped onion
1 oz. celery, finely chopped	25 g. celery, finely chopped	¼ cup finely chopped celery
1 tablespoon grated Parmesan cheese	1 tablespoon grated Parmesan cheese	1 tablespoon grated Parmesan cheese
salt and pepper to taste	salt and pepper to taste	salt and pepper to taste
1 tablespoon mild Continental mustard	1 tablespoon mild Continental mustard	1 tablespoon mild Continental mustard
2 tablespoons natural yogurt	2 tablespoons natural yogurt	3 tablespoons natural yogurt
4 medium tomatoes	4 medium tomatoes	4 medium tomatoes

Mix together the onion, celery, Parmesan cheese, seasonings and yogurt. Slice the tomatoes, place in a greased baking dish and spoon the topping over. Bake in a moderately hot oven (400°F, 200°C, Gas Mark 6) for about 15 minutes. Serve half – cool remainder; pack in shaped foil container and freeze. Defrost and reheat by placing in a moderately hot oven for 20 minutes.

Spicy cauliflower salad

IMPERIAL	METRIC	AMERICAN
1 lb. cooked cauliflower florets	450 g. cooked cauliflower florets	1 lb. cooked cauliflower florets
2 tablespoons chopped green pepper	2 tablespoons chopped green pepper	3 tablespoons chopped green sweet pepper
1 tablespoon clear honey	1 tablespoon clear honey	1 tablespoon clear honey
1 tablespoon wine vinegar	1 tablespoon wine vinegar	1 tablespoon wine vinegar
2 tablespoons lemon juice	2 tablespoons lemon juice	3 tablespoons lemon juice
1 teaspoon Worcestershire sauce	1 teaspoon Worcestershire sauce	1 teaspoon Worcestershire sauce
½ teaspoon onion salt	½ teaspoon onion salt	½ teaspoon onion salt
½ teaspoon paprika pepper	½ teaspoon paprika pepper	½ teaspoon paprika pepper
dash Tabasco sauce	dash Tabasco sauce	dash Tabasco sauce

Place the florets in a shallow bowl, mix together the remaining ingredients and pour over the cauliflower. Cover and leave overnight in the refrigerator. Serve half – place remainder in plastic container and freeze. Allow to defrost at room temperature and serve cold.

Pineapple salad

IMPERIAL	METRIC	AMERICAN
2 rings canned pineapple, chopped	2 rings canned pineapple, chopped	2 rings canned pineapple, chopped
little lemon juice	little lemon juice	little lemon juice
2 canned red pimentos, chopped	2 canned red pimentos, chopped	2 canned red pimientos, chopped
8 oz. cottage cheese	225 g. cottage cheese	1 cup cottage cheese
2 spring onions, chopped	2 spring onions, chopped	2 scallions, chopped

Gently combine all the ingredients. Chill and serve half. Place remainder in plastic container and freeze. Allow to defrost at room temperature and serve cold.

Provençale topping for fish

IMPERIAL	METRIC	AMERICAN
½ oz. butter	15 g. butter	1 tablespoon butter
1 small green pepper, chopped	1 small green pepper, chopped	1 small green sweet pepper, chopped
1 medium onion, chopped	1 medium onion, chopped	1 medium onion, chopped
½ oz. flour	15 g. flour	2 tablespoons all-purpose flour
¼ pint water	1½ dl. water	⅔ cup water
3 tablespoons tomato purée	3 tablespoons tomato purée	4 tablespoons tomato paste
1 teaspoon clear honey	1 teaspoon clear honey	1 teaspoon clear honey
1 teaspoon vinegar	1 teaspoon vinegar	1 teaspoon vinegar

Melt the butter and fry the green pepper and onion in it until limp but not coloured. Stir in the flour and cook gently for 2 minutes. Gradually add the water, tomato purée, honey and vinegar and bring to the boil, stirring constantly. Cool. Divide between three small plastic containers and freeze. Defrost 1 portion and reheat gently in saucepan and spoon over steamed fish.

Braised celery

Place cleaned and quartered celery heads in an ovenproof dish, add 1 teaspoon vegetable extract to the water, cover and braise until tender.

Minced beef à la grecque

IMPERIAL	METRIC	AMERICAN
1 tablespoon oil	1 tablespoon oil	1 tablespoon oil
1 onion, chopped	1 onion, chopped	1 onion, chopped
8 oz. lean minced beef	225 g. lean minced beef	1 cup lean ground beef
1 5-oz. can tomatoes	1 150-g. can tomatoes	1 5-oz. can tomatoes
1 teaspoon dried mixed herbs	1 teaspoon dried mixed herbs	1 teaspoon dried mixed herbs
salt and pepper to taste	salt and pepper to taste	salt and pepper to taste
1 carton natural yogurt	1 carton natural yogurt	1 carton natural yogurt
2 egg yolks	2 egg yolks	2 egg yolks

Heat the oil in a small frying pan and cook the onion gently until soft. Add the minced beef and stir over medium heat until it changes colour. Add the tomatoes, herbs and seasoning, bring to the boil, cover and simmer gently for about 20 minutes. Divide the mixture between two ovenproof dishes. Mix together the yogurt and egg yolks, season to taste and spoon evenly over the two meat dishes. Bake in a moderate oven (350°F, 180°C, Gas Mark 4) for about 15 minutes until topping is lightly set. Serve 1 portion – cool remaining one; cover with foil and freeze. Defrost in the refrigerator overnight and reheat in a moderate oven for about 30 minutes.

Spinach mousse

IMPERIAL	METRIC	AMERICAN
4 oz. frozen spinach purée	100 g. frozen spinach purée	$\frac{1}{4}$ lb. frozen spinach purée
salt and pepper to taste	salt and pepper to taste	salt and pepper to taste
good pinch nutmeg	good pinch nutmeg	good pinch nutmeg
1 egg, separated	1 egg, separated	1 egg, separated

Defrost spinach over moderate heat, beat in the seasonings and remove from the heat. Lightly beat the egg yolk, stir into the spinach and return to the heat for 1 minute, stirring constantly. Remove from the heat again and fold in the stiffly beaten egg white. Serve hot.

Pork stuffed with prunes, instant lamb casserole (see page 198).

Chinese chicken portions

IMPERIAL	METRIC	AMERICAN
2 tablespoons orange juice	2 tablespoons orange juice	3 tablespoons orange juice
2 tablespoons soy sauce	2 tablespoons soy sauce	3 tablespoons soy sauce
2 tablespoons water	2 tablespoons water	3 tablespoons water
1 tablespoon clear honey	1 tablespoon clear honey	1 tablespoon clear honey
$\frac{1}{2}$ teaspoon Worcestershire sauce	$\frac{1}{2}$ teaspoon Worcestershire sauce	$\frac{1}{2}$ teaspoon Worcestershire sauce
2 small chicken portions	2 small chicken portions	2 small chicken joints
4 oz. canned bamboo shoots or button mushrooms	100 g. canned bamboo shoots or button mushrooms	1 cup canned bamboo shoots or button mushrooms

Mix together the orange juice, soy sauce, water, honey and Worcestershire sauce. Place the chicken portions in an ovenproof dish, pour over the marinade and leave in a cool place for a few hours, or overnight. Cook in the centre of a moderate oven (325°F, 170°C, Gas Mark 3) for about $1\frac{1}{2}$ hours. Drain and slice the bamboo shoots finely or slice the mushrooms. Add these to the chicken 15 minutes before the end of cooking time. Serve 1 portion – cool remainder; pack in a shaped foil container with the sauce and vegetables and freeze. Defrost and reheat in a moderately hot oven (400°F, 200°C, Gas Mark 6) for 30 minutes.

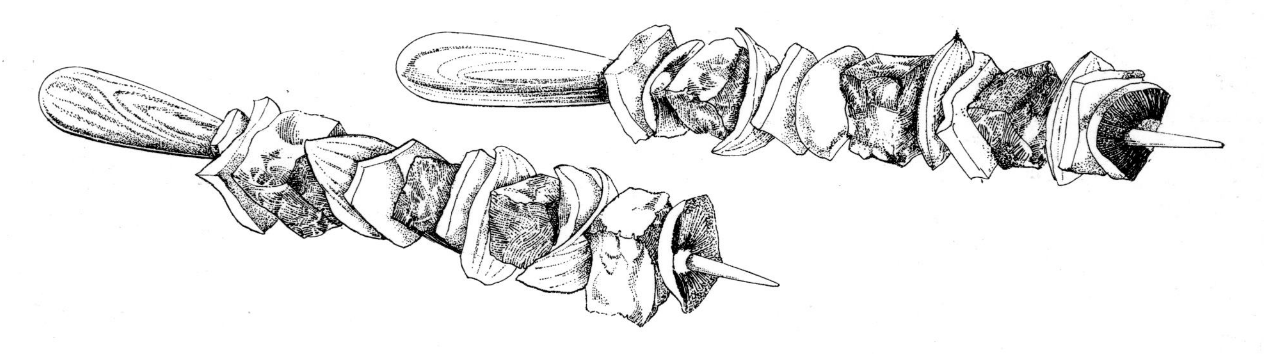

Lamb kebabs

IMPERIAL	METRIC	AMERICAN
8 oz. lean leg of lamb	225 g. lean leg of lamb	$\frac{1}{2}$ lb. boneless lean lamb leg
1 green pepper	1 green pepper	1 green sweet pepper
1 small onion, quartered	1 small onion, quartered	1 small onion, quartered
8 button mushrooms, halved	8 button mushrooms, halved	8 button mushrooms, halved
1 tablespoon olive oil	1 tablespoon olive oil	1 tablespoon olive oil
1 tablespoon wine vinegar	1 tablespoon wine vinegar	1 tablespoon wine vinegar
salt and pepper to taste	salt and pepper to taste	salt and pepper to taste
$\frac{1}{2}$ clove garlic, crushed	$\frac{1}{2}$ clove garlic, crushed	$\frac{1}{2}$ clove garlic, crushed
$\frac{1}{2}$ teaspoon ground coriander	$\frac{1}{2}$ teaspoon ground coriander	$\frac{1}{2}$ teaspoon ground coriander

Cut the lamb into neat cubes. Deseed the pepper and cut the flesh into squares. Place the meat, pepper squares, onion quarters and mushroom halves in a shallow dish. Mix together the remaining ingredients and pour over. Allow to marinate for at least 2 hours. Thread half the ingredients onto a skewer and grill under medium heat for about 20 minutes until cooked, basting with the marinade. Pack remainder, with the marinade, in a plastic container and freeze. Defrost at room temperature and cook as above.

Tournedos aux coeurs d'artichauts, boeuf en croûte à la reine Marie (see pages 227, 228).

French bean salad

IMPERIAL	METRIC	AMERICAN
8 oz. cooked French beans	225 g. cooked French beans	½ lb. cooked green beans
1 4-oz. pickled cucumber, chopped	1 100-g. pickled cucumber, chopped	1 ¼-lb. pickled cucumber, chopped
2 tablespoons vinaigrette dressing	2 tablespoons vinaigrette dressing	3 tablespoons vinaigrette dressing

Gently combine all the ingredients. Serve half – pack remainder in plastic container and freeze. Defrost and serve chilled.

Cauliflower and black olives

IMPERIAL	METRIC	AMERICAN
4 oz. frozen cauliflower florets	100 g. frozen cauliflower florets	¼ lb. frozen cauliflower florets
1 oz. black olives	25 g. black olives	scant ¼ cup ripe olives
1 teaspoon capers	1 teaspoon capers	1 teaspoon capers

Cook the frozen cauliflower until tender. Stone and chop the olives and stir into the drained cauliflower with the capers. Serve hot.

Spanish salad

IMPERIAL	METRIC	AMERICAN
1 large Spanish onion	1 large Spanish onion	1 large Spanish onion
1 red pepper	1 red pepper	1 red sweet pepper
1 green pepper	1 green pepper	1 green sweet pepper
few white peppercorns	few white peppercorns	few white peppercorns
salt to taste	salt to taste	salt to taste
2 tablespoons wine vinegar	2 tablespoons wine vinegar	3 tablespoons wine vinegar

Slice the onion very finely and deseed and chop the peppers. Toss together with the pepper, salt and vinegar. Chill and serve half – pack remainder in plastic container and freeze. Defrost and serve chilled.

Golden guide number

to parcels

The Continental method of cooking *en papillote* has always been a great favourite with me because it conserves the flavour, aroma and juices of the most delicate foods and at the same time eliminates the necessity to clean cooking utensils or the oven. The fact that it enables one to cook main meal dishes without additional fat is a delightful bonus for dieters. Chicken portions, turkey portions, veal and pork chops, all make excellent parcels for oven-baking with the addition of herbs, seasoning, onion or sliced tomato. The parcels are frozen and transferred, unopened, straight to a baking tray and popped in the oven when required.

Steak in a parcel

Imperial	Metric	American
4 lean rump steaks	4 lean rump steaks	4 lean sirloin steaks
1 small cooking apple	1 small cooking apple	1 small baking apple
1 leek	1 leek	1 leek
2 tablespoons hot stock	2 tablespoons hot stock	3 tablespoons hot stock
2 tablespoons tomato purée	2 tablespoons tomato purée	3 tablespoons tomato paste
$\frac{1}{4}$ teaspoon ground ginger	$\frac{1}{4}$ teaspoon ground ginger	$\frac{1}{4}$ teaspoon ground ginger
salt and pepper to taste	salt and pepper to taste	salt and pepper to taste
4 oz. mushrooms, sliced	100 g. mushrooms, sliced	1 cup sliced mushrooms

Remove any fat from steaks. Peel, core and chop the apple, wash and slice leek into rings. Mix together the hot stock, tomato purée, ginger and seasoning. Stir in the mushrooms, apple and leeks. Divide the vegetable mixture between four large squares of freezer foil. Place a steak on each. Seal the foil completely to make four airtight parcels. Place one parcel in a moderately hot oven (375°F, 190°C, Gas Mark 5) for 45 minutes. Freeze remaining parcels. Defrost and cook as above.

Suitable alternatives for steak are pork fillet slices, lamb chump chops and lamb's liver.

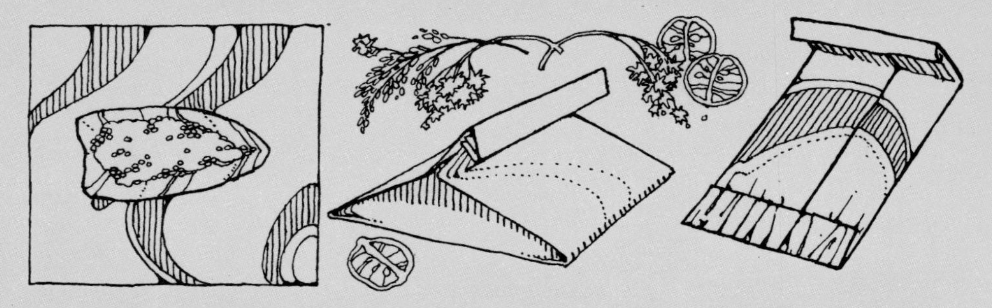

Haddock in a parcel

Imperial	Metric	American
4 small haddock fillets	4 small haddock fillets	4 small haddock fillets
4 tablespoons parsley and thyme stuffing mix	4 tablespoons parsley and thyme stuffing mix	5 tablespoons parsley and thyme stuffing mix
finely grated zest of 1 small lemon	finely grated zest of 1 small lemon	finely grated zest of 1 small lemon
4 tablespoons very hot water	4 tablespoons very hot water	$\frac{1}{3}$ cup very hot water
salt and pepper to taste	salt and pepper to taste	salt and pepper to taste
beaten egg to bind	beaten egg to bind	beaten egg to bind

Place fish fillets on four large squares of freezer foil. Mix together stuffing mix, lemon zest, hot water and seasoning with sufficient beaten egg to bind. Divide mixture into four and spread evenly over the fillets. Seal the foil completely to make four airtight parcels. Place one parcel in a moderately hot oven (375°F, 190°C, Gas Mark 5) for 45 minutes. Freeze remaining parcels, defrost and cook as above. Suitable alternatives for haddock are cod fillet, halibut steak or plaice fillet.

Note For method of sealing parcels, see drawing above.

Stuffed tomatoes

Imperial	Metric	American
8 oz. minced cooked lean lamb or beef	225 g. minced cooked lean lamb or beef	1 cup ground cooked lean lamb or beef
1 medium onion, minced	1 medium onion, minced	1 medium onion, minced
4 oz. mushrooms, minced	100 g. mushrooms, minced	$\frac{1}{4}$ lb. mushrooms, minced
8 large tomatoes	8 large tomatoes	8 large tomatoes
2 teaspoons Worcestershire sauce	2 teaspoons Worcestershire sauce	2 teaspoons Worcestershire sauce
$\frac{1}{4}$ pint thick brown gravy	$1\frac{1}{2}$ dl. thick brown gravy	$\frac{2}{3}$ cup thick brown gravy
salt and pepper to taste	salt and pepper to taste	salt and pepper to taste

Mince together the meat, onion and mushrooms twice. Slice the tops off the tomatoes and scoop out the centres. Sieve and mix the tomato purée with the minced mixture, Worcestershire sauce, gravy and seasoning to taste. Pack filling into the tomato cases and replace the tops. Take four large squares of freezer foil, place two stuffed tomatoes on each and seal to make airtight parcels. Place one parcel in a moderately hot oven (375°F, 190°C, Gas Mark 5) for 45 minutes. Freeze remaining parcels. To serve, defrost and cook as above.

Note This recipe is given for a larger quantity because the preparation is fairly complicated.

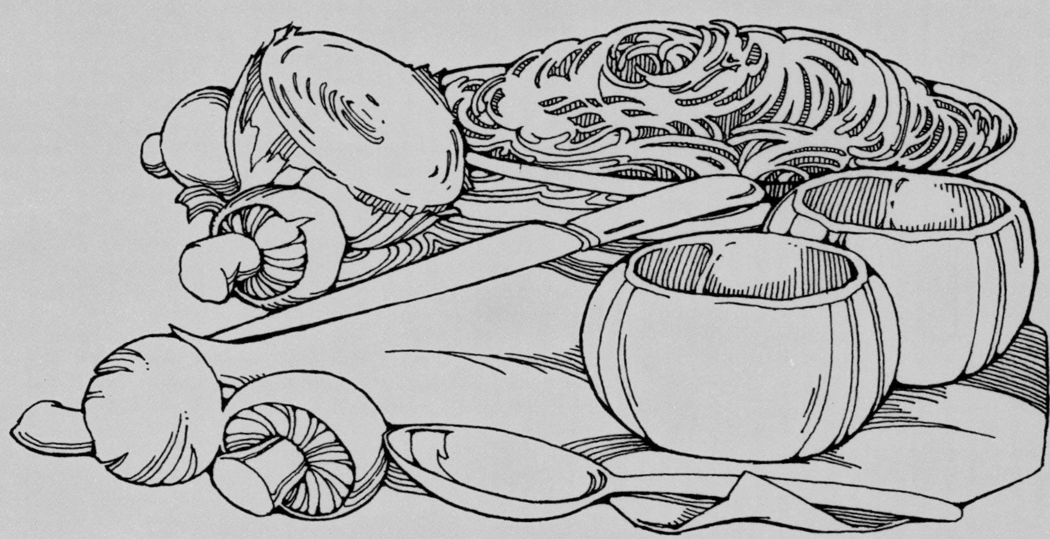

Herb-stuffed hearts

Imperial	Metric	American
2 lambs' hearts	2 lambs' hearts	2 lamb hearts
stuffing	*stuffing*	*stuffing*
2 oz. lean bacon	50 g. lean bacon	3 bacon slices
1 oz. seedless raisins	25 g. seedless raisins	3 tablespoons seedless raisins
2 teaspoons grated lemon zest	2 teaspoons grated lemon zest	2 teaspoons grated lemon zest
2 oz. soft white breadcrumbs	50 g. soft white breadcrumbs	1 cup fresh soft bread crumbs
1 teaspoon chopped parsley	1 teaspoon chopped parsley	1 teaspoon chopped parsley
pinch dried mixed herbs	pinch dried mixed herbs	pinch dried mixed herbs
salt and pepper to taste	salt and pepper to taste	salt and pepper to taste
$\frac{1}{2}$ oz. butter	15 g. butter	1 tablespoon butter
beaten egg to bind	beaten egg to bind	beaten egg to bind
1 tablespoon seasoned flour	1 tablespoon seasoned flour	1 tablespoon seasoned flour
1 tablespoon oil	1 tablespoon oil	1 tablespoon oil

Wash and trim the hearts. De-rind the bacon and chop it finely. Mix together the bacon, raisins, lemon zest, breadcrumbs, herbs and seasoning. Melt the butter and stir into the stuffing with enough beaten egg to bind. Divide the stuffing between the hearts and either sew up the opening or close with a skewer or wooden cocktail stick. Coat the hearts in seasoned flour. Heat the oil in a pan and use to fry the hearts until golden brown all over. Cool, wrap each heart in freezer foil and freeze. Defrost, still wrapped, place in a small ovenproof casserole and cook in a moderate oven (350°F, 180°C, Gas Mark 4) for about $1\frac{1}{2}$ hours. Remove thread or skewers and serve sliced.

Cordoba meat rolls

IMPERIAL	METRIC	AMERICAN
2 4-oz. slices topside, silverside or minute steaks	2 100-g. slices topside, silverside or minute steaks	2 $\frac{1}{4}$-lb. slices top round or rump steaks or Swiss steaks
2 rashers bacon	2 rashers bacon	2 bacon slices
1 oz. butter	25 g. butter	2 tablespoons butter
1 tablespoon oil	1 tablespoon oil	1 tablespoon oil
1 medium onion, chopped	1 medium onion, chopped	1 medium onion, chopped
1 oz. cheese, grated	25 g. cheese, grated	$\frac{1}{4}$ cup grated cheese
$\frac{1}{2}$ teaspoon dried marjoram	$\frac{1}{2}$ teaspoon dried marjoram	$\frac{1}{2}$ teaspoon dried marjoram
1 oz. sultanas	25 g. sultanas	3 tablespoons seedless white raisins
salt and pepper to taste	salt and pepper to taste	salt and pepper to taste
1 tablespoon tomato purée	1 tablespoon tomato purée	1 tablespoon tomato paste
$\frac{1}{2}$ pint stock	3 dl. stock	$1\frac{1}{4}$ cups stock

Place the meat slices between two sheets of greaseproof paper and bat out thinly. De-rind and grill the bacon and put a rasher on each slice of meat. Heat the butter and half the oil in a frying pan and use to fry the onion gently until softened but not browned. Drain and divide fried onion in half. Mix together one portion of onion with the cheese, herbs, sultanas and seasoning with a little of the fat from the pan to moisten. Divide this stuffing between the meat slices, roll up tightly and secure with thread or wooden cocktail sticks. Heat the remaining oil in the frying pan and use to brown the meat rolls all over. Place the meat rolls in two small shaped foil dishes. To the juices in the frying pan add the remaining onion, tomato purée, stock and seasoning and bring to boiling point. Pour over the meat rolls and cool. Cover, seal and freeze. Defrost one portion and cook covered in a cool oven (300°F, 150°C, Gas Mark 2) for about 45 minutes. Remove thread or cocktail sticks before serving.

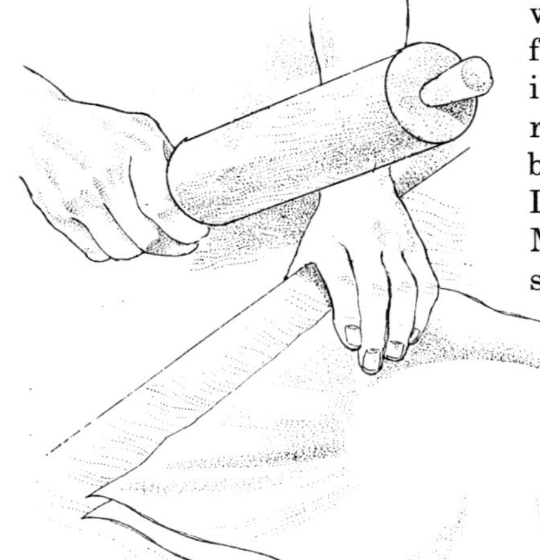

South sea kebabs

IMPERIAL	METRIC	AMERICAN
1 tablespoon soy sauce	1 tablespoon soy sauce	1 tablespoon soy sauce
1 tablespoon wine or wine vinegar	1 tablespoon wine or wine vinegar	1 tablespoon wine or wine vinegar
1 tablespoon oil	1 tablespoon oil	1 tablespoon oil
1 clove garlic, crushed	1 clove garlic, crushed	1 clove garlic, crushed
black pepper to taste	black pepper to taste	black pepper to taste
8 oz. rump steak	225 g. rump steak	$\frac{1}{2}$ lb. sirloin steak

Mix together the soy sauce, wine, oil, garlic and pepper. Remove any fat from meat and cut into neat cubes. Divide the meat between two small polythene containers, pour over the marinade and allow to stand for about 2 hours. Cover, seal and freeze. Defrost one portion, thread meat on a skewer with a few drained water chestnuts, button mushrooms or squares of red or green pepper. Brush with marinade and grill for about 15 minutes, turning to cook evenly.

Leek and liver casserole

IMPERIAL	METRIC	AMERICAN
2 leeks	2 leeks	2 leeks
8 oz. lamb's liver	225 g. lamb's liver	½ lb. lamb liver
½ oz. lard	15 g. lard	1 tablespoon lard
1 medium onion, chopped	1 medium onion, chopped	1 medium onion, chopped
1 carrot, chopped	1 carrot, chopped	1 carrot, chopped
salt and pepper to taste	salt and pepper to taste	salt and pepper to taste
¼ teaspoon dried mixed herbs	¼ teaspoon dried mixed herbs	¼ teaspoon dried mixed herbs
½ pint stock	3 dl. stock	1¼ cups stock

Wash and slice the leeks into rings. Cut liver into small dice. Melt the lard in a saucepan and use to fry the liver for 1 minute to seal. Add the remaining ingredients and bring to the boil, stirring constantly. Cover tightly and simmer very gently for about 15 minutes. Cool. Divide between two shaped foil dishes, cover, seal and label. Place 1 portion, still wrapped and frozen, in a moderate oven (350°F, 180°C, Gas Mark 4) for 25–30 minutes.

Kidneys in yogurt sauce

IMPERIAL	METRIC	AMERICAN
4 lambs' kidneys	4 lambs' kidneys	4 lamb kidneys
4 oz. mushrooms, sliced	100 g. mushrooms, sliced	1 cup sliced mushrooms
¼ pint beef stock	1½ dl. beef stock	⅔ cup beef stock
½ tablespoon tomato purée	½ tablespoon tomato purée	½ tablespoon tomato paste
1 teaspoon made mustard	1 teaspoon made mustard	1 teaspoon made mustard
4 tablespoons natural yogurt	4 tablespoons natural yogurt	5 tablespoons natural yogurt
salt and pepper to taste	salt and pepper to taste	salt and pepper to taste

Skin the kidneys, halve them and remove cores. Place kidneys in a small saucepan with the mushrooms, stock and tomato purée. Bring to the boil, cover and simmer gently for about 20 minutes. Remove kidneys and place in two shaped foil dishes. Add the remaining ingredients to the sauce and bring to boiling point, stirring constantly. Adjust seasoning and pour over the kidneys. Cool. Cover, seal and freeze. Reheat from the frozen state in a moderate oven (350°F, 180°C, Gas Mark 4) for 25–30 minutes.

Split devilled kidneys

IMPERIAL	METRIC	AMERICAN
4 lambs' kidneys	4 lambs' kidneys	4 lamb kidneys
1 oz. butter	25 g. butter	2 tablespoons butter
salt and pepper to taste	salt and pepper to taste	salt and pepper to taste
½ teaspoon curry paste	½ teaspoon curry paste	½ teaspoon curry paste
½ teaspoon Worcestershire sauce	½ teaspoon Worcestershire sauce	½ teaspoon Worcestershire sauce

Skin the kidneys, split them almost in half and remove the cores. Thread a wooden cocktail stick through each kidney to keep it open flat. Melt the butter and mix with the other ingredients. Place kidneys in grill pan, pour in the sauce and turn kidneys in this. Grill under medium heat until cooked, about 15 minutes, turning to cook evenly. Cool and remove cocktail sticks. Pack kidneys in two small shaped foil dishes and pour over any remaining sauce. Cool. Cover containers, seal and freeze. Defrost and reheat in a moderate oven (350°F, 180°C, Gas Mark 4) for 20 minutes.

Fruity frankfurters

IMPERIAL	METRIC	AMERICAN
2 tablespoons wine vinegar	2 tablespoons wine vinegar	3 tablespoons wine vinegar
2 tablespoons lemon juice	2 tablespoons lemon juice	3 tablespoons lemon juice
pinch ground ginger	pinch ground ginger	pinch ground ginger
pinch grated nutmeg	pinch grated nutmeg	pinch grated nutmeg
1 teaspoon grated lemon zest	1 teaspoon grated lemon zest	1 teaspoon grated lemon zest
salt and pepper to taste	salt and pepper to taste	salt and pepper to taste
2 tablespoons mango chutney	2 tablespoons mango chutney	3 tablespoons mango chutney
1 dessert apple	1 dessert apple	1 dessert apple
2 oz. seedless raisins	50 g. seedless raisins	generous ½ cup seedless raisins
6 frankfurter sausages	6 frankfurter sausages	6 frankfurter sausages

Mix together the vinegar, lemon juice, spices, lemon zest and seasoning in a small saucepan. Chop any large lumps in the mango chutney, peel core and chop the apple and add both to the pan with the raisins. Stir well and bring the sauce to boiling point. Cool. Divide the frankfurter sausages between two small polythene containers and pour the sauce over. Seal and freeze. Turn the contents of one container into a saucepan and reheat gently, stirring to prevent burning. Bring to boiling point and simmer for about 7 minutes.

Curried cod with mushrooms

IMPERIAL	METRIC	AMERICAN
2 tablespoons oil	2 tablespoons oil	3 tablespoons oil
1 large onion, chopped	1 large onion, chopped	1 large onion, chopped
4 oz. button mushrooms, sliced	100 g. button mushrooms, sliced	1 cup sliced mushrooms
$1\frac{1}{2}$ teaspoons curry powder	$1\frac{1}{2}$ teaspoons curry powder	$1\frac{1}{2}$ teaspoons curry powder
$\frac{1}{2}$ teaspoon curry paste	$\frac{1}{2}$ teaspoon curry paste	$\frac{1}{2}$ teaspoon curry paste
1 chicken stock cube	1 chicken stock cube	1 chicken bouillon cube
1 tablespoon tomato purée	1 tablespoon tomato purée	1 tablespoon tomato paste
$\frac{1}{2}$ pint boiling water	3 dl. boiling water	$1\frac{1}{4}$ cups boiling water
12 oz. cod fillet	350 g. cod fillet	$\frac{3}{4}$ lb. cod fillet
1 oz. sultanas	25 g. sultanas	3 tablespoons seedless white raisins

Heat the oil and fry the onion until softened but not browned. Add the mushrooms and cook for 2 minutes. Stir in the curry powder and paste and cook, for a further 5 minutes. Dissolve the stock cube and tomato purée in the water and gradually add to the pan. Bring to the boil, stirring constantly. Cut the cod into chunks and add to the sauce with the sultanas. Bring back to boiling point, cover and simmer for about 30 minutes. Check seasoning and cool. Divide between two polythene containers, seal and freeze. Turn 1 portion into a saucepan and reheat gently to boiling point, stirring to prevent burning.

Fruit snows

IMPERIAL	METRIC	AMERICAN
$\frac{1}{4}$ pint canned or frozen unsweetened apple purée	$1\frac{1}{2}$ dl. canned or frozen unsweetened apple purée	$\frac{2}{3}$ cup canned or frozen unsweetened applesauce
4 tablespoons unsweetened strawberry purée	4 tablespoons unsweetened strawberry purée	5 tablespoons unsweetened strawberry purée
artificial liquid sweetener to taste	artificial liquid sweetener to taste	artificial liquid sweetener to taste
2 egg whites	2 egg whites	2 egg whites

Combine the fruit purées and sweeten to taste with artificial sweetener. Beat the egg whites until stiff and fold evenly into the purée. Divide between two individual serving dishes.

Apple purée makes a good basis and combines with other fruit besides strawberries. Try gooseberries, apricots or raspberries.

Variety for the vegetarian

So many people today shop at health food stores besides the convinced vegetarian (who believes eating meat is wrong or bad for his health). So a corner must be found in the freezer for dishes to enhance a vegetarian diet and this means a little more than making up special packs of peanut butter sandwiches.

Many of the recipes in this section could be improved by adding wheatgerm and brewer's yeast; for example, when the bean soup is defrosted and reheated. See the note attached to this recipe.

To turn any of the dishes into a satisfying meal, they can be eaten with wholemeal bread or rolls, hard-boiled eggs, generous portions of cheese and fresh salads.

Bean soup

Makes 4 servings.
To freeze Pack in polythene containers leaving headspace. Seal and label.
To serve Turn frozen soup into a saucepan and reheat gently to boiling point. To each ½ pint (3 dl.) of soup add 1 teaspoon wheatgerm and ½ teaspoon brewer's yeast. Mix it with a little warm soup and add to the pan. Stir and continue heating gently for a few minutes.
Storage time 4–6 months.

IMPERIAL	METRIC	AMERICAN
8 oz. haricot beans	225 g. haricot beans	generous 1 cup navy beans
2 pints cold water	generous 1 litre water	5 cups cold water
6 tablespoons oil	6 tablespoons oil	½ cup oil
1 large onion, chopped	1 large onion, chopped	1 large onion, chopped
1 teaspoon yeast extract	1 teaspoon yeast extract	1 teaspoon yeast extract
salt and pepper to taste	salt and pepper to taste	salt and pepper to taste

Soak the beans in the cold water overnight. Cook in the same water until tender. In a second pan heat the oil and use to fry the onion until softened and golden but not brown. Add to the beans with the yeast extract and seasoning to taste. Stir well, cool and liquidise or sieve.

Corn and tomato chowder

Makes 6–8 servings.
To freeze Pack in polythene containers leaving headspace. Seal and label.
To serve Turn frozen chowder into a saucepan, add a little milk and reheat gently almost to boiling point, stirring frequently to prevent burning.
Storage time 4–6 months.

IMPERIAL	METRIC	AMERICAN
2 tablespoons corn oil	2 tablespoons corn oil	3 tablespoons corn oil
2 onions, chopped	2 onions, chopped	2 onions, chopped
8 oz. potatoes, diced	225 g. potatoes, diced	½ lb. potatoes, diced
1 15-oz. can tomato soup	1 425-g. can tomato soup	1 15-oz. can tomato soup
½ pint water	3 dl. water	1¼ cups water
1 teaspoon yeast extract	1 teaspoon yeast extract	1 teaspoon yeast extract
½ teaspoon dried mixed herbs	½ teaspoon dried mixed herbs	½ teaspoon dried mixed herbs
1 teaspoon sugar	1 teaspoon sugar	1 teaspoon sugar
6 oz. frozen or canned sweetcorn kernels	175 g. frozen or canned sweetcorn kernels	1 cup frozen or canned kernel corn
½ pint milk	3 dl. milk	1¼ cups milk
salt and pepper to taste	salt and pepper to taste	salt and pepper to taste

Heat the oil and use to fry the onion until softened but not browned. Add the potatoes and cook for 5 minutes, stirring occasionally. Stir in the soup, water, yeast extract, herbs, sugar and sweetcorn and bring to the boil. Cover and simmer for about 20 minutes. Stir in the milk and season to taste. Cool.

Asparagus flan

Makes 4–6 servings.
To freeze Open freeze until solid, remove flan ring and wrap in freezer foil or polythene bag. Seal and label.
To serve Defrost in the refrigerator for 6 hours. If required hot, place in a moderate oven (350°F, 180°C, Gas Mark 4) for 20 minutes.
Storage time 4–6 months.

IMPERIAL	METRIC	AMERICAN
1 8-inch frozen pastry flan case, baked	1 20-cm. frozen pastry flan case, baked	1 8-inch frozen pie shell, baked
1 10-oz. can asparagus tips, drained	1 275-g. can asparagus tips, drained	1 10-oz. can asparagus tips, drained
2 tomatoes, chopped	2 tomatoes, chopped	2 tomatoes, chopped
2 oz. button mushrooms, sliced	50 g. button mushrooms, sliced	½ cup sliced button mushrooms
2 eggs	2 eggs	2 eggs
¼ pint natural yogurt	1½ dl. natural yogurt	⅔ cup natural yogurt
¼ pint milk	1½ dl. milk	⅔ cup milk
salt and pepper to taste	salt and pepper to taste	salt and pepper to taste
2 tablespoons dried white breadcrumbs	2 tablespoons dried white breadcrumbs	3 tablespoons fine dry bread crumbs

Place the baked flan case on a baking tray in the flan ring. Chop the asparagus and mix with the tomatoes and mushrooms. Spread this mixture in the base of the flan case. Beat together the eggs, yogurt, milk and seasoning and pour into the flan case. Sprinkle with the breadcrumbs and bake in a moderately hot oven (400°F, 200°C, Gas Mark 6) for 20 minutes. Reduce heat to 350°F, 180°C, Gas Mark 4 for a further 20–30 minutes, until set. Cool.

Nut-stuffed courgettes

Makes 3–4 servings.
To freeze Label.
To serve Place still frozen and sealed in a moderately hot oven (400°F, 200°C, Gas Mark 6) for 30 minutes.
Storage time 3 months.

IMPERIAL	METRIC	AMERICAN
2 tablespoons oil	2 tablespoons oil	3 tablespoons oil
1 medium onion, chopped	1 medium onion, chopped	1 medium onion, chopped
8 oz. mixed nuts, chopped	225 g. mixed nuts, chopped	2 cups chopped mixed nuts
4 oz. cooked brown rice	100 g. cooked brown rice	scant 1 cup cooked brown rice
1 small can tomato purée	1 small can tomato purée	1 small can tomato paste
1 teaspoon mild Continental mustard	1 teaspoon mild Continental mustard	1 teaspoon mild Continental mustard
3 tablespoons red wine	3 tablespoons red wine	scant $\frac{1}{4}$ cup red wine
salt and pepper to taste	salt and pepper to taste	salt and pepper to taste
1 lb. courgettes	450 g. courgettes	1 lb. small zucchini

Heat the oil and use to fry the onion until softened and just beginning to brown. Remove from the heat and add the nuts, rice, tomato purée, mustard, wine and seasoning to taste. Top and tail the courgettes and cut a deep V-shaped wedge from the length of each one. Stuff courgettes with the nut mixture and place close together in a shaped foil dish. Cover dish with freezer foil, crimp edges together to make an airtight seal and place in a moderate oven (350°F, 180°C, Gas Mark 4) for 35 minutes. Cool.

Nut croquettes

Makes Approximately 8 croquettes.
To freeze Pack croquettes in a rigid-based polythene container with foil dividers.
To serve Deep fry from the frozen state in hot oil for about 10 minutes, until golden brown all over. Drain on absorbent paper.
Storage time 6 months.

IMPERIAL	METRIC	AMERICAN
8 oz. shelled mixed nuts	225 g. shelled mixed nuts	2 cups shelled mixed nuts
3 oz. margarine	75 g. margarine	6 tablespoons margarine
3 oz. soft wholemeal breadcrumbs	75 g. soft wholemeal breadcrumbs	$1\frac{1}{2}$ cups soft wholewheat bread crumbs
2 onions, chopped	2 onions, chopped	2 onions, chopped
1 teaspoon dried mixed herbs	1 teaspoon dried mixed herbs	1 teaspoon dried mixed herbs
salt and pepper to taste	salt and pepper to taste	salt and pepper to taste
2 eggs, beaten	2 eggs, beaten	2 eggs, beaten
dried breadcrumbs for coating	dried breadcrumbs for coating	dry bread crumbs for coating

Mince the nuts or chop them finely in an electric blender. Place in a bowl with the margarine, soft breadcrumbs, chopped onion, herbs and seasoning. Stir in sufficient beaten egg to bind and mix thoroughly. Form the mixture into small croquettes with floured hands. Dip croquettes in the remaining beaten egg and coat with dried breadcrumbs. Place on a baking tray and chill until firm.

Beetroot bake

Makes 6 servings.
To freeze Pack beetroot in small polythene containers and pour the sauce over. Seal and label.
To serve Turn contents into a saucepan and reheat gently, stirring to prevent burning. Serve with buttered wholemeal rolls.
Storage time 4–6 months.

IMPERIAL	METRIC	AMERICAN
sauce	*sauce*	*sauce*
1 medium onion, chopped	1 medium onion, chopped	1 medium onion, chopped
4 tablespoons vinegar	4 tablespoons vinegar	$\frac{1}{3}$ cup vinegar
1 tablespoon tomato purée	1 tablespoon tomato purée	1 tablespoon tomato paste
1 oz. brown sugar	25 g. brown sugar	2 tablespoons brown sugar
1 tablespoon Worcestershire sauce	1 tablespoon Worcestershire sauce	1 tablespoon Worcestershire sauce
$\frac{1}{2}$ teaspoon mustard	$\frac{1}{2}$ teaspoon mustard	$\frac{1}{2}$ teaspoon mustard
pinch cayenne pepper	pinch cayenne pepper	pinch cayenne pepper
$\frac{3}{4}$ pint stock	scant $\frac{1}{2}$ litre stock	scant 2 cups stock
2 tablespoons cornflour	2 tablespoons cornflour	3 tablespoons cornstarch
2 lb. small beetroot, cooked	1 kg. small beetroot, cooked	2 lb. small beets, cooked

To make the sauce, place the chopped onion in a saucepan with the vinegar and boil to reduce liquid by three-quarters. Stir in the tomato purée, sugar, Worcestershire sauce and seasonings then add the stock. Moisten the cornflour with a little cold water, stir into the sauce and bring to the boil, stirring constantly until smooth and thickened. Cool. Peel the beetroots and cut into quarters.

Curried cauliflower

Makes 2 servings.
To freeze Pack into polythene containers leaving headspace. Seal and label.
To serve Turn contents into a saucepan and reheat gently to boiling point. Stir in 2 oz. (50 g.) small raw cauliflower florets, 2 oz. (50 g.) thinly sliced raw carrot and 1 oz. (25 g.) sultanas and stir over moderate heat for 3 minutes.
Storage time 4–6 months.

Note Other finely sliced raw vegetables can be substituted for the cauliflower and carrot but the texture should still be crisp without being unpleasantly hard.

IMPERIAL	METRIC	AMERICAN
$1\frac{1}{2}$ pints water	scant 1 litre water	$3\frac{3}{4}$ cups water
8 oz. lentils	225 g. lentils	1 cup lentils
large pinch salt	large pinch salt	large pinch salt
1 tablespoon curry powder	1 tablespoon curry powder	1 tablespoon curry powder
3 large onions	3 large onions	3 large onions
4 tablespoons corn oil	4 tablespoons corn oil	$\frac{1}{3}$ cup corn oil
3 large tomatoes, sliced	3 large tomatoes, sliced	3 large tomatoes, sliced

Put the water into a saucepan and bring to the boil. Add the lentils, salt and curry powder and return to boiling point. Cover and simmer for about 1 hour, until lentils are tender and mixture is very thick. Slice the onions very thinly. Heat the oil in a frying pan and use to fry the onions and tomatoes until softened but not browned. Combine the two mixtures and cool.

Roast potatoes

(Illustrated on page 18)

To freeze Pack in polythene bags. Seal and label.
To serve Place still frozen in a roasting tin containing a very little hot oil. Turn the potatoes in the oil and place in a moderately hot oven (400°F, 200°C, Gas Mark 6) for 30 minutes.
Storage time 4–6 months.

IMPERIAL	METRIC	AMERICAN
¼ pint oil	1½ dl. oil	⅔ cup oil
4 lb. potatoes	2 kg. potatoes	4 lb. potatoes
salt	salt	salt

Place oil in a roasting tin and preheat in a moderately hot oven (400°F, 200°C, Gas Mark 6). Peel and cut the potatoes into even-sized pieces, bring to the boil in salted water, strain off the water and shake dry. Add potatoes to roasting tin, turning to coat evenly with the oil. Sprinkle with salt, return to oven and roast for about 1 hour, turning once to brown evenly. Drain well and cool. For non-vegetarians use good meat dripping in place of the oil.

French fried potatoes

(Illustrated on page 18)

To freeze Pack in polythene bags. Seal and label.
To serve Defrost sufficiently to be able to separate the chips. Turn into a frying basket and plunge again into heated deep hot oil for a few minutes until golden brown.
Storage time 4–6 months.

Note The uncooked chips may also be blanched for 1 minute in boiling water and open frozen before packing to prevent them from sticking together. Cook from the frozen state in hot deep oil.

IMPERIAL	METRIC	AMERICAN
4 lb. potatoes	2 kg. potatoes	4 lb. potatoes
oil for frying	oil for frying	oil for frying

Peel the potatoes and cut into even-shaped chips. Wash and pat dry in a clean tea towel. Heat oil to 360°F, 185°C and deep fry chips in a basket until tender but not coloured. Remove, drain on absorbent kitchen paper and cool.

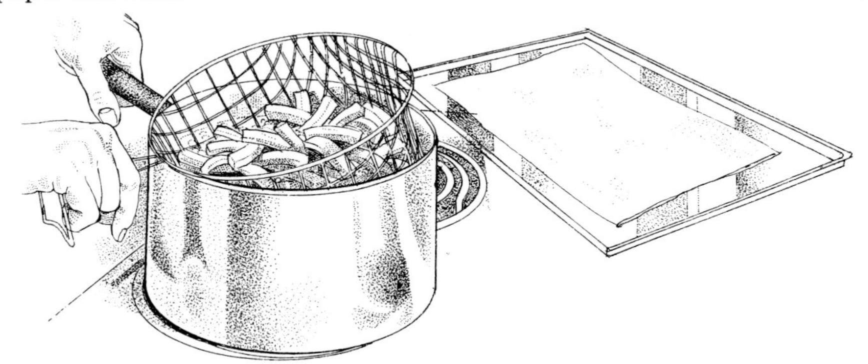

Duchesse potatoes

(Illustrated on page 18)

To freeze Open freeze until solid then pack in polythene containers in layers with foil sheets as dividers. Seal and label.
To serve Place still frozen on greased baking tray, brush with beaten egg and reheat in a moderately hot oven (400°F, 200°C, Gas Mark 6) for 25 minutes.
Storage time 4–6 months.

Note Large potatoes baked in their jackets can be frozen if the cooked potato is scooped out, mixed with a savoury ingredient such as grated cheese, packed back into the potato shells with foil dividers placed between the two halves. Mould the potatoes in foil for freezing.

IMPERIAL	METRIC	AMERICAN
4 lb. potatoes	2 kg. potatoes	4 lb. potatoes
4 oz. butter	100 g. butter	½ cup butter
2 eggs	2 eggs	2 eggs
½ teaspoon nutmeg	½ teaspoon nutmeg	½ teaspoon nutmeg
salt and pepper to taste	salt and pepper to taste	salt and pepper to taste

Peel and slice the potatoes. Cook in boiling salted water until tender then drain well and mash thoroughly. Beat in the butter, eggs and seasonings. Pipe the potato mixture into pyramids onto lightly greased baking trays (or onto trays lined with foil).

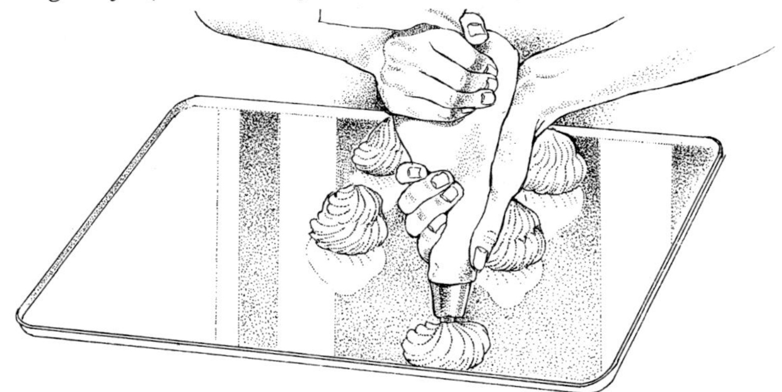

Three ways with frozen vegetables

1. With mushrooms

IMPERIAL	METRIC	AMERICAN
8 oz. frozen peas	225 g. frozen peas	½ lb. frozen peas
2 teaspoons corn oil	2 teaspoons corn oil	2 teaspoons corn oil
2 oz. mushrooms, sliced	50 g. mushrooms, sliced	½ cup sliced mushrooms
2 tablespoons soured cream	2 tablespoons soured cream	3 tablespoons sour cream
salt and pepper to taste	salt and pepper to taste	salt and pepper to taste
chopped parsley to garnish	chopped parsley to garnish	chopped parsley to garnish

Cook the peas in a very little lightly salted water until tender, drain and keep hot. Heat the oil and cook mushrooms until tender, about 3–5 minutes. Add the peas, stir in the cream and season well. Reheat gently but do not allow to boil. Turn into a warm serving dish and sprinkle with parsley.

2. With tomato purée

1 tablespoon olive oil	1 tablespoon olive oil	1 tablespoon olive oil
1 medium onion, chopped	1 medium onion, chopped	1 medium onion, chopped
2 tablespoons tomato purée	2 tablespoons tomato purée	3 tablespoons tomato paste
½ teaspoon sugar	½ teaspoon sugar	½ teaspoon sugar
salt and pepper to taste	salt and pepper to taste	salt and pepper to taste
8 oz. frozen broad beans	225 g. frozen broad beans	½ lb. frozen broad or fava beans

Heat the oil and fry the onion gently until limp but not coloured. Add the tomato purée, sugar and seasoning. Stir well and add beans. Cover and simmer gently for 15–20 minutes, stirring occasionally. Adjust seasoning before serving.

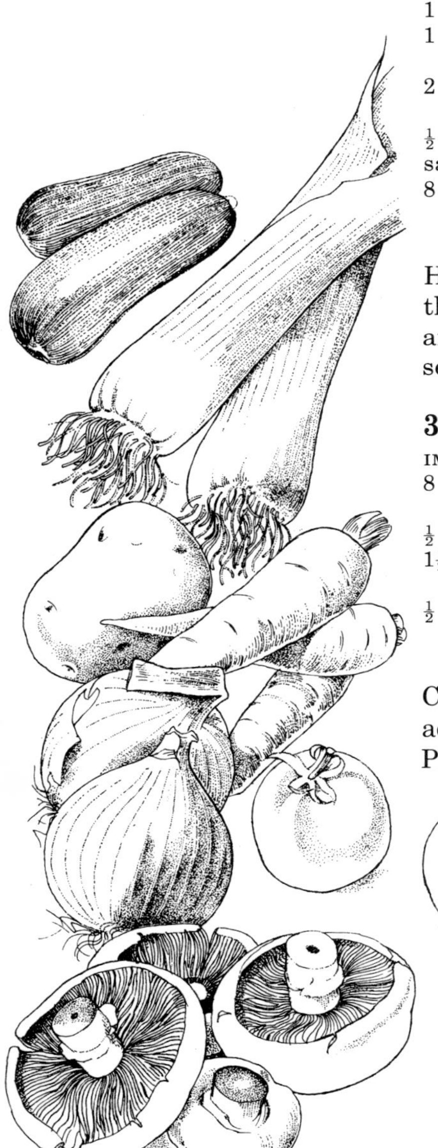

3. With almonds

IMPERIAL	METRIC	AMERICAN
8 oz. frozen French beans	225 g. frozen French beans	½ lb. frozen green beans
½ oz. butter	15 g. butter	1 tablespoon butter
1½ teaspoons lemon juice	1½ teaspoons lemon juice	1½ teaspoons lemon juice
½ oz. flaked almonds, toasted	15 g. flaked almonds, toasted	about ¼ cup slivered almonds, toasted

Cook the beans in a very little lightly salted water until tender. Drain, add the butter and lemon juice and stir until beans are well coated. Place in a warm serving dish and sprinkle with the almonds.

Piquant avocado dip

(Illustrated on page 108)

Makes 6 servings.
To freeze Pack into a suitable polythene container. Seal and label.
To serve Defrost in the refrigerator overnight, stir thoroughly and serve chilled. If required quickly, hold container under warm water until dip can be broken up and mixed with a fork until smooth. Garnish with lemon and sprigs of parsley. Serve with potato crisps and small savoury biscuits.
Storage time 4–6 months.

IMPERIAL	METRIC	AMERICAN
2 medium avocados	2 medium avocados	2 medium avocados
2 tablespoons lemon juice	2 tablespoons lemon juice	3 tablespoons lemon juice
6 oz. cream cheese	175 g. cream cheese	¾ cup cream cheese
1 teaspoon salt	1 teaspoon salt	1 teaspoon salt
¼ teaspoon pepper	¼ teaspoon pepper	¼ teaspoon pepper
dash Tabasco sauce	dash Tabasco sauce	dash Tabasco sauce
1 tablespoon finely grated onion	1 tablespoon finely grated onion	1 tablespoon finely grated onion

Stone the avocados and mash or sieve the flesh until smooth. Add the remaining ingredients and mix well.

Hazelnut cheese spread

Makes 4 servings.
To freeze Pack into small polythene containers. Seal and label.
To serve Defrost at room temperature for 2 hours and stir until smooth. Serve spread on crispbread or wholemeal bread.
Storage time 4–6 months.

IMPERIAL	METRIC	AMERICAN
4 oz. cream cheese	100 g. cream cheese	½ cup cream cheese
2 oz. soft margarine	50 g. soft margarine	¼ cup soft margarine
¼ teaspoon made mustard	¼ teaspoon made mustard	¼ teaspoon made mustard
2 oz. hazelnuts, chopped	50 g. hazelnuts, chopped	½ cup chopped hazelnuts
3 gherkins, chopped	3 gherkins, chopped	3 sweet dill pickle, chopped
salt and pepper to taste	salt and pepper to taste	salt and pepper to taste
¼ teaspoon paprika pepper	¼ teaspoon paprika pepper	¼ teaspoon paprika pepper

Cream the cheese and margarine until smooth. Gradually beat in the mustard, hazelnuts and gherkins. Add the paprika pepper and adjust seasoning.

Herbed celery scones

Makes about 18 scones.
To freeze Pack in polythene bags. Seal and label.
To serve Place on a baking tray and reheat in a moderate oven (325°F, 170°C, Gas Mark 3) for 20 minutes. Serve split and buttered.
Storage time 4–6 months.

IMPERIAL	METRIC	AMERICAN
4 oz. wholemeal flour	100 g. wholemeal flour	1 cup wholewheat flour
4 oz. plain flour	125 g. plain flour	1 cup all-purpose flour
½ teaspoon bicarbonate soda	½ teaspoon bicarbonate soda	½ teaspoon baking soda
1 teaspoon baking powder	1 teaspoon baking powder	1 teaspoon baking powder
½ teaspoon salt	½ teaspoon salt	½ teaspoon salt
2 oz. butter	50 g. butter	¼ cup butter
3 sticks celery	3 sticks celery	3 stalks celery
1 teaspoon dried mixed herbs	1 teaspoon dried mixed herbs	1 teaspoon dried mixed herbs
1 carton natural yogurt	1 carton natural yogurt	⅔ cup natural yogurt

Mix together the flours, bicarbonate of soda, baking powder and salt and rub in the butter. Chop the celery finely and stir into the mixture with the remaining ingredients and knead to form a soft dough. Roll out lightly and cut into 2-inch (5-cm.) rounds. Place on greased baking trays and bake in a moderately hot oven (400°F, 200°C, Gas Mark 6) for 10–15 minutes, until golden brown. Cool.

Honey fruit scones

Makes approximately 18 scones.
To freeze Pack in polythene bags.
To serve Defrost at room temperature for 30 minutes. If required hot, place on a baking tray and reheat in a moderate oven (350°F, 180°C, Gas Mark 4) for 15 minutes.
Storage time 4–6 months.

IMPERIAL	METRIC	AMERICAN
4 oz. wholemeal flour	100 g. wholemeal flour	1 cup wholewheat flour
4 oz. plain flour	125 g. plain flour	1 cup all-purpose flour
2 teaspoons baking powder	2 teaspoons baking powder	2 teaspoons baking powder
pinch salt	pinch salt	pinch salt
2 oz. margarine	50 g. margarine	$\frac{1}{4}$ cup margarine
1 oz. mixed dried fruit	25 g. mixed dried fruit	3 tablespoons mixed dried fruit
1 tablespoon honey	1 tablespoon honey	1 tablespoon honey
$\frac{1}{4}$ pint milk	$1\frac{1}{2}$ dl. milk	$\frac{2}{3}$ cup milk

Mix together the flours, baking powder and salt. Rub in the margarine and stir in the fruit. Dissolve the honey in a little of the milk and add to the dry ingredients with the remaining milk. Mix to a soft dough, roll out on a floured surface and cut into rounds. Place on a greased baking tray, brush with milk and bake in a hot oven (450°F, 230°C, Gas Mark 8) for 10 minutes. Cool.

Health fruit bran loaf

(Illustrated opposite)

Makes 2 loaves.
To freeze Pack whole loaf in a polythene bag, or cut into slices, wrap individually in foil and pack altogether in a polythene bag. Seal and label.
To serve Allow to defrost, still wrapped, at room temperature. Whole loaf – 4 hours, individual slices – 1 hour. Serve slices buttered.
Storage time 4–6 months.

IMPERIAL	METRIC	AMERICAN
8 oz. All-bran	225 g. All-bran	$\frac{1}{2}$ lb. All-bran
10 oz. castor sugar	250 g. castor sugar	$1\frac{1}{4}$ cups granulated sugar
8 oz. seedless raisins	225 g. seedless raisins	$1\frac{1}{3}$ cups seedless raisins
8 oz. dates, chopped	225 g. dates, chopped	$1\frac{1}{4}$ cups chopped dates
8 oz. dried bananas, chopped	225 g. dried bananas, chopped	$\frac{1}{2}$ lb. dried bananas, chopped
1 pint milk	generous $\frac{1}{2}$ litre milk	$2\frac{1}{2}$ cups milk
8 oz. self-raising flour	225 g. self-raising flour	2 cups all-purpose flour sifted with 2 teaspoons baking powder

Put the All-bran, sugar and fruit into a basin and mix them well together. Stir in the milk and leave to stand for about 30 minutes. Sieve in the flour, mix well and divide the mixture between two well-greased 2-lb. (1-kg.) loaf tins. Bake in a moderate oven (350°F, 180°C, Gas Mark 4) for about 1 hour. Turn onto a wire tray and allow to cool.

Chocolate fruit jiffies, spice cakes, health fruit bran loaf, orange bran cake (see pages 155, 156, 182, 187).

Sponge fruit flan, peach flan (see pages 209, 210).

Beef with herbed cobbler (see page 199).

1 Making the apple purée by cooking sliced apples with sugar and lemon juice to a pulp.

2 The apple purée being spooned into Tupperware containers ready for freezing.

3 Adding a few drops of green food colouring from the end of a skewer to the defrosted apple purée.

1

2

3

Swiss apple trifle (see opposite).

Orange bran cake

(Illustrated on page 183)

Makes 2 cakes.
To freeze Pack whole cake in a polythene bag, or cut into wedges, wrap individually in foil and pack altogether in a polythene bag. Seal and label.
To serve Unwrap cake and allow to defrost at room temperature for about 4 hours. Unwrap slices and defrost for about 1 hour.
Storage time 4–6 months.

IMPERIAL	METRIC	AMERICAN
8 oz. butter or margarine	225 g. butter or margarine	1 cup butter or margarine
8 oz. castor sugar	225 g. castor sugar	1 cup granulated sugar
4 eggs, beaten	4 eggs, beaten	4 eggs, beaten
grated zest of 2 oranges	grated zest of 2 oranges	grated zest of 2 oranges
¼ pint orange juice	1½ dl. orange juice	⅔ cup orange juice
12 oz. self-raising flour	350 g. self-raising flour	3 cups all-purpose flour sifted with 3 teaspoons baking powder
4 oz. All-bran	100 g. All-bran	¼ lb. All-bran
2 oz. chopped mixed peel	50 g. chopped mixed peel	generous ½ cup chopped mixed peel
8 oz. icing sugar	225 g. icing sugar	2 cups sifted confectioners' sugar
water to mix	water to mix	water to mix
crushed All-bran to decorate	crushed All-bran to decorate	crushed All-bran to decorate

Cream butter and sugar together until light and fluffy. Gradually add eggs and beat the mixture thoroughly. Add orange rind and juice. Sieve flour into mixture and fold in All-bran and mixed peel. Divide the mixture between two 6-inch (15-cm.) greased cake tins and bake in a moderate oven (350°F, 180°C, Gas Mark 4) for about 1 hour. Leave in the tins for 5 minutes then turn out and cool on a wire tray. Sieve the icing sugar into a small basin and mix in sufficient cold water to give a coating consistency. Pour over the cakes and decorate with crushed All-bran.

Swiss apple trifle

(Illustrated opposite)

Makes 4–6 servings.

IMPERIAL	METRIC	AMERICAN
¼ pint double cream	1½ dl. double cream	⅔ cup whipping cream
1 30-oz. container sweetened frozen apple purée, defrosted	1 850-g. container sweetened frozen apple purée, defrosted	1 30-oz. container sweetened frozen applesauce, defrosted
½ oz. gelatine	15 g. gelatine	2 envelopes gelatin
3 tablespoons hot water	3 tablespoons hot water	scant ¼ cup hot water
2 egg whites	2 egg whites	2 egg whites
few drops green food colouring	few drops green food colouring	few drops green food coloring
1 lemon jam Swiss roll	1 lemon jam Swiss roll	1 jam-filled jelly roll
1 red-skinned dessert apple	1 red-skinned dessert apple	1 red-skinned dessert apple
lemon juice	lemon juice	lemon juice

Whip the cream, set aside half for the decoration and stir the rest into the apple purée. Dissolve the gelatine in the hot water and stir into the apple mixture with the stiffly beaten egg whites. Add a few drops green food colouring, if liked. Line a glass serving dish with slices of Swiss roll and pour in the apple mousse. When set, decorate with small cream rosettes and quartered slices of unpeeled apple, dipped in lemon juice.

Orange, prune and apple ice cream

Makes 8 servings.
To freeze Pack into polythene containers. Seal and label.
To serve Place scoops of the ice cream in individual glasses and decorate with orange slices, apple slices and wafer biscuits.
Storage time 4–6 months.

IMPERIAL	METRIC	AMERICAN
2 egg whites	2 egg whites	2 egg whites
8 oz. castor sugar	225 g. castor sugar	1 cup granulated sugar
1 15-oz. can prunes, sieved	1 425-g. can prunes, sieved	1 15-oz. can prunes, sieved
8 oz. frozen apple purée, defrosted	225 g. frozen apple purée, defrosted	1 cup frozen applesauce, defrosted
4 tablespoons lemon juice	4 tablespoons lemon juice	$\frac{1}{3}$ cup lemon juice
$\frac{1}{2}$ 6-oz. can frozen unsweetened orange juice concentrate	$\frac{1}{2}$ 175-g. can frozen unsweetened orange juice concentrate	$\frac{1}{2}$ 6-oz. can frozen unsweetened orange juice concentrate
grated zest of 1 orange	grated zest of 1 orange	grated zest of 1 orange
$\frac{3}{4}$ pint double cream	$4\frac{1}{2}$ dl. double cream	$1\frac{1}{4}$ cups whipping cream plus $\frac{2}{3}$ cup heavy cream

Stiffly whip the egg whites then whisk in half the sugar. Fold in remaining sugar and stir in the prunes, apple purée, fruit juices and zest and mix well. Lightly whip the cream and fold into the mixture.

Blackcurrant cheesecake

Makes 1 cheesecake, 10 servings.
To freeze Pack whole cheesecake in a polythene bag or cut into portions and wrap individually in foil then altogether in a polythene bag. Seal and label.
To serve Unwrap cake, place on serving dish and allow to defrost at room temperature for 4 hours. Unwrap portions and defrost for 1 hour.
Storage time 4–6 months.

Note To insure a firm set when using canned blackcurrants, add 1 envelope (1 tablespoon) powdered gelatine to the blackcurrant and jelly mixture while dissolved.

IMPERIAL	METRIC	AMERICAN
1 15-oz. can blackcurrants or 1 lb. lightly stewed blackcurrants	1 425-g. can blackcurrants or 450 g. lightly stewed blackcurrants	1 15-oz. can black currants or 1 lb. lightly stewed black currants
1 blackcurrant jelly	1 blackcurrant jelly	1 package black currant-flavored gelatin
1 lb. cottage cheese	450 g. cottage cheese	2 cups cottage cheese
1 oz. castor sugar	25 g. castor sugar	2 tablespoons granulated sugar
$\frac{1}{4}$ pint double cream	$1\frac{1}{2}$ dl. double cream	$\frac{2}{3}$ cup whipping cream
crust	*crust*	*crust*
8 digestive biscuits	8 digestive biscuits	10 graham crackers
1 oz. demerara sugar	25 g. demerara sugar	2 tablespoons brown sugar
2 oz. butter, melted	50 g. butter, melted	$\frac{1}{4}$ cup butter, melted

Drain the blackcurrants and sieve, or blend and then sieve. Place the blackcurrant syrup and the jelly in a saucepan. Heat gently until dissolved, then cool. Sieve the cottage cheese and combine with the blackcurrant purée, sugar and cooled jelly. Lightly whip the cream and fold into the mixture when it thickens but before it sets. Pour the mixture into a lightly oiled 8-inch (20-cm.) round cake tin and chill until set. To make the crust, put the digestive biscuits into a polythene bag and crush finely with a rolling pin. Combine crumbs with the sugar and butter, sprinkle over the cheesecake mixture and press down lightly. Chill until firm then unmould.

Making your freezer a party treasure chest

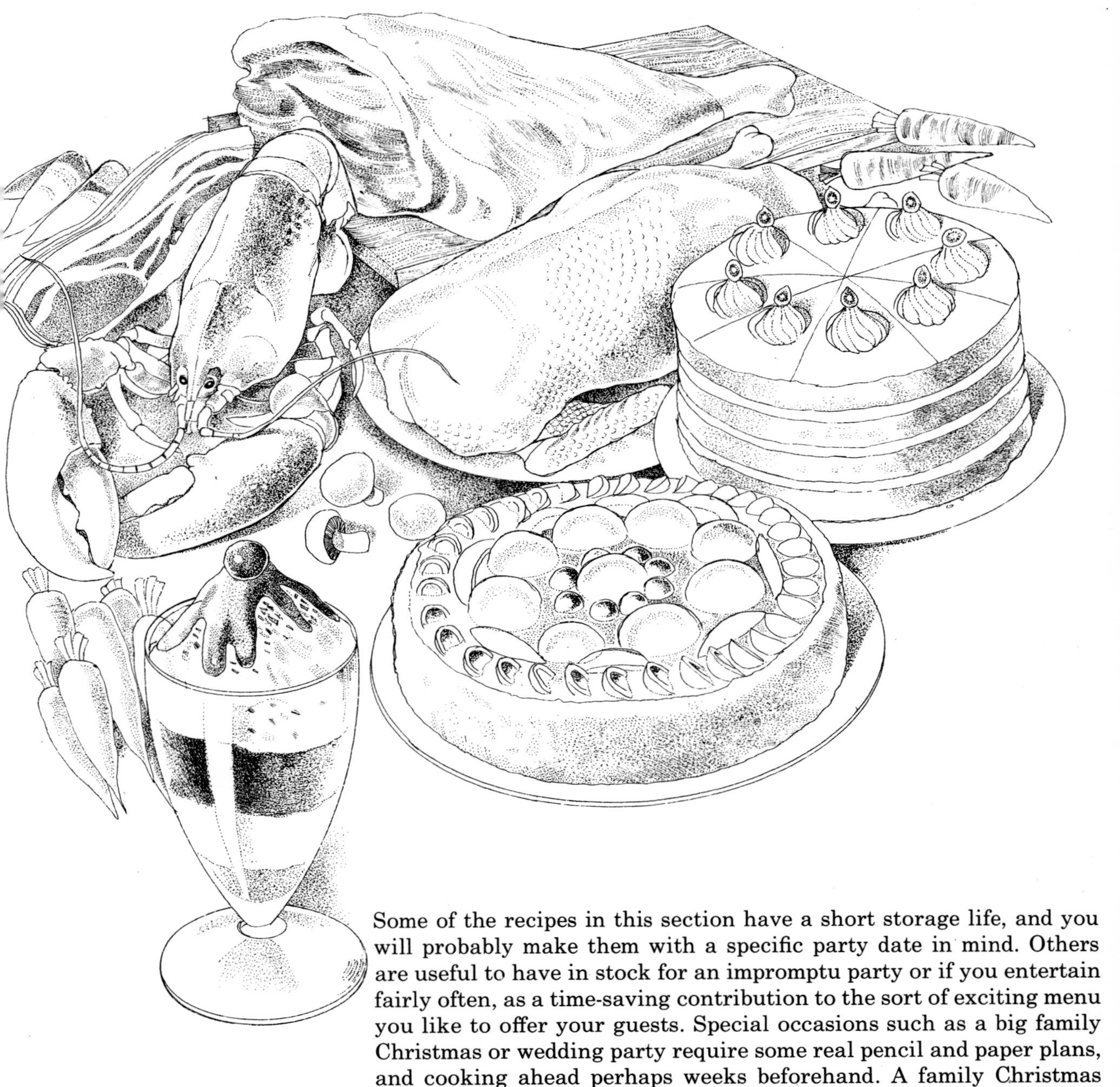

Some of the recipes in this section have a short storage life, and you will probably make them with a specific party date in mind. Others are useful to have in stock for an impromptu party or if you entertain fairly often, as a time-saving contribution to the sort of exciting menu you like to offer your guests. Special occasions such as a big family Christmas or wedding party require some real pencil and paper plans, and cooking ahead perhaps weeks beforehand. A family Christmas deserves a Golden Guide all to itself and there are suggestions on page 238 to help you cater for a wedding buffet. Exotic, yet delicately-flavoured soups for instance, transform an entire meal – try the lemon avocado soup recipe overleaf.

Lemon avocado soup

(Illustrated on page 108)

Makes 4 servings.
To freeze Pour into tumbler-shaped polythene containers and seal or line containers with polythene bags, pour in soup, partially freeze then remove containers and seal bags. Label.
To serve Defrost completely, add $\frac{1}{4}$ pint ($1\frac{1}{2}$ dl.) single cream, blend well and pour into a tureen. Float thin avocado slices on the surface of the soup. Brush them with lemon juice to prevent discolouration. If required hot, turn soup into a saucepan and add about $\frac{1}{4}$ pint ($1\frac{1}{2}$ dl.) milk. Reheat gently, blending well. Re-adjust seasoning and pour a swirl of single cream over the surface of each plate of soup.
Storage time 4–6 months.

IMPERIAL	METRIC	AMERICAN
1 oz. butter	25 g. butter	2 tablespoons butter
1 tablespoon grated onion	1 tablespoon grated onion	1 tablespoon grated onion
1 pint chicken stock	generous $\frac{1}{2}$ litre chicken stock	$2\frac{1}{2}$ cups chicken stock
2 medium avocados	2 medium avocados	2 medium avocados
2 tablespoons lemon juice	2 tablespoons lemon juice	3 tablespoons lemon juice
salt and pepper to taste	salt and pepper to taste	salt and pepper to taste

Melt the butter and use to cook the onion gently until transparent. Add the stock and bring just to boiling point. Stone the avocados, dice the flesh and add it to the lemon juice to prevent discolouration. Add the diced avocado to the pan, cover and simmer for 10 minutes. Cool slightly, sieve or liquidise in an electric blender. Adjust seasoning to taste. Cool. If liked add a few drops of green food colouring to make this a pleasant colour.

Devilled lobster bisque

Makes 6 servings.
To freeze Pour into tumbler-shaped polythene containers and seal or line containers with polythene bags, pour in soup, partially freeze then remove containers and seal bags. Label.
To serve Turn contents into a saucepan and reheat gently to boiling point, stirring frequently to prevent burning. Add 2 tablespoons dry sherry and 4 tablespoons single cream and reheat but do not allow to boil.
Storage time 3 months.

IMPERIAL	METRIC	AMERICAN
1 teaspoon curry paste	1 teaspoon curry paste	1 teaspoon curry paste
pinch ground nutmeg	pinch ground nutmeg	pinch ground nutmeg
1 teaspoon dried dill weed	1 teaspoon dried dill weed	1 teaspoon dried dill weed
$\frac{1}{4}$ teaspoon pepper	$\frac{1}{4}$ teaspoon pepper	$\frac{1}{4}$ teaspoon pepper
1 teaspoon salt	1 teaspoon salt	1 teaspoon salt
1 pint chicken stock	generous $\frac{1}{2}$ litre chicken stock	$2\frac{1}{2}$ cups chicken stock
8 oz. lobster meat	225 g. lobster meat	$\frac{1}{2}$ lb. lobster meat
1 oz. butter	25 g. butter	2 tablespoons butter
1 mild onion, chopped	1 mild onion, chopped	1 mild onion, chopped
4 oz. mushrooms, sliced	100 g. mushrooms, sliced	1 cup sliced mushrooms
1 oz. flour	25 g. flour	$\frac{1}{4}$ cup all-purpose flour
$\frac{1}{4}$ pint milk	$1\frac{1}{2}$ dl. milk	$\frac{2}{3}$ cup milk

Mix the curry paste, nutmeg, dill and other seasonings with a little of the stock, stir in the lobster meat and remaining stock. If liked, liquidise in an electric blender to make a smooth-textured soup. Heat the butter and use to sauté the onion and mushrooms until limp but not turning brown. Stir in the flour, then the milk and gradually add the lobster and stock mixture. Bring to the boil, cover and simmer, stirring occasionally, for 5 minutes. Cool.

Golden guide number

to Christmas catering

All the following recipes can be made and frozen up to three months before Christmas. Other items for the Christmas feast which can be frozen in advance are – roast potatoes, bread sauce, Brussels sprouts, green beans, stuffings and the brandy butter.

WORK PLAN FOR A WELL-ORGANISED CHRISTMAS PARTY

Two days before
Start defrosting turkey, goose, duck, pheasant, etc. at room temperature as follows:
18 lb. (9 kg.) and over – 48 hours
12–18 lb. (6–9 kg.) – 36 hours
under 12 lb. (6 kg.) – 24 hours

One day before
Defrost stuffing and sausagemeat.
Stuff and truss turkey.
Defrost Christmas jewel cake and decorate if necessary.

On the day
Defrost bread sauce, mincemeat shortbread* and brandy butter.
Roast turkey; cook roast potatoes. Brussels sprouts and green beans from the frozen state.
Bake mincemeat and apple plait* after turkey has been removed from oven.
Serve iced Christmas pudding and mincemeat and apple plait with brandy butter to follow the turkey; the Christmas jewel cake and mincemeat shortbread for tea.
*These are pleasant alternatives to the more usual mince pies.

Iced Christmas pudding

Makes 4–6 servings.
To freeze Cover with moulded foil and smooth down edges. Seal and label.
To serve Unmould, defrost very slightly at room temperature and serve with brandy- or liqueur-flavoured cream.
Storage time 4–6 months.

Imperial	Metric	American
2 oz. raisins, chopped	50 g. raisins, chopped	generous ½ cup chopped raisins
2 oz. currants	50 g. currants	generous ½ cup currants
½ oz. candied peel, chopped	15 g. candied peel, chopped	1½ tablespoons chopped candied peel
1 oz. glacé cherries, chopped	25 g. glacé cherries, chopped	scant ¼ cup chopped candied cherries
3 tablespoons rum or brandy	3 tablespoons rum or brandy	scant ¼ cup rum or brandy
1 pint vanilla ice cream	generous ½ litre vanilla ice cream	2½ cups vanilla ice cream

Soak the fruits in the spirit overnight. Slightly soften the ice cream and fold in the fruits. Pack into a 1-pint (generous ½-litre) mould.

Mincemeat and apple plait

(Illustrated on page 204)

Makes 1 plait, 4–6 servings.
To freeze Pack in a polythene bag. Seal and label.
To serve Place, still frozen, on a greased baking tray and place in a moderately hot oven for about 20 minutes to defrost and reheat. Serve hot or cold.
Storage time 4–6 months.

Imperial	Metric	American
1 8-oz. packet short pastry mix or 8 oz. frozen shortcrust pastry, defrosted	1 225-g. packet short pastry mix or 225 g. frozen shortcrust pastry, defrosted	1 ½-lb. package basic pie dough mix or ½ lb. frozen basic pie dough, defrosted
1 large cooking apple	1 large cooking apple	1 large baking apple
2 oz. icing sugar, sieved	50 g. icing sugar, sieved	½ cup sifted confectioners' sugar
8 oz. mincemeat	225 g. mincemeat	1 cup mincemeat

Make up the pastry mix, following instructions on the pack. Roll out to a rectangle 8 inches (20 cm.) by 14 inches (36 cm.) and trim the edges. Lift onto a baking tray.

Peel, core and slice the apple. Arrange a layer of apple slices, 4 inches (10 cm.) wide, down the centre of the pastry, leaving 5 inches (13 cm). uncovered on either side. Sprinkle with icing sugar and top with the mincemeat. Cut the borders obliquely in ½-inch (1-cm.) strips, removing alternate strips. Dampen edges and carefully plait remaining strips of the pastry over the filling. Place on a greased baking tray and brush with beaten egg. Bake in a moderately hot oven (400°F, 200°C, Gas Mark 6) for about 30 minutes. Cool.

My brandy butter

Beat together 4 oz. (100 g.) unsalted butter and 4 oz. (100 g.) sieved icing sugar until smooth and creamy. Add 4 tablespoons brandy, a little at a time, alternately with another 2 oz. (50 g.) icing sugar. This method is very quick and easy. Pack in suitable pots or small polythene containers and freeze.

Christmas jewel cake

(Illustrated on page 205)

Makes 1 cake.
To freeze Pack in polythene bag. Seal and label.
To serve Unwrap, place on serving dish and allow to defrost for about 10 hours at room temperature.
Storage time 4–6 months.

Imperial	Metric	American
8 oz. butter or margarine	225 g. butter or margarine	1 cup butter or margarine
8 oz. castor sugar	225 g. castor sugar	1 cup granulated sugar
½ teaspoon almond essence	½ teaspoon almond essence	½ teaspoon almond extract
3 eggs	3 eggs	3 eggs
8 oz. plain flour	225 g. plain flour	2 cups all-purpose flour
pinch salt	pinch salt	pinch salt
4 oz. ground almonds	100 g. ground almonds	1 cup ground almonds
4 oz. glacé cherries, quartered	100 g. glacé cherries, quartered	½ cup quartered candied cherries
8 oz. glacé pineapple, chopped	225 g. glacé pineapple, chopped	1 cup chopped candied pineapple
2 oz. angelica, chopped	50 g. angelica, chopped	¼ cup chopped candied angelica
1 oz. crystallised ginger, chopped	25 g. crystallised ginger, chopped	scant ¼ cup chopped candied ginger
2 oz. mixed peel, chopped	50 g. mixed peel, chopped	generous ½ cup chopped mixed peel
to decorate	*to decorate*	*to decorate*
2 tablespoons sieved apricot jam	2 tablespoons sieved apricot jam	3 tablespoons sieved apricot jam
2 tablespoons castor sugar	2 tablespoons castor sugar	3 tablespoons granulated sugar
8 oz. marzipan	225 g. marzipan	½ lb. almond paste
assorted glacé cherries and other glacé fruits	assorted glacé cherries and other glacé fruits	assorted candied cherries and other candied fruits

Cream the fat and sugar until light and fluffy, add almond essence and beat in the eggs, one at a time, adding a tablespoon of flour with each. Fold in remaining flour, salt, ground almonds and prepared fruit. Turn into a greased and bottom-lined 8-inch (20-cm) round or 7-inch (18-cm.) square cake tin and smooth top. Bake in the centre of a cool oven (275°F, 140°C, Gas Mark 1) for 3½ hours. Turn out, cool on a wire tray, and remove paper.

To decorate, brush top of cake with apricot jam. Sprinkle sugar on greaseproof paper and on this roll out marzipan to fit top of cake. Lift and press firmly into place. Arrange glacé cherries and sliced glacé fruits on top of the marzipan to give the jewelled effect. Store cake in an airtight tin to mature for 1 week before freezing.

Note If preferred, freeze cake plain and decorate when required to serve.

Mincemeat shortbread

Makes 2 shortbreads, 16 servings.
To freeze Remove from tins and wrap in freezer foil. Seal and label.
To serve Defrost, still wrapped at room temperature for 4 hours. Dust with icing sugar and serve with cream.
Storage time 4–6 months.

Imperial	Metric	American
8 oz. butter	225 g. butter	1 cup butter
6 oz. sugar	175 g. sugar	¾ cup sugar
2 eggs, beaten	2 eggs, beaten	2 eggs, beaten
12 oz. self-raising flour	350 g. self-raising flour	3 cups all-purpose flour sifted with 3 teaspoons baking powder
1 lb. mincemeat	450 g. mincemeat	2 cups mincemeat

Cream the butter and sugar until soft, stir in the eggs and flour to form a firm dough, divide into four equal portions and chill for 45 minutes. Grease two 7-inch (18-cm) sandwich tins and line the bases with greaseproof paper. Roll out two portions of the shortbread to fit the tins exactly, place in the tins. Divide the mincemeat between the tins, spreading evenly, then roll out the remaining two portions and place on top, prick lightly with a fork, mark into eight portions and pinch the edges. Bake in a moderate oven (325°F, 170°C, Gas Mark 3) for 1 hour, or until light golden brown. Cool on a wire tray.

Potted shrimps

Makes 4 servings.
To freeze Cover each ramekin dish with a cap of foil and smooth down edges. Label.
To serve Defrost in the refrigerator overnight and serve garnished with lemon wedges and accompanied by hot toast.
Storage time 4–6 months.

IMPERIAL	METRIC	AMERICAN
4 oz. butter	100 g. butter	$\frac{1}{2}$ cup butter
8 oz. shelled shrimps	225 g. shelled shrimps	$1\frac{1}{2}$ cups shelled shrimp
juice of $\frac{1}{2}$ lemon	juice of $\frac{1}{2}$ lemon	juice of $\frac{1}{2}$ lemon
$\frac{1}{2}$ teaspoon black pepper	$\frac{1}{2}$ teaspoon black pepper	$\frac{1}{2}$ teaspoon black pepper
pinch cayenne pepper	pinch cayenne pepper	pinch cayenne pepper
$\frac{1}{2}$ teaspoon ground mace	$\frac{1}{2}$ teaspoon ground mace	$\frac{1}{2}$ teaspoon ground mace

Melt one-quarter of the butter in a saucepan and when hot add the shrimps, lemon juice and seasonings. Stir well, taste and add more lemon juice if preferred. Divide mixture between four small ramekin dishes and smooth tops. Melt the remaining butter until frothy, skim and pour over the shrimps. Leave to set.

Terrine of game

Makes 8 servings.
To freeze Cover dish with freezer foil and smooth down edges or remove from dish and wrap in freezer foil. Seal and label.
To serve Defrost, still wrapped, in the refrigerator overnight or at room temperature for 6 hours.
Storage time 3 months.

IMPERIAL	METRIC	AMERICAN
$1\frac{1}{2}$ lb. rabbit or hare	$\frac{3}{4}$ kg. rabbit or hare	$1\frac{1}{2}$ lb. rabbit or hare
8 oz. streaky bacon	225 g. streaky bacon	$\frac{1}{2}$ lb. bacon slices
8 oz. minced pork	225 g. minced pork	1 cup ground pork
3 tablespoons brandy	3 tablespoons brandy	4 tablespoons brandy
1 egg	1 egg	1 egg
salt and pepper to taste	salt and pepper to taste	salt and pepper to taste
pinch ground nutmeg	pinch ground nutmeg	pinch ground nutmeg
1 small bunch parsley, chopped	1 small bunch parsley, chopped	1 small bunch parsley, chopped

Remove rabbit flesh from bone and cut into small pieces. De-rind and chop the bacon and combine with the pork, rabbit and brandy. Allow to stand for 2 hours. Put the mixture through the mincer with all the remaining ingredients. Lightly grease a terrine and press the mixture into it firmly. Cover tightly and stand terrine in a bain-marie. Place in a moderately hot oven (375°F, 190°C, Gas Mark 5) for $1\frac{1}{4}$ hours. Cool.

Duck pâté

Makes 12 servings.
To freeze Wrap in freezer foil or place dish inside polythene bag. Seal and label.
To serve Uncover and defrost at room temperature for 4 hours or turn out while frozen and defrost on serving dish.
Storage time 3 months.

IMPERIAL	METRIC	AMERICAN
1 3–4-lb. duck	1 $1\frac{1}{2}$–2-kg. duck	1 3–4-lb. duck
8 oz. lean pork	225 g. lean pork	$\frac{1}{2}$ lb. lean pork
2 tablespoons dry sherry	2 tablespoons dry sherry	3 tablespoons dry sherry
2 tablespoons dry white wine	2 tablespoons dry white wine	3 tablespoons dry white wine
pinch ground mace	pinch ground mace	pinch ground mace
pinch ground nutmeg	pinch ground nutmeg	pinch ground nutmeg
salt and pepper to taste	salt and pepper to taste	salt and pepper to taste
4 oz. streaky bacon	100 g. streaky bacon	$\frac{1}{4}$ lb. bacon slices

Roast duck in the normal way, cool slightly and remove meat from the bones. Mince the duck, duck liver and pork and add the sherry, wine and seasonings. Line a terrine or loaf tin with the de-rinded bacon rashers and press in the duck mixture. Cover with lid or foil, place the dish in a bain-marie half-filled with water and bake in a cool oven (300°F, 150°C, Gas Mark 2) for about $1\frac{1}{2}$ hours, until the mixture shrinks slightly from the sides of the dish. Cool.

Skipper pâté

(Illustrated on page 128)

Makes 8 servings.
To freeze Cover tops with freezer film and pack together in a rigid-based polythene container. Seal and label.
To serve Defrost covered until it is easy to remove the freezer film. Garnish with parsley and halved black olives and serve semi-frozen.
Storage time 4–6 months.

IMPERIAL	METRIC	AMERICAN
4 large lemons, halved	4 large lemons, halved	4 large lemons, halved
2 3¾-oz. cans skippers in oil, drained	2 95-g. cans skippers in oil, drained	2 3¾-oz. cans skippers in oil, drained
4 oz. butter, softened	100 g. butter, softened	½ cup softened butter
1 egg, separated	1 egg, separated	1 egg, separated
salt and freshly ground black pepper to taste	salt and freshly ground black pepper to taste	salt and freshly ground black pepper to taste
1 tablespoon chopped parsley	1 tablespoon chopped parsley	1 tablespoon chopped parsley

Squeeze the juice from the lemons, and scoop out pith. Mix the juice with the mashed skippers and butter to make a soft mixture. Add the egg yolk, seasoning and parsley. Beat the egg white until stiff and fold into the mixture. Divide between the eight lemon halves.

Brisling mousse

(Illustrated on page 128)

Makes 4 servings.
To freeze Open freeze until solid, place in a large polythene bag, seal. Then put in a rigid-based container if available as the frozen soufflé is easily damaged.
To serve Remove container and allow to defrost for 5–6 hours. Holding a knife blade against the side of the dish, peel off the paper. Press chopped parsley against the exposed edges and garnish with lemon slices.
Storage time 4–6 months.

IMPERIAL	METRIC	AMERICAN
2 3¾-oz. cans skippers in oil	2 95-g. cans skippers in oil	2 3¾-oz. cans skippers in oil
¼ pint mayonnaise	1½ dl. mayonnaise	⅔ cup mayonnaise
½ oz. gelatine	15 g. gelatine	2 envelopes gelatin
3 tablespoons water	3 tablespoons water	scant ¼ cup water
¼ pint double cream	1½ dl. double cream	⅔ cup whipping cream
salt and pepper to taste	salt and pepper to taste	salt and pepper to taste
juice of ½ lemon	juice of ½ lemon	juice of ½ lemon
½ teaspoon Worcestershire sauce	½ teaspoon Worcestershire sauce	½ teaspoon Worcestershire sauce
3 eggs whites	3 egg whites	3 egg whites
1 tablespoon tomato purée	1 tablespoon tomato purée	1 tablespoon tomato paste

Prepare a 5-inch (13-cm.) soufflé dish by fastening a collar of doubled greaseproof paper, lightly oiled on the inner side, around the top, standing well up above the rim. Secure tightly with freezer tape. Drain the skippers, mash well and mix with the mayonnaise. Place the gelatine and water in a small basin over hot water and leave until dissolved, then stir into the skipper mixture. Lightly whip the cream and fold in. Season well with salt, pepper, lemon juice and Worcestershire sauce. When on the point of setting, beat the egg whites until stiff and fold in, with purée. Pour the mixture into the prepared case and allow to set.

Stuffed fish turbans with oriental sauce

(Illustrated on page 148)

Makes 3 servings.

IMPERIAL	METRIC	AMERICAN
3 frozen haddock fillets (tail pieces)	3 frozen haddock fillets (tail pieces)	3 frozen haddock fillets (tail pieces)
1 oz. butter	25 g. butter	2 tablespoons butter
salt and pepper to taste	salt and pepper to taste	salt and pepper to taste
stuffing	*stuffing*	*stuffing*
4 oz. frozen chopped spinach	100 g. frozen chopped spinach	about ½ cup frozen chopped spinach
4 oz. creamed potato	100 g. creamed potato	½ cup creamed potato
1 small egg	1 small egg	1 egg
¼ teaspoon nutmeg	¼ teaspoon nutmeg	¼ teaspoon nutmeg
sauce	*sauce*	*sauce*
1 7-oz. can red pimentos	1 200-g. can red pimentos	1 7-oz. can red pimientos
¼ pint tomato juice	1½ dl. tomato juice	⅔ cup tomato juice
1 teaspoon sugar	1 teaspoon sugar	1 teaspoon sugar
2 teaspoons cornflour	2 teaspoons cornflour	2 teaspoons cornstarch

Gently heat and defrost the spinach, draining off all surplus liquid. Beat in the creamed potato, then the egg and seasoning, including a very little salt and pepper. The mixture must be just stiff enough to pipe. Skin the fish and cut each fillet in half lengthways. Roll the fillets with the tail end inside, secure with wooden cocktail sticks and place upright in a buttered baking dish. Pipe the stuffing into the centres of the fillets. Melt the butter, pour over the stuffed fillets and sprinkle lightly with salt and pepper. Cover with foil and bake in a moderately hot oven (375°F, 190°C, Gas Mark 5) for 20 minutes. Meanwhile make the sauce. Dice the pimentos and cook them in the tomato juice until the mixture can be put through a sieve or liquidised. Season and add the sugar. Moisten the cornflour with a little cold water, add to the sauce and bring to the boil, stirring constantly until thickened. Pour the sauce around the cooked fish, allowing it to blend with the buttery juices in the baking dish.

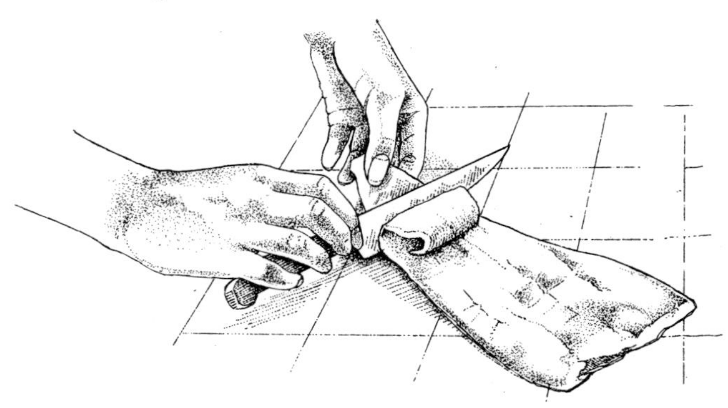

Apricot-baked bacon

(Illustrated on page 107)

Makes 12 servings (6 hot and 6 cold).
To freeze Wrap joint in freezer foil. Seal and label.
To serve Defrost completely, still wrapped, and place in a moderately hot oven (375°F, 190°C, Gas Mark 5) for 30 minutes. Open foil, take the apricots from an 8-oz. (225-g.) can and place around the joint. Moisten joint with syrup from the can and sprinkle with demerara sugar. Replace in oven for 15–20 minutes to brown.
Storage time 4–6 months.

IMPERIAL	METRIC	AMERICAN
$3\frac{1}{2}$–4-lb. joint middle cut bacon	$1\frac{3}{4}$–2-kg. joint middle cut bacon	$3\frac{1}{2}$–4-lb. piece ham butt
1 bay leaf	1 bay leaf	1 bay leaf
6 peppercorns	6 peppercorns	6 peppercorns
3 cloves	3 cloves	3 cloves
4 tablespoons apricot jam	4 tablespoons apricot jam	5 tablespoons apricot jam
stuffing	*stuffing*	*stuffing*
2 oz. dried apricots	50 g. dried apricots	$\frac{1}{3}$ cup dried apricots
3 oz. soft white breadcrumbs	75 g. soft white breadcrumbs	$1\frac{1}{2}$ cups fresh soft bread crumbs
$\frac{1}{2}$ teaspoon mixed spice	$\frac{1}{2}$ teaspoon mixed spice	$\frac{1}{2}$ teaspoon mixed spice
1 tablespoon lemon juice	1 tablespoon lemon juice	1 tablespoon lemon juice
1 small onion, grated	1 small onion, grated	1 small onion, grated
1 egg yolk	1 egg yolk	1 egg yolk

Soak the bacon joint in cold water overnight. Drain joint and roll it around a piece of crumpled foil or a jam jar and secure with string. Place in a saucepan with the bay leaf, peppercorns and cloves and sufficient fresh cold water to cover. Bring to the boil, cover and simmer for 20 minutes per lb. ($\frac{1}{2}$ kg.) and 20 minutes over. Soak the dried apricots for the stuffing in cold water for 2 hours then bring to the boil and simmer gently until tender. Chop the apricots and mix with the remaining stuffing ingredients. When joint is cooked, remove pan from heat and leave in the liquid until quite cold. Drain and remove string and foil or jam jar. Place stuffing in joint, spread all over with apricot jam and place on a large square of freezer foil. Bring edges together over the top and seal to make an airtight parcel.

Bacon grill fritters

(Illustrated on page 130)

Makes 4 servings.
To freeze Pack fritters in a polythene container with foil dividers. Seal and label.
To serve Fry fritters, still frozen, in hot deep fat for about 8 minutes, until reheated and golden brown. Drain well. Serve with jacket potatoes, grilled tomatoes and mushrooms.
Storage time 4–6 months.

IMPERIAL	METRIC	AMERICAN
4 oz. plain flour	100 g. plain flour	1 cup all-purpose flour
salt and pepper to taste	salt and pepper to taste	salt and pepper to taste
2 tablespoons oil	2 tablespoons oil	3 tablespoons oil
2 eggs, separated	2 eggs, separated	2 eggs, separated
about ¼ pint water	about 1½ dl. water	about ⅔ cup water
1 10-oz. can bacon grill	1 275-g. can bacon grill	1 10-oz. can bacon grill or corned beef

Sieve the flour into a bowl, season and beat in the oil, egg yolks, and enough water to make a smooth batter. Beat the egg whites stiffly and fold into the mixture. Cut the meat into neat slices, dip in the batter and fry in hot deep fat until batter is crisp but not browned. Drain on absorbent kitchen paper and cool.

Hot dog rolls

(Illustrated on page 130)

Makes 40.
To freeze Pack the rolls in a shallow shaped foil tray. Cover with lid or sheet of foil. Seal and label.
To serve Uncover, brush generously with melted butter and place, still frozen, in a moderately hot oven (400°F, 200°C, Gas Mark 6) for about 25 minutes. Garnish with tomato quarters and sprigs of parsley.
Storage time 4–6 months.

IMPERIAL	METRIC	AMERICAN
1 large can hot dog sausages	1 large can hot dog sausages	1 large can hot dog sausages
3 tablespoons made mustard	3 tablespoons made mustard	4 tablespoons made mustard
3 oz. butter	75 g. butter	6 tablespoons butter
1 large white loaf	1 large white loaf	1 large white loaf

Drain the sausages and cut each one in half. Blend together the mustard and butter. Remove crusts from bread and spread each slice with the mustard mixture. Cut the slices in half and wrap each piece around a halved sausage, securing with a wooden cocktail stick. Place rolls on a baking tray and open freeze until solid. Remove sticks.

Country house chicken

(Illustrated on page 147)

Makes 4 servings.
To freeze Pack in polythene containers. Seal and label.
To serve Defrost, turn into a saucepan, reheat gently to boiling point and simmer for 10 minutes. Stir frequently to prevent burning.
Storage time 4–6 months.

IMPERIAL	METRIC	AMERICAN
1 3–4-lb. roasting chicken	1 1½–2-kg. roasting chicken	1 3–4-lb. roasting chicken
2 tablespoons seasoned flour	2 tablespoons seasoned flour	3 tablespoons seasoned flour
2 oz. butter	50 g. butter	¼ cup butter
8 oz. button mushrooms	225 g. button mushrooms	2 cups button mushrooms
2 chicken stock cubes	2 chicken stock cubes	2 chicken bouillon cubes
¾ pint boiling water	scant ½ litre boiling water	scant 2 cups boiling water
4 tablespoons dry sherry	4 tablespoons dry sherry	⅓ cup dry sherry
2 teaspoons cornflour	2 teaspoons cornflour	2 teaspoons cornstarch
1 tablespoon orange jelly marmalade	1 tablespoon orange jelly marmalade	1 tablespoon orange jelly marmalade

Cut the chicken into eight pieces and turn in the seasoned flour. Heat half the butter in a flameproof casserole and use to sauté the chicken pieces gently until golden brown all over. Cover and continue cooking gently, turning frequently, for 20 minutes. In a separate pan, sauté the mushrooms in the remaining butter for 3 minutes, slicing the larger ones. Dissolve the stock cubes in the boiling water, stir in sherry, cornflour moistened with a little cold water and the marmalade. Add the mushrooms to the chicken, strain the liquid over them. Bring to the boil, cover and simmer for a further 5 minutes. Cool.

Note This is an excellent dish made with frozen chicken for refreezing or for serving immediately.

Raisin-stuffed duck with vermouth

Makes 4 servings.

IMPERIAL	METRIC	AMERICAN
1 packet sage and onion stuffing	1 packet sage and onion stuffing	1 package sage and onion stuffing
2 oz. seedless raisins	50 g. seedless raisins	generous ½ cup seedless raisins
2 eggs, separated	2 eggs, separated	2 eggs, separated
1 5-lb. frozen duck, defrosted	1 2½-kg. frozen duck, defrosted	1 5-lb. frozen duck, defrosted
¼ pint strong chicken stock	1½ dl. strong chicken stock	⅔ cup strong chicken stock
4 tablespoons vermouth	4 tablespoons vermouth	⅓ cup vermouth
1½ teaspoons cornflour	1½ teaspoons cornflour	1½ teaspoons cornstarch
4 tablespoons double cream	4 tablespoons double cream	5 tablespoons whipping cream
salt and pepper to taste	salt and pepper to taste	salt and pepper to taste

Make up the stuffing with the raisins, lightly beaten egg whites and sufficient hot water to moisten. Use to stuff the duck. Stand duck on a trivet in a roasting tin. Pour the stock and vermouth over the duck and place in a moderately hot oven (400°F, 200°C, Gas Mark 6) for 30 minutes. Reduce heat to 375°F, 190°C, Gas mark 5 and continue roasting for a further 1¼ hours, basting frequently and skimming off surplus fat. Then turn duck over on trivet and roast for a further 10 minutes to brown the under side. Place duck on serving dish and keep hot. Remove as much fat as possible from the juices in the pan, moisten cornflour with a little cold water, stir into the juices, cooking over gentle heat until smooth and clear. Beat together the egg yolks and cream in a basin, gradually strain in the hot sauce, stirring all the time. Reheat in a saucepan and season to taste. Pour sauce into a sauceboat and hand separately.

Pork chops baked in foil

(Illustrated on page 129)

To freeze Label.
To serve Defrost and place in a moderately hot oven (375°F, 190°C, Gas Mark 5) for 1 hour. Serve garnished with fresh tomato.
Storage time 4–6 months.

IMPERIAL	METRIC	AMERICAN
4 large pork chops	4 large pork chops	4 large pork chops
4 oz. long-grain rice	100 g. long-grain rice	generous ½ cup long-grain rice
1 oz. butter, melted	25 g. butter, melted	2 tablespoons melted butter
1 small pickled cucumber, chopped	1 small pickled cucumber, chopped	1 sweet dill pickle, chopped
2 tablespoons chopped red and green pepper	2 tablespoons chopped red and green pepper	3 tablespoons chopped red and green sweet pepper
salt and pepper to taste	salt and pepper to taste	salt and pepper to taste
4 tablespoons chicken stock	4 tablespoons chicken stock	⅓ cup chicken stock

Trim most of the fat from the chops and place each on a large square of double thickness foil. Cook the rice, drain and mix with the melted butter, pickled cucumber, peppers and seasoning. Cover chops with the rice mixture and add a spoonful of stock to each. Bring the edges of the foil together over the filling and seal to make airtight parcels.

Pork stuffed with prunes

(Illustrated on page 165)

Makes 6–8 servings.
To freeze Wrap stuffed joint in freezer foil; pack stuffing balls in a shallow polythene container or in layers with foil dividers. Seal and label.
To serve Defrost completely. Rub skin with salt and place in a roasting tin. Roast in a moderately hot oven (375°F, 190°C, Gas Mark 5) for 40 minutes then reduce oven heat to 350°F, 180°C, Gas Mark 4 for a further 2–2½ hours. Add the frozen stuffing balls for the last 35 minutes. Serve garnished with orange wedges and a sprig of watercress.
Storage time 4–6 months.

IMPERIAL	METRIC	AMERICAN
4-lb. joint belly of pork	2-kg. joint belly of pork	4-lb. piece fresh picnic shoulder
1 tablespoon lemon juice	1 tablespoon lemon juice	1 tablespoon lemon juice
salt and pepper to taste	salt and pepper to taste	salt and pepper to taste
stuffing	*stuffing*	*stuffing*
12 oz. cooking apples	350 g. cooking apples	¾ lb. baking apples
2 oz. soft white breadcrumbs	50 g. soft white breadcrumbs	1 cup fresh soft bread crumbs
grated zest and juice of 1 orange	grated zest and juice of 1 orange	grated zest and juice of 1 orange
8 oz. prunes, soaked, stoned and chopped	225 g. prunes, soaked, stoned and chopped	½ lb. prunes, soaked, pitted and chopped
1 tablespoon chopped parsley	1 tablespoon chopped parsley	1 tablespoon chopped parsley
1 teaspoon ground cinnamon	1 teaspoon ground cinnamon	1 teaspoon ground cinnamon
½ teaspoon Tabasco sauce	½ teaspoon Tabasco sauce	½ teaspoon Tabasco sauce
1 egg	1 egg	1 egg

Bone the joint and remove any excess fat. Sprinkle with lemon juice and seasoning. To make the stuffing, peel, core and chop the cooking apples and place in a bowl with the breadcrumbs, orange zest and juice, chopped prunes, salt to taste, parsley and cinnamon and mix well. Beat together the Tabasco and egg and use to bind the stuffing ingredients together. Spread some of the stuffing evenly over the boned side of the meat, roll up and tie tightly with string. Form the remaining stuffing into balls.

Instant lamb casserole

(Illustrated on page 165)

Makes 4 servings.

IMPERIAL	METRIC	AMERICAN
2 tablespoons oil	2 tablespoons oil	3 tablespoons oil
8 frozen lamb cutlets	8 frozen lamb cutlets	8 frozen rib chops
1 teaspoon Tabasco sauce	1 teaspoon Tabasco sauce	1 teaspoon Tabasco sauce
salt and freshly ground black pepper to taste	salt and freshly ground black pepper to taste	salt and freshly ground black pepper to taste
2 beef stock cubes	2 beef stock cubes	2 beef bouillon cubes
2 pints water	generous 1 litre water	5 cups water
8 oz. frozen peas	225 g. frozen peas	½ lb. frozen peas
1 lb. frozen carrots	450 g. frozen carrots	1 lb. frozen carrots
1 lb. frozen new potatoes	450 g. frozen new potatoes	1 lb. frozen new potatoes

Heat the oil in a frying pan and use to brown the cutlets on all sides. Sprinkle with half the Tabasco sauce and seasoning to taste while cooking. Remove cutlets to a flameproof casserole. Make up the stock cubes with the meat juices and water. Pour over the cutlets and bring to the boil. Cover and simmer very gently for 20 minutes. Add the frozen vegetables, bring back to the boil and simmer for a further 10 minutes.

Note If preferred, the stock may be thickened with 1 tablespoon cornflour, moistened with a little water.

Beef with herbed cobbler

(Illustrated on page 185)

Makes 6 servings.
To freeze Cover casserole with air-tight lid or sheet of moulded foil and smooth down edges. Pack cobbler topping in polythene bag. Seal and label.
To serve Defrost casserole and reheat in a moderately hot oven (375°F, 190°C, Gas Mark 5) for 25 minutes. Place frozen cobbler topping over the casserole, overlapping. Brush with beaten egg and sprinkle with 'Pinch of herbs'. Replace in oven for a further 15 minutes to reheat and brown the topping.
Storage time 4–6 months.

IMPERIAL	METRIC	AMERICAN
1 oz. flour	25 g. flour	¼ cup all-purpose flour
salt	salt	salt
1½ lb. stewing beef, cubed	¾ kg. stewing beef, cubed	1½ lb. beef stew meat, cubed
2 tablespoons oil	2 tablespoons oil	3 tablespoons oil
2 medium onions, sliced	2 medium onions, sliced	2 medium onions, sliced
1 lb. carrots, sliced	450 g. carrots, sliced	1 lb. carrots, sliced
1 15-oz. can tomatoes	1 425-g. can tomatoes	1 15-oz. can tomatoes
2 tablespoons tomato purée	2 tablespoons tomato purée	3 tablespoons tomato paste
½ teaspoon seasoned pepper	½ teaspoon seasoned pepper	½ teaspoon seasoned pepper
cobbler	*cobbler*	*cobbler*
8 oz. self-raising flour	225 g. self-raising flour	2 cups all-purpose flour sifted with 2 teaspoons baking powder
pinch salt	pinch salt	pinch salt
2 oz. margarine	50 g. margarine	¼ cup margarine
1 teaspoon 'Pinch of herbs'	1 teaspoon 'Pinch of herbs'	1 teaspoon 'Pinch of herbs'
¼ pint milk	1½ dl. milk	⅔ cup milk

Season the flour with salt and use to coat the meat. Heat the oil and in it fry the meat until browned on all sides. Add the remaining ingredients, stir well and turn into an ovenproof casserole. Cover and cook in a moderate oven (350°F, 180°C, Gas Mark 4) for 1½–2 hours. Cool. Meanwhile make the cobbler topping. Sieve flour and salt into a bowl and rub in fat. Add the herbs and sufficient milk to make a soft dough. Roll out on a floured surface and cut into rounds. Place on a greased baking tray, brush with milk and cook in a hot oven (450°F, 230°C, Gas Mark 8) for 10 minutes. Cool.

Liver with cream sauce

Makes 4 servings.
To freeze Pack in a shallow shaped foil container and cover with lid or sheet of foil. Seal and label.
To serve Uncover and place in a moderate oven (350°F, 180°C, Gas Mark 4) for 20 minutes. Pour over ¼ pint (1½ dl.) single cream and replace in oven for a further 15 minutes. Serve liver on a bed of fluffy boiled rice with the cream sauce poured over. Garnish with grilled bacon rolls and sautéed mushrooms.
Storage time 4–6 months.

IMPERIAL	METRIC	AMERICAN
2 tablespoons oil or butter	2 tablespoons oil or butter	3 tablespoons oil or butter
1 onion, chopped	1 onion, chopped	1 onion, chopped
1½ lb. calf's liver, sliced	¾ kg. calf's liver, sliced	1½ lb. calf liver, sliced
1 oz. seasoned flour	25 g. seasoned flour	¼ cup seasoned flour
½ teaspoon dried mixed herbs	½ teaspoon dried mixed herbs	½ teaspoon dried mixed herbs
¼ pint stock	1½ dl. stock	⅔ cup stock
salt and pepper to taste	salt and pepper to taste	salt and pepper to taste
juice of ½ lemon	juice of ½ lemon	juice of ½ lemon

Heat the oil in a frying pan and use to sauté the onion until softened but not browned. Dip the liver in the seasoned flour and add to the pan. Fry for about 5 minutes on each side then add the remaining ingredients to the pan. Bring to the boil and simmer for 5 minutes. Cool.

Glazed beef roll

Makes 4 servings.
To freeze Wrap in freezer foil. Seal and label.
To serve Unwrap and defrost in the refrigerator overnight. Dissolve 2 teaspoons gelatine in ¼ pint (1½ dl.) hot beef consommé and allow to cool. Take the segments from a large orange and arrange on top of the roll. When gelatine is on the point of setting, spoon carefully over the roll to glaze; allow to set.
Storage time 3 months.

IMPERIAL	METRIC	AMERICAN
½ oz. dripping or lard	15 g. dripping or lard	1 tablespoon drippings or lard
2 medium onions, chopped	2 medium onions, chopped	2 medium onions, chopped
1 clove garlic, crushed	1 clove garlic, crushed	1 clove garlic, crushed
1 lb. minced beef	450 g. minced beef	1 lb. ground beef
3 oz. fresh brown breadcrumbs	75 g. fresh brown breadcrumbs	1½ cups fresh brown bread crumbs
1 tablespoon lemon juice	1 tablespoon lemon juice	1 tablespoon lemon juice
1 teaspoon Worcestershire sauce	1 teaspoon Worcestershire sauce	1 teaspoon Worcestershire sauce
½ teaspoon ground nutmeg	½ teaspoon ground nutmeg	½ teaspoon ground nutmeg
1 teaspoon mixed herbs	1 teaspoon mixed herbs	1 teaspoon mixed herbs
½ teaspoon black pepper	½ teaspoon black pepper	½ teaspoon black pepper
2 tablespoons tomato ketchup	2 tablespoons tomato ketchup	3 tablespoons tomato catsup
pinch salt	pinch salt	pinch salt
1 egg, beaten	1 egg, beaten	1 egg, beaten
¼ pint strong beef stock	1½ dl. strong beef stock	⅔ cup strong beef stock

Melt the dripping in a frying pan and use to sauté the onions and garlic until softened but not browned. Stir in all remaining ingredients except the stock and mix well. Add sufficient of the stock to bind the mixture, and shape it with floured hands into a roll. Place on a baking tray and cook in a moderately hot oven (375°F, 190°C, Gas Mark 5) for 1 hour. Cool.

Farmhouse galantine

Makes 2 galantines, 8–12 servings.
To freeze Wrap each galantine in freezer foil. Seal and label.
To serve Defrost, still wrapped, in the refrigerator overnight. Unwrap and coat all over with toasted breadcrumbs.
Storage time 4–6 months.

IMPERIAL	METRIC	AMERICAN
8 oz. streaky bacon	225 g. streaky bacon	½ lb. bacon slices
1 lb. chuck steak, minced	450 g. chuck steak, minced	1 lb. chuck steak, ground
8 oz. soft white bread-crumbs	225 g. soft white bread-crumbs	4 cups fresh soft bread crumbs
1 medium onion, grated	1 medium onion, grated	1 medium onion, grated
2 teaspoons chopped thyme	2 teaspoons chopped thyme	2 teaspoons chopped thyme
2 teaspoons chopped parsley	2 teaspoons chopped parsley	2 teaspoons chopped parsley
salt and black pepper to taste	salt and black pepper to taste	salt and black pepper to taste
4 eggs, beaten	4 eggs, beaten	4 eggs, beaten

De-rind and chop the bacon and mix in a bowl with the meat, bread-crumbs, onion, herbs and seasoning. Mix in sufficient beaten egg to bind the mixture, divide it between two greased 1-lb. (½-kg.) loaf tins and smooth the tops. Place the loaf tins in a shallow roasting tin containing a little cold water. Cover completely with a sheet of foil and crimp the edges under those of the roasting tin. Place in a moderate oven (350°F, 180°C, Gas Mark 4) for 1½ hours. Cool and remove from loaf tins.

Game pie

Makes 6 servings.
To freeze Wrap in freezer foil. Seal and label.
To serve Place, still frozen and wrapped, in a moderately hot oven (400°F, 200°C, Gas Mark 6) for 40 minutes.
Storage time 3 months.

IMPERIAL	METRIC	AMERICAN
8 oz. pheasant	225 g. pheasant	½ lb. pheasant
1½ lb. hare or rabbit	¾ kg. hare or rabbit	1½ lb. hare or rabbit
8 oz. calf's or chicken liver	225 g. calf's or chicken liver	½ lb. calf's or chicken liver
1 oz. seasoned flour	25 g. seasoned flour	¼ cup seasoned flour
2 tablespoons oil	2 tablespoons oil	3 tablespoons oil
1 pint brown stock	generous ½ litre brown stock	2½ cups brown stock
4 oz. mushrooms	100 g. mushrooms	¼ lb. mushrooms
2 bay leaves	2 bay leaves	2 bay leaves
12 oz. shortcrust pastry	350 g. shortcrust pastry	¾ lb. basic pie dough

Remove meat from bones and cut up neatly. Wash the liver and cut into small pieces. Toss all these in seasoned flour. Heat the oil in a large frying pan and use to fry the meat until lightly browned on all sides. Turn into a large deep pie dish. Pour a little of the stock into the frying pan and heat, scraping up the sediment. Pour this over the meat with enough stock almost to cover. Add the mushrooms, only slicing if they are large, and the bay leaves. Leave to cool. Roll out the pastry and use to make a lid. Brush with beaten egg and decorate with leaves made from pastry trimmings. Bake in a moderate oven (350°F, 180°C, Gas Mark 4) for 1½ hours. Cool.

Lamb and lemon curry

Makes 6 servings.
To freeze Use a polythene bag to line a saucepan, pour in the curry and partially freeze. Remove from freezer, seal bag and remove saucepan. Label.
To serve Remove polythene bag while curry is still frozen, replace in saucepan and allow to defrost. Reheat gently to boiling point and simmer for 2 minutes. Garnish with raw onion rings and serve with dahl, basmati rice, fried poppadums, mango chutney and sambals (see page 202).
Storage time 4–6 months.

IMPERIAL	METRIC	AMERICAN
6 oz. desiccated coconut	175 g. desiccated coconut	2 cups shredded coconut
2½ lb. boned shoulder or leg of lamb, diced	1¼ kg. boned shoulder or leg of lamb, diced	2½ lb. boneless lamb shoulder or leg, diced
seasoned flour	seasoned flour	seasoned flour
3 oz. butter	75 g. butter	6 tablespoons butter
2 large onions, chopped	2 large onions, chopped	2 large onions, chopped
2 cloves garlic, chopped	2 cloves garlic, chopped	2 cloves garlic, chopped
2 tablespoons curry powder	2 tablespoons curry powder	3 tablespoons curry powder
2 teaspoons curry paste	2 teaspoons curry paste	2 teaspoons curry paste
2 oz. flour	50 g. flour	½ cup all-purpose flour
¾ pint chicken stock	scant ½ litre chicken stock	scant 2 cups chicken stock
2 green dessert apples	2 green dessert apples	2 green dessert apples
2 tablespoons sultanas	2 tablespoons sultanas	3 tablespoons seedless white raisins
2 tablespoons apricot jam	2 tablespoons apricot jam	3 tablespoons apricot jam
2 lemons, halved	2 lemons, halved	2 lemons, halved
2 sticks cinnamon and 2 or 3 red chilli peppers tied in muslin	2 sticks cinnamon and 2 or 3 red chilli peppers tied in muslin	2 sticks cinnamon and 2 or 3 red chili peppers tied in muslin

Pour ¾ pint (scant ½ litre) boiling water over the coconut and leave to stand for at least 1 hour, then strain off the coconut milk and reserve. Coat the lamb lightly in seasoned flour. Heat the butter and use to fry the onion, garlic and lamb for 4 minutes, turning to seal the surfaces of the meat. Stir in the curry powder, curry paste and flour and cook over gentle heat for a further 3 minutes, stirring constantly. Gradually add the stock and coconut milk, bring to the boil and stir until thickened. Core and chop the apple and add to the pan with the sultanas, apricot jam, spices tied in muslin and the lemon halves. Cover and simmer for 1–1½ hours, until meat is tender. Remove spice bag and lemon halves and adjust seasoning. Cool.

Dahl

Makes 6–8 servings.
To freeze Pack into polythene container leaving headspace. Seal and label.
To serve Turn contents into a saucepan and reheat gently to boiling point.
Storage time 4–6 months.

IMPERIAL	METRIC	AMERICAN
8 oz. brown lentils	225 g. brown lentils	1 cup brown lentils
1 oz. butter	25 g. butter	2 tablespoons butter
1 onion, chopped	1 onion, chopped	1 onion, chopped
½ teaspoon turmeric	½ teaspoon turmeric	½ teaspoon turmeric
1 teaspoon salt	1 teaspoon salt	1 teaspoon salt
1 teaspoon dry mustard	1 teaspoon dry mustard	1 teaspoon dry mustard
1 teaspoon ground coriander	1 teaspoon ground coriander	1 teaspoon ground coriander

Cover the lentils with cold water and leave to stand for 1 hour. Drain. Melt the butter in a saucepan, sauté the chopped onion until limp. Add the drained lentils, turmeric and enough water to cover. Bring to the boil and simmer until the lentils are soft. Add the other seasonings and mix well. The mixture can be a thin soup or reduced to the consistency of a thick soup, or that of pease pudding. Cool.

Rice to serve with curries Basmati rice has a distinctive nutty flavour and requires to be washed before cooking, but good Patna rice is an acceptable substitute and will not require washing. Drain and add to a large pan of fast-boiling salted water, cook at a rolling boil until one or two grains, cooled and pressed between thumb and forefinger, feel tender. This will take from 12–18 minutes, according to the type of rice. Turn into a fine colander or large sieve to drain, pour fresh hot water through to separate the grains, and shake gently to remove surplus water.

Note The cooled cooked rice may be frozen and reheated when required. To keep grains separate, partially freeze and squeeze the bag or break up contents of a container with a fork, then complete freezing process. To serve, allow to thaw at room temperature for 1 hour, then reheat gently in a greased shallow baking tray, dotted with butter and covered with foil.

To cook poppadums Allow 2 poppadums per person. Heat about ½ inch (1 cm.) oil in a strong frying pan slightly larger than the size of the poppadums, as they expand during cooking. To test the heat of the oil, break a small piece from a poppadum and drop into the oil. If it begins to frizzle immediately, the oil is hot enough. Slide in a poppadum, hold lightly under surface of the oil with a fish slice until it expands, turn carefully and cook for a further few seconds until golden brown, then remove and drain on soft kitchen paper.

Yogurt and mint sambal

IMPERIAL	METRIC	AMERICAN
2 cartons natural yogurt	2 cartons natural yogurt	2 cartons natural yogurt
3 tablespoons chopped fresh mint	3 tablespoons chopped fresh mint	4 tablespoons chopped fresh mint

Stir the yogurt with a fork until smooth, add the mint and mix in until evenly dispersed.

Banana and nut sambal

IMPERIAL	METRIC	AMERICAN
2 bananas	2 bananas	2 bananas
lemon juice	lemon juice	lemon juice
2 oz. salted nuts, chopped	50 g. salted nuts, chopped	½ cup chopped salted nuts

Peel the bananas and slice them. Dip the slices in lemon juice and roll in the salted nuts.

1 Mixing defrosted raspberry purée with whisked egg yolks and sugar before adding the dissolved gelatine, egg whites and cream.

2 Spooning the soufflé mixture into Tupperware fancy moulds.

3 Tupperware Jel'n'serve, with the seal removed, inverted on a serving dish so that the frozen soufflé slips out when the top seal is removed. Extra whipped cream is piped onto trays for open freezing.

1

2

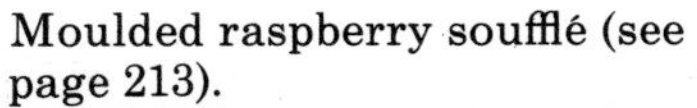

Moulded raspberry soufflé (see page 213).

3

1 Making up an 8-oz. (225-g.) packet shortcrust pastry mix for a quick Christmas special as an alternative to mince pies.

2 Folding in the obliquely-cut strips of pastry over the apple and mincemeat filling – the alternate strips have been cut out and removed.

3 Putting the baked plait, packed in polythene with the air removed, into the freezer.

1

2

3

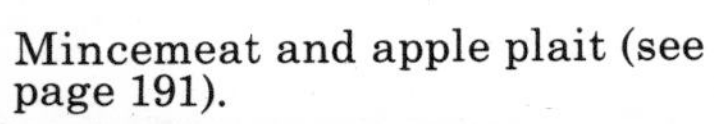

Mincemeat and apple plait (see page 191).

Opposite Christmas jewel cake (see page 192).

1 Crushing the digestive biscuits, placed in a polythene bag, with a rolling pin.

2 Mixing the condensed milk and cream together for the filling.

3 The cream pie, placed in a polythene bag, the air withdrawn through a straw, ready for freezing.

Orange and lemon cream pie (see page 213).

1

2

3

Golden guide number

to party gâteaux

Preparing a spectacular party sweet is a time-consuming task. One needs patience, a certain amount of artistry and the rest of the family well out of the way. Given these requisites one is amply rewarded by the gasps of pleasure from your family or friends when you bring the finished gâteau to the table.

Chocolate walnut flan

Makes 8 servings.
To freeze Open freeze until solid. Pack in a polythene bag. Seal and label.
To serve Unwrap while still frozen, place on serving dish and allow to defrost at room temperature for 4 hours.
Storage time 4–6 months.

Note The same filling can also be served in one of the following flan cases.

Imperial	Metric	American
½ pint milk	3 dl. milk	1¼ cups milk
1 teaspoon vanilla essence	1 teaspoon vanilla essence	1 teaspoon vanilla extract
30 marshmallows	30 marshmallows	30 marshmallows
½ pint double cream	3 dl. double cream	1¼ cups whipping cream
3 oz. plain chocolate, grated	75 g. plain chocolate, grated	3 squares semi-sweet chocolate pieces
2 oz. walnuts, chopped	50 g. walnuts, chopped	½ cup chopped walnuts
1 9-inch baked flan case	1 23-cm. baked flan case	1 9-inch baked pie shell
grated chocolate and chopped walnuts to decorate	grated chocolate and chopped walnuts to decorate	grated chocolate and chopped walnuts to decorate

Place the milk, vanilla essence and marshmallows in the top of a double boiler and stir over hot water until melted. Cool. Half whip the cream and fold in with the chocolate and walnuts. Pour into the flan case and sprinkle with more chocolate and walnuts. Allow to set.

Fancy flan cases

Crumb crust A nutty-textured crumb crust to fill with mousses or creams can be frozen if baked in a moderate oven for 7–8 minutes to set. Combine 4 oz. (100 g.) wheatmeal biscuit crumbs, 2 tablespoons toasted flaked almonds, 2 tablespoons castor sugar and 4 tablespoons melted butter. If liked, add a few drops almond flavouring.
Crispy crust Crisp rice cereal makes a case that requires no cooking. Melt 2 tablespoons butter with 3 tablespoons golden syrup and 2 oz. (50 g.) crisp rice cereal. Stir gently until well mixed. Press into a foil pie plate and cool until firm. To make a chocolate crust, mix the butter and syrup with 3 oz. (75 g.) plain chocolate melted, and 4 oz. (100 g.) cornflakes.
Freezing and storage All these fancy flan cases can be made in quantity, open frozen until hard, then removed from the foil plates and stacked, with dividers, then wrapped in a parcel. However, as foil plates are expendable, it is less trouble to freeze them on the plates rather than risk damage handling them too much when defrosting. Allow to defrost at room temperature, unwrapped (about 1 hour) or in the refrigerator (wrapped) for about 2½ hours.

No-bake chocolate cake

Makes 4–6 servings.
To freeze Cover with lid or sheet of foil. Seal and label.
To serve Unwrap and defrost at room temperature for 4 hours. Melt 2 oz. (50 g.) plain chocolate, spread over the cake and decorate with cherries and walnuts. Serve cut into fingers.
Storage time 4–6 months.

Imperial	Metric	American
4 oz. butter	100 g. butter	½ cup butter
1 oz. sugar	25 g. sugar	2 tablespoons sugar
2 oz. cocoa powder	50 g. cocoa powder	¼ cup unsweetened cocoa powder
1 tablespoon golden syrup	1 tablespoon golden syrup	1 tablespoon maple syrup
8 oz. digestive biscuits	225 g. digestive biscuits	½ lb. graham crackers
1 oz. glacé cherries, chopped	25 g. glacé cherries, chopped	scant ¼ cup chopped candied cherries

Cream butter and sugar, add cocoa powder and syrup and mix well. Place the biscuits in a polythene bag and crush with a rolling pin. Add to the creamed mixture and stir in the cherries. Press into a shaped foil tray about 1 inch (½ cm.) deep.

Frosted coffee gâteau

Makes 6–8 servings.
To freeze Pack in a polythene bag. Seal and label.
To serve Unwrap while still frozen and place on serving dish. Allow to defrost for 4 hours before serving.
Storage time 4–6 months.

Imperial	Metric	American
4 oz. butter	100 g. butter	½ cup butter
4 oz. castor sugar	100 g. castor sugar	½ cup sugar
2 eggs	2 standard eggs	2 eggs
1 tablespoon coffee essence or very strong black coffee	1 tablespoon coffee essence or very strong black coffee	1 tablespoon very strong black coffee
3 tablespoons brandy or sherry	3 tablespoons brandy or sherry	4 tablespoons brandy or sherry
2 7-inch baked sponge layers (see opposite)	2 18-cm. baked sponge layers (see opposite)	2 7-inch baked sponge layers (see opposite)
to decorate	*to decorate*	*to decorate*
3 oz. butter	75 g. butter	¾ cup butter
6 oz. icing sugar, sieved	175 g. icing sugar, sieved	1⅓ cups sifted confectioners' sugar
1 tablespoon strong black coffee	1 tablespoon strong black coffee	1 tablespoon strong black coffee
¼ pint double cream, whipped	1½ dl. double cream, whipped	⅔ cup whipping cream, whipped
2 oz. walnuts, finely chopped	50 g. walnuts, finely chopped	½ cup finely chopped walnuts
few halved walnuts	few halved walnuts	few halved walnuts

Cream butter and sugar together until light and fluffy. Beat in the eggs and gradually add the coffee essence and brandy. Carefully cut each sponge cake into three layers. Place one layer in the base of a 7-inch (18-cm.) cake tin and spread with about one-fifth of the cream mixture. Continue with layers of cake and cream, ending with a layer of cake. Place a second 7-inch (18-cm.) cake tin on top and weight down. Place in the freezer until solid. Remove from cake tin. To decorate the cake, beat together the butter and icing sugar and gradually beat in the coffee. Fold in the whipped cream. Spread a thin coating of icing around the sides of the cake and roll the sides in the chopped walnuts. Spread top of cake with icing and decorate with piped rosettes with the remaining icing. Decorate with walnut halves.

Note The decoration on this cake is easily damaged even in the frozen state. If preferred, freeze the cake plain and decorate when required to serve.

Pineapple cheesecake

(Illustrated on page 71)

Makes 1 cheesecake, 8 servings.
To freeze Place on a serving dish and pack in a polythene bag. Seal and label.
To serve Unwrap and allow to defrost at room temperature for 4 hours. Pipe whipped cream around the edge and decorate with pieces of canned pineapple.
Storage time 4–6 months.

Imperial	Metric	American
1 8-oz. can pineapple pieces	1 225-g. can pineapple pieces	1 8-oz. can pineapple pieces
½ oz. powdered gelatine	15 g. powdered gelatine	2 envelopes gelatin
1 lb. cottage cheese	450 g. cottage cheese	2 cups cottage cheese
finely grated zest of 1 lemon	finely grated zest of 1 lemon	finely grated zest of 1 lemon
4 oz. castor sugar	100 g. castor sugar	½ cup granulated sugar
2 eggs, separated	2 eggs, separated	2 eggs, separated
¼ pint double cream	1½ dl. double cream	⅔ cup whipping cream
crust	*crust*	*crust*
6 digestive biscuits	6 digestive biscuits	9 graham crackers
1 tablespoon castor sugar	1 tablespoon castor sugar	1 tablespoon granulated sugar
1½ oz. butter, melted	40 g. butter, melted	3 tablespoons butter, melted

Drain pineapple syrup into a basin and sprinkle in the gelatine. Sieve the cottage cheese into a basin. Chop the pineapple and add to the cheese with the lemon zest and juice. Place the pineapple syrup and gelatine over a pan of hot water and allow to dissolve. Stir in the sugar and egg yolks and continue stirring over the hot water until mixture is the consistency of pouring cream. Cool. When thick but not set blend into the cheese mixture then whip the cream and stir in. Beat the egg whites stiffly, fold in and turn into a well-oiled 7-inch (18-cm.) cake tin and chill until set. To make the crust, put the digestive biscuits into a polythene bag and crush finely with a rolling pin. Combine crumbs with the sugar and butter, sprinkle over the cheesecake mixture and press down lightly. Chill until firm then unmould.

Basic sponge flan cases

Makes 2 flan cases.
To freeze Pack in a polythene bag. Seal and label.
To serve Unwrap, place on serving dish and allow to defrost for 2 hours and fill as desired.
Storage time 4–6 months.

IMPERIAL	METRIC	AMERICAN
6 eggs	6 eggs	6 eggs
6 oz. castor sugar	175 g. castor sugar	$\frac{3}{4}$ cup granulated sugar
6 oz. plain flour	175 g. plain flour	$1\frac{1}{2}$ cups all-purpose flour
2 oz. butter, melted	50 g. butter, melted	$\frac{1}{4}$ cup butter, melted

Whisk the eggs and sugar together in a bowl over a pan of hot water until thick and whisk leaves a trail when lifted out of the mixture. Remove from the heat and whisk steadily until mixture is quite cold. Sieve the flour and fold evenly into the mixture with the melted butter. Divide the mixture between two greased and floured 8-inch (20-cm.) flan tins and bake in a moderately hot oven (375°F, 190°C, Gas Mark 5) for 20 minutes. Turn out onto a wire tray and cool.

Sponge fruit flan

(Illustrated on page 184)

IMPERIAL	METRIC	AMERICAN
1 sponge flan case, defrosted	1 sponge flan case, defrosted	1 sponge flan case, defrosted
1 15-oz. can pear halves or pineapple rings	1 425-g. can pear halves or pineapple rings	1 15-oz. can pear halves or pineapple rings
1 teaspoon arrowroot	1 teaspoon arrowroot	1 teaspoon arrowroot flour
1 tablespoon cherry brandy	1 tablespoon cherry brandy	1 tablespoon cherry brandy
1 oz. flaked almonds, toasted	25 g. flaked almonds, toasted	$\frac{1}{4}$ cup almonds, toasted

Place flan case on a serving dish. Drain the fruit (slice pears) and arrange in flan case. Moisten the arrowroot with a little fruit syrup and put the rest into a saucepan and heat. Add the blended arrowroot and stir constantly until clear and thick. Stir in the cherry brandy; cool slightly and pour over the fruit and scatter the almonds over the top.

Apricot and cherry flan

IMPERIAL	METRIC	AMERICAN
1 sponge flan case, defrosted	1 sponge flan case, defrosted	1 sponge flan case, defrosted
1 15-oz. can apricot halves	1 425-g. can apricot halves	1 15-oz. can apricot halves
1 teaspoon arrowroot	1 teaspoon arrowroot	1 teaspoon arrowroot flour
1 tablespoon apricot brandy	1 tablespoon apricot brandy	1 tablespoon apricot brandy
8 glacé cherries, halved	8 glacé cherries, halved	8 candied cherries, halved

Place flan case on a serving dish. Drain the apricots and arrange them, cut-side-down, in flan case. Moisten the arrowroot with a little apricot syrup and put the rest into a saucepan and heat. Add the blended arrowroot and stir constantly until clear and thick. Stir in the apricot brandy, cool slightly and pour over the apricots and decorate with the cherry halves.

Peach flan

(Illustrated on page 184)

IMPERIAL	METRIC	AMERICAN
1 frozen baked 9-inch pastry flan case	1 frozen baked 23-cm. flan case	1 frozen baked 9-inch pie shell
pastry cream	*pastry cream*	*pastry cream*
1 egg	1 egg	1 egg
½ oz. flour	15 g. flour	2 tablespoons all-purpose flour
1 oz. castor sugar	25 g. castor sugar	2 tablespoons granulated sugar
¼ pint milk	1½ dl. milk	⅔ cup milk
1 oz. butter	25 g. butter	2 tablespoons butter
few drops vanilla essence	few drops vanilla essence	few drops vanilla extract
filling	*filling*	*filling*
1 1¾-lb. can peach slices	1 800-g. can peach slices	1 1¾-lb. can peach slices
2 tablespoons apricot jam	2 tablespoons apricot jam	3 tablespoons apricot jam

Place the flan case on a serving dish and allow to defrost. To make the pastry cream, beat the egg, flour and sugar together. Heat the milk to boiling point and pour over the mixture, stirring well. Return the mixture to the saucepan, bring slowly just to the boil, stirring constantly, until mixture is smooth and thick. Remove from the heat and beat in the butter and vanilla essence. Cool quickly and pour into the flan case. Drain the peaches and cut each slice in half. Arrange slices, overlapping, in concentric circles over the pastry cream. Heat together the jam and 4 tablespoons of the peach syrup and strain over the fruit to glaze.

Note The pastry cream can be made up in a larger quantity and frozen in several small polythene containers or polythene bags.

Pear condé

(Illustrated on page 36)

Makes 8 servings.

IMPERIAL	METRIC	AMERICAN
1 15-oz. can creamed rice pudding	1 425-g. can creamed rice pudding	1 15-oz. can creamed rice pudding
1 30-oz. container pear halves in sugar syrup	1 850-g. container pear halves in sugar syrup	1 30-oz. container pear halves in sugar syrup
2 tablespoons apricot jam	2 tablespoons apricot jam	3 tablespoons apricot jam
¼ pint double cream	1½ dl. double cream	⅔ cup whipping cream
8 chocolate leaves (see note)	8 chocolate leaves (see note)	8 chocolate leaves (see note)

Place creamed rice in base of a shallow serving dish. Defrost the container of pears in warm water. Drain pears well and arrange, cut-side-down, over the rice. Heat the pear syrup with the jam, allow to cool slightly and strain over the pears to glaze. Whip the cream, pipe a border of rosettes all around the dish and between the pear halves. Decorate with chocolate leaves.

Note To make chocolate leaves, melt plain chocolate in a small bowl over hot water. Choose even-size rose leaves and wash and dry carefully. Using the back of a teaspoon coat the underside of the leaves with melted chocolate. Allow to cool until quite hard then gently ease off the leaves.

Golden guide number

to ice cream desserts

Home-made ice cream can be varied in many ways to produce delicious flavours utterly different from any commercial ice cream. The same applies to sorbets and sherberts. Yet ice creams which you buy by the pack or in large containers can be layered with fruit and sponge and refrozen to produce your own individual speciality. Once you have tried your hand at this sort of making and mixing you will be able to think of endless new variations.

Pineapple sherbert

Makes 6–8 servings.
To freeze Pack in polythene containers. Seal and label.
To serve Serve scoops of the frozen sherbert in sundae glasses and decorate with glacé pineapple or frosted grapes.
Storage time 4–6 months.

Imperial	Metric	American
1 large can condensed milk	1 large can condensed milk	1 large can sweetened condensed milk
4 tablespoons lemon juice	4 tablespoons lemon juice	$\frac{1}{3}$ cup lemon juice
4 tablespoons melted butter	4 tablespoons melted butter	$\frac{1}{3}$ cup melted butter
1 $13\frac{1}{2}$-oz. can crushed pineapple	1 390-g. can crushed pineapple	1 $13\frac{1}{2}$-oz. can crushed pineapple
4 egg whites	4 egg whites	4 egg whites

Combine the condensed milk, lemon juice and melted butter until well blended. Add the pineapple and syrup and chill thoroughly. Beat the egg whites until stiff and fold into the chilled mixture. Partially freeze, then remove from freezer and beat until smooth.

Layered ice cream gâteau

Makes 8 servings.
To freeze Seal and label.
To serve Remove foil while still frozen, place on serving dish. Decorate with rosettes of whipped cream and crystallised rose petals. Replace in freezer and remove a few minutes only before serving.
Storage time 4–6 months.

Imperial	Metric	American
1 16-oz. container frozen strawberries or strawberry purée	1 450-g. container frozen strawberries or strawberry purée	1 16-oz. container frozen strawberries or strawberry purée
1 family brick rich strawberry ice cream	1 family brick rich strawberry ice cream	1 family brick rich strawberry ice cream
1 tablespoon cornflour	1 tablespoon cornflour	1 tablespoon cornstarch
2 tablespoons Kirsch	2 tablespoons Kirsch	3 tablespoons Kirsch
few drops red food colouring	few drops red food colouring	few drops red food coloring
1 quantity sponge flan mixture (see page 209)	1 quantity sponge flan mixture (see page 209)	1 quantity sponge flan mixture (see page 209)

Purée or sieve the strawberries and sweeten if necessary. Bring slowly to the boil in a small saucepan, stir in cornflour moistened with 2 tablespoons water. Cook, stirring constantly for 2 minutes, cool and add Kirsch. Colour pink. Chill in the freezer until thick (about 40 minutes). Meanwhile, divide the sponge flan mixture between two well greased and sugared 2-lb. (1-kg.) loaf tins. Cook in a moderate oven (350°F, 180°C, Gas Mark 4) for 30–35 minutes; turn out and cool on a wire tray. Slice one sponge loaf horizontally to make three layers. Cut the ice cream in half lengthwise and build up alternate layers of sponge and ice cream side by side on a large sheet of freezer foil. Wrap closely to make a neat oblong parcel, put into the freezer until coating mixture is ready. Remove ice cream and sponge parcel from freezer, open and coat with strawberry mixture. Open freeze until hard, then fold in foil again carefully.

Note Freeze the other sponge loaf until required. Different flavoured ice creams and fruit coating mixtures can be used.

Frozen orange syllabub

Makes 12 servings.
To freeze Cover open dishes with caps of freezer foil smoothed close to the sides. Seal moulds. Label.
To serve Top soufflé dishes with rosettes of whipped cream and transfer to the refrigerator cabinet 30 minutes before serving. Dip frozen moulds into hot water for 2 seconds to release and turn out. Decorate with halved glacé cherries.
Storage time 4–6 months.

Imperial	Metric	American
2 tablespoons lemon juice	2 tablespoons lemon juice	3 tablespoons lemon juice
finely grated zest and juice of 2 large oranges	finely grated zest and juice of 2 large oranges	finely grated zest and juice of 2 large oranges
¼ pint sweet sherry	1½ dl. sweet sherry	⅔ cup sweet sherry
4 tablespoons brandy	4 tablespoons brandy	⅓ cup brandy
6 oz. castor sugar	175 g. castor sugar	¾ cup granulated sugar
1 pint double cream	generous ½ litre double cream	2½ cups whipping cream

Put the lemon juice, orange zest and juice, sherry, brandy and sugar together in a bowl. Stir until the sugar has dissolved, leaving to stand if necessary. Whip the cream until just beginning to thicken. Gradually beat it into the mixture until very thick and foamy. Divide between 12 small individual dishes, such as miniature soufflé dishes, or fancy moulds such as Tupperware jellettes.

Orange and lemon sorbet

Makes 4–6 servings.
To freeze Pack in polythene containers. Seal and label.
To serve Serve sorbet in scoops while still frozen.
Storage time 4–6 months.

Imperial	Metric	American
2 teaspoons powdered gelatine	2 teaspoons powdered gelatine	2 teaspoons powdered gelatin
½ pint water	3 dl. water	1¼ cups water
6 oz. sugar	175 g. sugar	¾ cup sugar
½ pint fresh or frozen orange juice	3 dl. fresh or frozen orange juice	1¼ cups fresh or frozen orange juice
juice of 1 large lemon	juice of 1 large lemon	juice of 1 large lemon
2 egg whites	2 egg whites	2 egg whites

Soak the gelatine in a little of the water, heat the remaining water and sugar and boil rapidly for 5 minutes to reduce slightly. Add softened gelatine and stir to dissolve completely; cool. Add the orange and lemon juices. Partially freeze, then remove from freezer and mash until smooth. Whisk the egg whites until stiff and fold in evenly.

Basic ice cream

Makes 1½ pints (scant 1 litre).
To freeze Pack into polythene containers. Seal and label.
To serve Remove from freezer and allow to soften slightly in the refrigerator for 30 minutes before serving.
Storage time 4–6 months.

Imperial	Metric	American
½ pint creamy milk	3 dl. creamy milk	1¼ cups milk
2 eggs, separated	2 eggs, separated	2 eggs, separated
3 oz. sugar	75 g. sugar	6 tablespoons sugar
1 vanilla pod or 1 teaspoon vanilla essence	1 vanilla pod or 1 teaspoon vanilla essence	1 vanilla bean or 1 teaspoon vanilla extract
pinch salt	pinch salt	pinch salt
¾ pint double cream	scant ½ litre double cream	scant 2 cups whipping cream

Place the milk, egg yolks, sugar, vanilla and salt in a double boiler and stir over simmering water until mixture is thick enough to coat the back of a wooden spoon. Remove from heat, discard vanilla pod (if used) and cool. Whip the cream until thick and fold into the custard. Beat the egg whites stiffly and fold in. Place bowl in freezer until mixture is partially frozen, remove and beat until smooth.

Variations

These quantities are sufficient to flavour half the basic quantity of ice cream.
Caramel Melt half the sugar required for basic recipe in a saucepan until golden and add to the egg custard.
Praline Toast 4 oz. (100 g.) finely chopped almonds, mix into the caramelised sugar and add, as above.
Ginger Add 2 tablespoons finely chopped preserved ginger and 2 tablespoons ginger syrup from the jar.
Mocha Add 2 tablespoons very strong black coffee to the custard and 2 oz. (50 g.) grated plain chocolate after the final beating. Mix in thoroughly.
Fruit Add 4 oz. (100 g.) fruit purée to ice cream, before freezing.
Cassata Line a fancy 2-pint (generous 1-litre) mould, such as the Tupperware Jel'n'serve, with orange and lemon sorbet. Freeze until solid and fill centre with basic vanilla ice cream combined with chopped glacé cherries and angelica. Freeze until firm.

Orange and lemon cream pie

(Illustrated on page 206)

Makes 2 pies.
To freeze Open freeze until solid and pack in polythene bags. Seal and label.
To serve Remove wrapping while still frozen and allow to defrost at room temperature for 4 hours. Decorate with rosettes of whipped cream, topped with quartered lemon or orange slices, or jellied fruit slices.
Storage time 4–6 months.

IMPERIAL	METRIC	AMERICAN
8 oz. digestive biscuits	225 g. digestive biscuits	½ lb. graham crackers
4 oz. butter	100 g. butter	½ cup butter
2 tablespoons sugar	2 tablespoons sugar	3 tablespoons sugar
filling	*filling*	*filling*
1 large can sweetened condensed milk	1 large can sweetened condensed milk	1 large can sweetened condensed milk
1 10-oz. can cream	1 275-g. can cream	1 10-oz. can cream
grated zest and juice of 1 large lemon and 1 large orange	grated zest and juice of 1 large lemon and 1 large orange	grated zest and juice of 1 large lemon and 1 large orange

Place the digestive biscuits in a polythene bag and crush with a rolling pin. Melt the butter in a saucepan and stir in the crumbs and sugar. Divide this mixture between two small pie plates and press firmly in place with the back of a metal spoon. Chill until firm. To make the filling, mix together the condensed milk and cream until well blended. Stir in the grated fruit zests and juices. This will thicken the mixture. Divide the filling between the pie cases, smooth tops with a palette knife and leave until set.

Moulded raspberry soufflé

(Illustrated on page 203)

Makes 6 servings.
To freeze Seal and label.
To serve Defrost large moulds at room temperature for about 5 hours. Turn out onto serving dish. Decorate with shell piping of whipped cream around the base; place some partially defrosted raspberries on top. Small moulds can be turned out in the frozen state. Dip, while still sealed, into hot water for 2 seconds to release them. Decorate with cream and raspberries.
Storage time 3 months.

IMPERIAL	METRIC	AMERICAN
6 large eggs, separated	6 large eggs, separated	6 eggs, separated
8 oz. castor sugar	225 g. castor sugar	1 cup granulated sugar
1 30-oz. container frozen raspberry purée (1 pint)	1 850-g. container frozen raspberry purée (generous ½ litre)	1 30-oz. container frozen raspberry purée (2½ cups)
2 tablespoons Kirsch	2 tablespoons Kirsch	3 tablespoons Kirsch
1 oz. gelatine	25 g. gelatine	4 envelopes gelatin
6 tablespoons warm water	6 tablespoons warm water	½ cup warm water
½ pint double cream, lightly whipped	3 dl. double cream, lightly whipped	1¼ cups whipping cream, lightly whipped
few drops red food colouring (optional)	few drops red food colouring (optional)	few drops red food coloring (optional)

Rinse out a fancy 2-pint (generous 1-litre) mould, such as the Tupperware Jel'n'serve and six individual moulds such as Jellettes with cold water. Whisk the egg yolks and sugar in a basin over hot water until the mixture begins to thicken. Gradually whisk in the raspberry purée and the Kirsch. Dissolve the gelatine in the water, stir into the mixture, and leave to cool until it becomes fairly thick, stirring occasionally. Fold in the whipped cream and then the stiffly beaten egg whites. Add a few drops of red food colouring if liked. Pour into the containers, chill until set.

Cream-filled nut torte

Makes 6–8 servings.
To freeze Pack in polythene bag with foil dividers. Seal and label. (These layers will be fragile, even in the frozen state.)
To serve Defrost, still wrapped, for 1 hour. Whip ½ pint (3 dl.) double cream with 1 oz. (25 g.) icing sugar and use to sandwich the cakes together. Pipe rosettes of cream over top of cake and decorate with toasted flaked almonds or chocolate shapes (see page 240).
Storage time 4–6 months.

IMPERIAL	METRIC	AMERICAN
4 oz. digestive biscuits	100 g. digestive biscuits	1 cup crushed graham crackers
6 eggs	6 eggs	6 eggs
6 oz. castor sugar	175 g. castor sugar	¾ cup granulated sugar
1 teaspoon baking powder	1 teaspoon baking powder	1 teaspoon baking powder
1 teaspoon vanilla essence	1 teaspoon vanilla essence	1 teaspoon vanilla extract
1 teaspoon almond essence	1 teaspoon almond essence	1 teaspoon almond extract
4 oz. mixed nuts, chopped	100 g. mixed nuts, chopped	1 cup chopped mixed nuts

Place the digestive biscuits in a polythene bag and crush finely with a rolling pin. Line three 7-inch (18-cm.) cake tins and grease and flour the greaseproof paper. Separate the eggs, add the sugar to the yolks and whisk until very light and fluffy. Fold in the baking powder and essences. Whisk the egg whites until stiff and fold into the yolk mixture with the biscuit crumbs and nuts. Divide the mixture between the prepared tins and bake in a moderate oven (325°F, 170°C, Gas Mark 3) for about 30 minutes. Cool for at least 10 minutes before removing from the tins.

Strawberry ice cream sundaes

(Illustrated on page 224)

Makes 48 servings.

IMPERIAL	METRIC	AMERICAN
2 pints double cream, whipped	generous 1 litre double cream, whipped	5 cups whipping cream, whipped
4 oz. pistachio nuts, blanched and chopped	100 g. pistachio nuts, blanched and chopped	1 cup blanched and chopped pistachio nuts
½ gallon rich vanilla ice cream	2¼ litres rich vanilla ice cream	10 cups rich vanilla ice cream
4 lb. strawberries, fresh or partially defrosted	2 kg. strawberries, fresh or partially defrosted	4 lb. strawberries, fresh or partially defrosted
½ gallon lemon and lime ice cream	2¼ litres lemon and lime ice cream	10 cups lemon and lime ice cream
hazelnut fingers to serve	hazelnut fingers to serve	ladyfingers to serve

This quantity would make about 48 sundaes, for a big party. However, if fewer are needed, as much ice cream as is required can be taken from the containers, the exposed surfaces of the ice cream covered with freezer film, and the containers returned to the freezer. Or both flavours can be bought in smaller quantities and the other ingredients used in proportion.

Have ready the whipped cream in a piping bag and a saucer of blanched and chopped pistachio nuts. Allow two full scoops of both flavours to make one sundae. Fill the bottom half of each sundae glass with rich vanilla ice cream. Arrange a few cut strawberries around the sides, pressing the cut surfaces against the glass. Fill the glass with small half-scoops of lemon and lime ice cream. Pipe a swirl of cream on this, sprinkle with pistachio nuts and top with a whole strawberry. If using frozen strawberries, pick out the best ones for the decoration. Serve at once, while fruit is still firm, with hazelnut fingers.

Great dishes of Europe

Many happy holiday memories can be revived by cooking, at home, the dishes you have enjoyed abroad, or others you have heard your friends describe with enthusiasm. If the ingredients are not always easy to come by, and the preparation is rather lengthy, take advantage of your friendly freezer. Cook these dishes when the opportunity presents itself and produce them for a festive occasion, or just when you feel a more exciting meal than usual is called for. Some of the recipes are for dishes which can take a freezer dive and emerge with the authentic, national flavour unimpaired. Others, like the two which follow for scampi, are intended to make good use of items you buy at the freezer food centre.

Pâté de luxe
(France)

Makes 4 servings.
To freeze Pack into small polythene containers. Seal and freeze.
To serve Defrost for 3–4 hours and serve with crusty French bread.
Storage time 6 weeks.

IMPERIAL	METRIC	AMERICAN
1 oz. butter	25 g. butter	2 tablespoons butter
8 oz. chicken livers	225 g. chicken livers	½ lb. chicken livers
1 tablespoon corn or olive oil	1 tablespoon corn or olive oil	1 tablespoon corn or olive oil
1 clove garlic, crushed	1 clove garlic, crushed	1 clove garlic, crushed
4 shallots, chopped	4 shallots, chopped	4 shallots, chopped
1 teaspoon chopped parsley	1 teaspoon chopped parsley	1 teaspoon chopped parsley
1 teaspoon dried thyme or mixed sweet herbs	1 teaspoon dried thyme or mixed sweet herbs	1 teaspoon dried thyme or mixed sweet herbs
salt and pepper to taste	salt and pepper to taste	salt and pepper to taste
¼ pint double cream	1½ dl. double cream	⅔ cup whipping cream
2 tablespoons brandy	2 tablespoons brandy	3 tablespoons brandy

Melt the butter in a small pan, add the chicken livers and cook them gently for 4 minutes. Remove livers and add the oil, garlic and chopped shallots to the pan. Fry gently until shallots are softened but not browned then stir in the parsley, herbs and seasoning. Liquidise the livers and the contents of the pan in an electric blender, or pass through a sieve. Allow to become quite cold then stir in the lightly whipped cream and brandy thoroughly to make a smooth mixture.

Crème Dubarry
(France)

Makes 8 servings.
To freeze Pack into polythene containers leaving headspace. Seal and label.
To serve Turn contents into a saucepan, reheat to boiling point adding sufficient additional milk or chicken stock to thin down to required consistency. Serve garnished with fried croûtons.
Storage time 4–6 months.

IMPERIAL	METRIC	AMERICAN
3 oz. butter	75 g. butter	6 tablespoons butter
8 oz. onions, chopped	225 g. onions, chopped	2 cups chopped onion
8 oz. potatoes, diced	225 g. potatoes, diced	½ lb. potatoes, diced
1 tablespoon flour	1 tablespoon flour	1 tablespoon all-purpose flour
2 lb. cauliflower florets (fresh or frozen)	1 kg. cauliflower florets (fresh or frozen)	2 lb. cauliflower florets (fresh or frozen)
2 pints milk	generous 1 litre milk	5 cups milk
2 bay leaves	2 bay leaves	2 bay leaves
salt and black pepper to taste	salt and black pepper to taste	salt and black pepper to taste

Melt the butter in a large saucepan, add the onion and potato and cook for 1 minute. Stir in the flour. Meanwhile, defrost frozen cauliflower or blanch fresh cauliflower for 2 minutes in boiling water. Add drained florets to pan with milk, bay leaves and salt and pepper to taste. Cover and simmer until vegetables are tender, about 15 minutes. Remove bay leaves, liquidise in an electric blender or sieve. Adjust seasoning and cool.

Pierre's vichyssoise
(France)

Makes 8 servings.
To freeze Pack in polythene containers leaving headspace. Seal and label.
To serve Defrost, add ¼ pint (1½ dl.) single cream and ¼ pint (1½ dl.) soured cream and stir thoroughly. Serve chilled and sprinkle each plate of soup with chopped chives.
Storage time 4–6 months.

IMPERIAL	METRIC	AMERICAN
4 small leeks	4 small leeks	4 small leeks
2 oz. butter	50 g. butter	¼ cup butter
4 small onions, chopped	4 small onions, chopped	4 small onions, chopped
2 pints chicken stock	generous 1 litre chicken stock	5 cups chicken stock
8 small potatoes, diced	8 small potatoes, diced	8 small potatoes, diced
1 tablespoon chopped parsley	1 tablespoon chopped parsley	1 tablespoon chopped parsley
salt and pepper to taste	salt and pepper to taste	salt and pepper to taste

Trim the leeks and chop finely. Melt the butter in a large saucepan and cook the leeks and onions until softened but not browned. Add the stock, potatoes, parsley and seasoning and bring to the boil. Cover and simmer gently for about 30 minutes. Sieve or liquidise in an electric blender and then sieve. Cool.

Fried scampi alla Gennaro

(Italy)

Makes 6 servings.

IMPERIAL	METRIC	AMERICAN
24 large frozen scampi, defrosted	24 large frozen scampi, defrosted	24 large frozen jumbo shrimp, defrosted
3 tablespoons seasoned flour	3 tablespoons seasoned flour	4 tablespoons seasoned flour
olive oil for shallow frying	olive oil for shallow frying	olive oil for shallow frying
1 oz. butter	25 g. butter	2 tablespoons butter
1 lb. courgettes, sliced	450 g. courgettes, sliced	1 lb. small zucchini, sliced
1 tablespoon tomato purée	1 tablespoon tomato purée	1 tablespoon tomato paste
2 lemons	2 lemons	2 lemons
salt and freshly ground black pepper to taste	salt and freshly ground black pepper to taste	salt and freshly ground black pepper to taste

Dry the scampi on absorbent kitchen paper and coat lightly in the seasoned flour. Heat the oil and use to fry the scampi over a high heat, turning frequently, for about 5 minutes, until golden brown. Drain on absorbent paper. Meanwhile, melt the butter in a small saucepan, add the courgettes, tomato purée, juice of 1 lemon and seasoning to taste. Cover and cook over low heat, stirring occasionally, until the courgettes are tender, about 10 minutes. Arrange scampi in the centre of a hot serving dish and surround with the courgettes. Quarter the remaining lemon and use to garnish the dish.

Note French fried potatoes or croquettes are the perfect accompaniment to this dish.

Scampi capriccio di Sophia

(Italy)

Makes 6 servings.

IMPERIAL	METRIC	AMERICAN
24 large frozen scampi, defrosted	24 large frozen scampi, defrosted	24 large frozen jumbo shrimp, defrosted
salt to taste	salt to taste	salt to taste
1½ oz. butter	40 g. butter	3 tablespoons butter
¼ pint dry sherry	1½ dl. dry sherry	⅔ cup dry sherry
4 tablespoons double cream	4 tablespoons double cream	5 tablespoons whipping cream
4 tablespoons grated Parmesan cheese	4 tablespoons grated Parmesan cheese	5 tablespoons grated Parmesan cheese
½ pint béchamel sauce (see page 100), defrosted	3 dl. béchamel sauce (see page 100), defrosted	1¼ cups béchamel sauce (see page 100), defrosted
parsley sprigs to garnish	parsley sprigs to garnish	parsley sprigs to garnish

Halve the scampi and season lightly with salt. Melt the butter in a frying pan and sauté the scampi until just golden. Drain and remove scampi and keep warm. Add the sherry to the juices in the pan and boil to reduce by two-thirds. Mix the cream and half the cheese into the béchamel sauce, add to the pan, stir and reheat without bringing to the boil. Spoon a little of the sauce into six scallop shells or individual ovenproof dishes. Divide the scampi between them, pour over the remaining sauce and sprinkle with the rest of the cheese. Place on a baking tray in a moderately hot oven (400°F, 200°C, Gas Mark 6) for 5–10 minutes, until lightly browned. Serve garnished with parsley sprigs.

Truites aux amandes

(France)

Makes 4 servings.

IMPERIAL	METRIC	AMERICAN
1 tablespoon flour	1 tablespoon flour	1 tablespoon all-purpose flour
salt and pepper to taste	salt and pepper to taste	salt and pepper to taste
4 frozen trout	4 frozen trout	4 frozen trout
3 oz. butter	75 g. butter	6 tablespoons butter
3 oz. flaked almonds	75 g. flaked almonds	½ cup almonds
1 tablespoon corn or olive oil	1 tablespoon corn or olive oil	1 tablespoon corn or olive oil
2 tablespoons lemon juice	2 tablespoons lemon juice	3 tablespoons lemon juice
parsley sprigs to garnish	parsley sprigs to garnish	parsley sprigs to garnish

Season the flour with salt and pepper and use to coat the trout. Melt one-third of the butter and use to fry the almonds until golden. Drain and keep hot. Add remaining butter and the oil to the pan and in it fry the trout gently for 5 minutes on each side until browned; drain on absorbent paper. Scatter the almonds over the fish, sprinkle with lemon juice and garnish with sprigs of parsley.

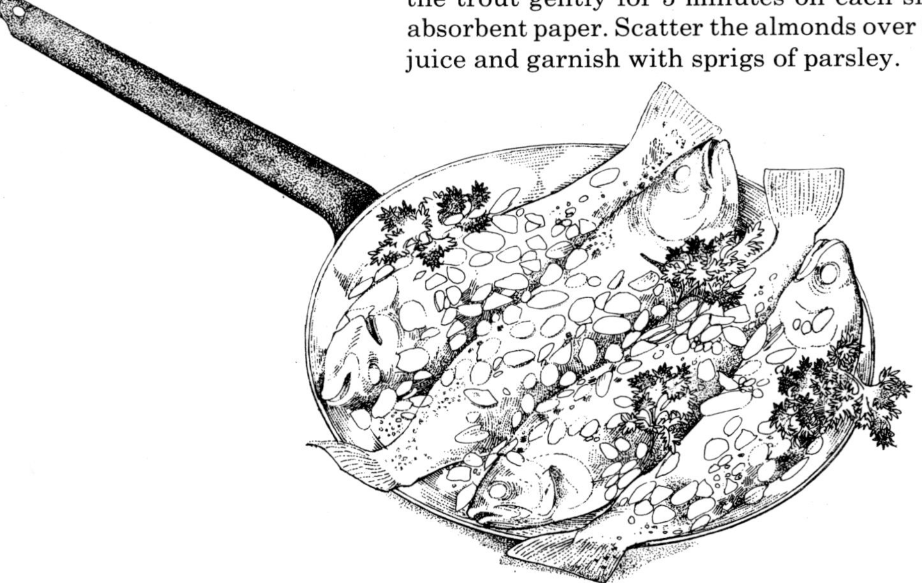

Haddock meunière

(France)

(Illustrated on page 110)

Makes 4 servings.

IMPERIAL	METRIC	AMERICAN
1 13-oz. packet frozen haddock fillets	1 375-g. packet frozen haddock fillets	1 13-oz. package frozen haddock fillets
2 tablespoons seasoned flour	2 tablespoons seasoned flour	3 tablespoons seasoned flour
about 4 oz. butter	about 100 g. butter	about ½ cup butter
juice of 1 large lemon	juice of 1 large lemon	juice of 1 large lemon
few lemon slices and chopped parsley to garnish	few lemon slices and chopped parsley to garnish	few lemon slices and chopped parsley to garnish

Defrost the fish fillets only sufficiently to separate them and coat in the seasoned flour. Heat half the butter in a heavy frying pan put in the fillets, skin side uppermost, and fry briskly for a few minutes until golden brown. Turn and fry on the other side. This ensures that the surfaces of the fish are a golden brown colour before the butter starts to darken. (If liked a little of the butter can be substituted by olive or corn oil which prevents the tendency of the butter to burn.) When the fish is cooked, drain well and remove to a hot serving dish. Add the rest of the butter to the pan with the lemon juice. Allow to reheat thoroughly, swirling round the pan, and pour over the fish. Garnish with lemon slices and chopped parsley.

Note There are a number of possible additions to the lemon and butter sauce; here are a few variations.
1. Add 1 tablespoon drained capers to the butter and lemon juice.
2. Add 1 tablespoon finely chopped parsley to the butter and lemon juice, or sprinkle the parsley over the dish to serve.
3. Brown 1 tablespoon of flaked almonds in the butter before adding the lemon juice. Add this when the almonds are fried a delicate golden brown and pour over the fish.

Roulade aux épinards

(France)

(Illustrated on page 110)

Makes 4 servings.

IMPERIAL	METRIC	AMERICAN
8 oz. frozen chopped spinach	225 g. frozen chopped spinach	½ lb. frozen chopped spinach
½ oz. butter	15 g. butter	1 tablespoon butter
salt and pepper to taste	salt and pepper to taste	salt and pepper to taste
1 tablespoon grated Parmesan cheese	1 tablespoon grated Parmesan cheese	1 tablespoon grated Parmesan cheese
4 eggs, separated	4 eggs, separated	4 eggs, separated
filling	*filling*	*filling*
½ oz. butter	15 g. butter	1 tablespoon butter
6 oz. button mushrooms, sliced	175 g. button mushrooms, sliced	1½ cups sliced button mushrooms
1 tablespoon flour	1 tablespoon flour	1 tablespoon all-purpose flour
¼ pint milk	1½ dl. milk	⅔ cup milk
pinch grated nutmeg	pinch grated nutmeg	pinch grated nutmeg
2 tablespoons double cream	2 tablespoons double cream	3 tablespoons whipping cream

Line a 12-inch (30-cm.) by 8-inch (20-cm.) Swiss roll tin with greaseproof paper, leaving a 1½-inch (3·5-cm.) overlap all round. Mitre the corners and secure each one with a paperclip to give the roulade a square edge. Brush inside the paper case with oil.

Cook the spinach and drain well. Put into a bowl with the butter, seasoning and cheese and mix well. Gradually beat in the egg yolks. Beat the egg whites until stiff, fold into the spinach mixture and turn into the prepared paper case. Bake in the top of a moderately hot oven (400°F, 200°C, Gas Mark 6) for about 10 minutes, until firm. Meanwhile, melt the butter and use to sauté the mushrooms lightly. Add the flour and stir until smooth. Gradually add the milk, nutmeg and salt and pepper to taste. Bring to the boil, stirring constantly until thick. Remove from the heat and stir in the cream. Turn the cooked roulade onto a sheet of greaseproof paper, trim if necessary and spread with some of the mushroom filling. Roll up and serve in a hot dish surrounded by the remaining mushroom mixture.

Quiche belle bretonne

(France)

Makes 2 flans, 8 servings.
To freeze Wrap in polythene bags. Seal and label.
To serve Unwrap, place on a baking tray and reheat in a moderate oven (350°F, 180°C, Gas Mark 4) for 25–30 minutes.
Storage time 4–6 months.

IMPERIAL	METRIC	AMERICAN
8 oz. frozen shortcrust pastry, defrosted	225 g. frozen shortcrust pastry, defrosted	$\frac{1}{2}$ lb. frozen basic pie dough, defrosted
2 tablespoons oil	2 tablespoons oil	3 tablespoons oil
2 small onions, chopped	2 small onions, chopped	2 small onions, chopped
4 oz. ham, chopped	100 g. ham, chopped	$\frac{1}{2}$ cup chopped cooked ham
6 oz. button mushrooms, sliced	175 g. button mushrooms, sliced	$1\frac{1}{2}$ cups sliced button mushrooms
3 eggs	3 eggs	3 eggs
$\frac{1}{4}$ pint single cream	$1\frac{1}{2}$ dl. single cream	$\frac{2}{3}$ cup coffee cream
$\frac{1}{4}$ pint milk	$1\frac{1}{2}$ dl. milk	$\frac{2}{3}$ cup milk
3 oz. mixed Gruyère and Parmesan cheese, grated	75 g. mixed Gruyère and Parmesan cheese, grated	$\frac{3}{4}$ cup mixed grated Gruyère and Parmesan cheese
salt and pepper to taste	salt and pepper to taste	salt and pepper to taste
pinch dried mixed herbs	pinch dried mixed herbs	pinch dried mixed herbs

Roll out the pastry and use to line two 7-inch (18-cm.) flan rings. Heat the oil in a frying pan and fry the onion in it gently until softened but not browned. Divide onions between the two flan cases and sprinkle over the ham and mushrooms. Beat the eggs, add the cream, milk, cheese and seasoning and pour into the flan cases. Sprinkle each with herbs and bake in a hot oven (400°F, 200°C, Gas Mark 6) for 15 minutes. Reduce heat to 350°F, 180°C, Gas Mark 4 and cook for a further 30 minutes, or until lightly browned and firm. Cool.

Pollo sauté alla cacciatora

(Italy)

Makes 8 servings.
To freeze Pack into two shaped foil dishes. Seal and label.
To serve Defrost and place in a moderate oven (350°F, 180°C, Gas Mark 4) for 30 minutes.
Storage time 4–6 months.

IMPERIAL	METRIC	AMERICAN
2 tablespoons olive oil	2 tablespoons olive oil	3 tablespoons olive oil
4 oz. butter	100 g. butter	$\frac{1}{2}$ cup butter
8 small chicken leg joints	8 small chicken leg joints	8 small chicken leg pieces
1 large onion, sliced	1 large onion, sliced	1 large onion, sliced
$1\frac{1}{2}$ lb. mushrooms, sliced	700 g. mushrooms, sliced	$1\frac{1}{2}$ lb. mushrooms, sliced
4 ripe tomatoes	4 ripe tomatoes	4 ripe tomatoes
1 teaspoon dried tarragon	1 teaspoon dried tarragon	1 teaspoon dried tarragon
salt and pepper to taste	salt and pepper to taste	salt and pepper to taste
6 tablespoons dry white wine	6 tablespoons dry white wine	$\frac{1}{2}$ cup dry white wine
2 oz. flour	50 g. flour	$\frac{1}{2}$ cup all-purpose flour
1 pint hot chicken stock	generous $\frac{1}{2}$ litre hot chicken stock	$2\frac{1}{2}$ cups hot chicken stock
2 tablespoons brandy	2 tablespoons brandy	3 tablespoons brandy

Heat together oil and half the butter and use to cook the chicken joints on all sides for about 10 minutes. Remove chicken to a large flameproof casserole. Add the onion to the remaining fat and cook gently until softened but not browned then add to the chicken. Melt the remaining butter and use to fry the mushrooms until golden brown then add to the casserole with the peeled and sliced tomatoes, tarragon, seasoning and wine. Cook over high heat for a few minutes. Remove from the heat, add the flour and stir until smooth. Gradually add the hot stock, and cook, stirring constantly until thickened. Heat the brandy, ignite and pour over the chicken. Cover and simmer for 10 minutes. Cool.

Coq au vin rouge
(France)

Makes 8 servings.
To freeze Divide the chicken between two shaped foil containers and pour a little of the sauce over, including most of the onions and mushrooms. Freeze any remaining sauce in a polythene container. Seal and label.
To serve Defrost, remove cover and reheat in a hot oven (425°F, 220°C, Gas Mark 7) for 40 minutes. Reheat the remaining sauce and spoon over the dish before serving.
Storage time With garlic 1 month; without garlic 4 months.

IMPERIAL	METRIC	AMERICAN
8 chicken portions	8 chicken portions	8 chicken pieces
6 thick rashers unsmoked streaky bacon	6 thick rashers unsmoked streaky bacon	6 thick bacon slices
2 oz. butter	50 g. butter	$\frac{1}{4}$ cup butter
4 tablespoons oil	4 tablespoons oil	$\frac{1}{3}$ cup oil
2 cloves garlic, crushed (optional)	2 cloves garlic, crushed (optional)	2 cloves garlic, crushed (optional)
24 small button onions	24 small button onions	24 small onions
8 oz. mushrooms	225 g. mushrooms	2 cups mushrooms
4 tablespoons brandy	4 tablespoons brandy	$\frac{1}{3}$ cup brandy
$\frac{1}{2}$–$\frac{3}{4}$ bottle red wine (preferably Burgundy)	$\frac{1}{2}$–$\frac{3}{4}$ bottle red wine (preferably Burgundy)	$\frac{1}{2}$–$\frac{3}{4}$ bottle red wine (preferably Burgundy)
beurre manié	*beurre manié*	*beurre manié*
$\frac{1}{2}$ oz. butter	15 g. butter	1 tablespoon butter
$\frac{1}{2}$ oz. flour	15 g. flour	2 tablespoons all-purpose flour
stock	*stock*	*stock*
1 onion, chopped	1 onion, chopped	1 onion, chopped
1 carrot, chopped	1 carrot, chopped	1 carrot, chopped
bouquet garni	bouquet garni	bouquet garni
1 chicken stock cube	1 chicken stock cube	1 chicken bouillon cube
2 bay leaves	2 bay leaves	2 bay leaves
about 1 pint water	about generous $\frac{1}{2}$ litre water	about $2\frac{1}{2}$ cups water

Make a stock with leg shanks and backbone trimmed from the chicken portions, giblets if available, onion and carrot, bouquet garni, stock cube, bay leaves and water. Simmer covered for at least 30 minutes. De-rind the bacon and cut into cubes. Heat the butter and oil in a wide casserole and use to sauté the bacon lightly. Add the chicken pieces, skin side downwards first, and cook until golden, then turn and cook for a further 2 minutes, adding the garlic, small onions and mushrooms. Pour brandy into a heated ladle, ignite and pour over the chicken in the pan. Shake the pan gently until flames die down. Add the strained stock and the wine. (If using the larger quantity of wine, add to the stock first and boil fiercely to reduce by $\frac{1}{4}$ pint ($1\frac{1}{2}$ dl.) to give a stronger wine flavour.) Cover and simmer over low heat for 1 hour. Remove chicken and cool. To make the beurre manié, soften the butter and gradually work in the flour. Add this to the pan, a little at a time, and stir over gentle heat until sauce is thick and smooth. Cool.

Canard à l'orange au Cointreau

(France)

(Illustrated opposite)

Makes 4 servings.

To freeze Pack duck into four shaped foil containers and spoon over a little of the sauce. Cover the irregular surface with freezer film to prevent air damage and cover each container with its own lid, or a sheet of freezer foil, crimping the edges together. Pack remaining sauce in a small polythene container. Seal and label.

To serve Uncover, remove freezer film and defrost. Reheat in a hot oven (425°F, 220°C, Gas Mark 7) for 30 minutes. Heat remaining sauce in a small saucepan and spoon a little extra sauce over each portion. Garnish with fresh orange segments, if liked, sprigs of parsley or watercress and serve with roast or boiled potatoes and peas.

Storage time 4–6 months.

IMPERIAL	METRIC	AMERICAN
2 oranges	2 oranges	2 oranges
1 4–4½-lb. frozen duck, defrosted	1 2–2½-kg. frozen duck, defrosted	1 4–4½-lb. frozen duck, defrosted
salt and pepper to taste	salt and pepper to taste	salt and pepper to taste
1 tablespoon cornflour	1 tablespoon cornflour	1 tablespoon cornstarch
3 tablespoons Cointreau	3 tablespoons Cointreau	scant ¼ cup Cointreau

Pare the zest thinly from 1 orange and cut into fine strips. Squeeze the juice from both oranges and reserve. Place the squeezed orange halves inside the cavity of the bird. Arrange on a trivet in a roasting tin without pricking, sprinkle with salt and roast in a moderately hot oven (400°F, 200°C, Gas Mark 6) for 30 minutes. Make stock by simmering the giblets in sufficient water to cover well, with a little salt and pepper. Brush duck with fat which has accumulated in the pan, lower heat to 350°F, 180°C, Gas Mark 4 and continue roasting for another hour. Skim off surplus fat from tin as often as is necessary to prevent duck from resting in the fat, each time brushing the skin with it. If liked, turn the duck over on its breast for a further 10 minutes to brown the skin underneath. Remove onto a warm serving dish and carve into portions, dividing the duck in half down the breastbone with poultry scissors, removing the backbone by cutting each side of it completely. This releases a lot of duck juices which should be added to the stock. Now divide the legs into thighs and drumsticks to facilitate packing.

To make the sauce, stir the cornflour into the orange juice and add to the pan juices. Stir constantly over a low heat, adding sufficient stock from the giblets to make a thick pouring sauce. Strain into a clean saucepan, add the orange zest strips, bring to the boil, taste and adjust the seasoning. (If a darker, richer flavoured sauce is preferred, substitute gravy powder for half of the cornflour.) Add the Cointreau, bring to the boil again and cook for 1 minute, then cool.

Enten weissauer

(Germany)

Makes 4 servings.

IMPERIAL	METRIC	AMERICAN
4 frozen duck portions, defrosted	4 frozen duck portions, defrosted	4 frozen duck portions, defrosted
1½ pints water	scant 1 litre water	3¾ cups water
1 teaspoon salt	1 teaspoon salt	1 teaspoon salt
2 cloves	2 cloves	2 cloves
3 peppercorns	3 peppercorns	3 peppercorns
1 carrot, quartered	1 carrot, quartered	1 carrot, quartered
1 bay leaf	1 bay leaf	1 bay leaf
1½ tablespoons wine vinegar	1½ tablespoons wine vinegar	2 tablespoons wine vinegar
¼ pint dry cider	1½ dl. dry cider	⅔ cup cider
½ oz. gelatine	15 g. gelatine	2 envelopes gelatin

Put the duck portions in a flameproof casserole. Bring water and all other ingredients except gelatine to the boil, pour over duck, cover and simmer for 1½ hours, or until tender. Remove duck portions, drain well and allow to get cold. Place in a shallow serving dish. Strain and cool the liquid and allow fat to solidify on top. Scoop off, making sure to remove all fat. Reheat 1 pint (generous ½ litre) of the liquid again (reduce if necessary) and mix the gelatine with a little of it when warm. Dissolve this in the remaining liquid and strain over the duck. Allow to get cold and set. Serve duck portions, garnished with gherkin fans, in the set jelly. Serve with a potato salad.

Canard à l'orange au Cointreau (see above).

40°
COINTREAU
LIQUEUR EXTRA DRY
ANGERS
FRANCE

1 Open freezing strawberries and packing in Tupperware square rounds or shaped foil containers, with foil dividers, for use in sweets.

2 Half filling sundae glasses with spoonfuls of vanilla ice cream, ready for covering with halved defrosted strawberries.

3 Sundaes topped with lemon and lime ice cream being piped with whipped cream ready for decorating with whole strawberries and chopped pistachio nuts.

1

2

3

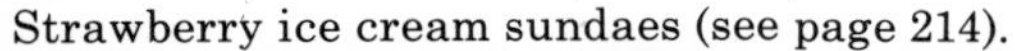

Strawberry ice cream sundaes (see page 214).

Lapin à la normande

(France)

Makes 4 servings.
To freeze Pack into a shaped foil dish. Seal and label.
To serve Place, still frozen, in a moderately hot oven (375°F, 190°C, Gas Mark 5) for 30 minutes.
Storage time 4–6 months.

IMPERIAL	METRIC	AMERICAN
1 rabbit	1 rabbit	1 rabbit
½ oz. flour	15 g. flour	2 tablespoons all-purpose flour
1 oz. butter	15 g. butter	2 tablespoons butter
1 cooking apple	1 cooking apple	1 baking apple
1 large carrot, chopped	1 large carrot, chopped	1 large carrot, chopped
1 large onion, chopped	1 large onion, chopped	1 large onion, chopped
¾ pint cider	scant ½ litre cider	scant 2 cups cider
bouquet garni	bouquet garni	bouquet garni
salt and pepper to taste	salt and pepper to taste	salt and pepper to taste

Cut the rabbit into serving pieces and coat in the flour. Melt the butter and use to brown the rabbit on all sides. Peel, core and chop the apple and add to the pan with the carrot and onion. Cook gently for a further 5 minutes. Add the cider, bouquet garni and seasoning and bring to the boil. Cover and simmer for about 1½ hours, until rabbit is tender. Remove bouquet garni and cool.

Faisan à la crème

(France)

Makes 4 servings.

IMPERIAL	METRIC	AMERICAN
3 oz. butter	75 g. butter	6 tablespoons butter
1 large frozen pheasant, defrosted	1 large frozen pheasant, defrosted	1 large frozen pheasant, defrosted
2 thick rashers streaky bacon	2 thick rashers streaky bacon	2 thick bacon slices
½ pint chicken stock	3 dl. chicken stock	1¼ cups chicken stock
1 tablespoon chopped parsley	1 tablespoon chopped parsley	1 tablespoon chopped parsley
3 tablespoons port	3 tablespoons port	scant ¼ cup port
1 large head celery	1 large head celery	1 large bunch celery
1 egg yolk	1 egg yolk	1 egg yolk
¼ pint double cream	1½ dl. double cream	⅔ cup whipping cream
salt and pepper to taste	salt and pepper to taste	salt and pepper to taste

Melt the butter and use to brown the pheasant all over. Remove pheasant and place in a deep ovenproof casserole. De-rind and dice the bacon and add to the casserole with the stock, parsley and port. Cover and place in a moderate oven (350°F, 180°C, Gas Mark 4) for 20 minutes. Clean and dice the celery and add to the casserole. Cook for a further hour. Place the pheasant on a warm serving dish, remove the celery with a slotted draining spoon and place in a hot vegetable dish. Beat together the egg yolk and cream and gradually stir into the hot stock. Bring almost to boiling point, remove from heat and season with salt and pepper to taste. Pour into a sauceboat.

Vitello milanese

(Italy)

Makes 8 servings.
To freeze Pack in large shaped foil containers, with a layer of meat surrounding the noodles (see page 56). Seal and label.
To serve Place covered in a moderately hot oven (375°F, 190°C, Gas Mark 5) for 30 minutes.
Storage time 4–6 months.

IMPERIAL	METRIC	AMERICAN
3 lb. shoulder or breast of veal, boned	1½ kg. shoulder or breast of veal, boned	3 lb. boneless veal shoulder or breast
2 oz. flour	50 g. flour	½ cup all-purpose flour
5 tablespoons olive oil	5 tablespoons olive oil	6 tablespoons olive oil
1 large onion, chopped	1 large onion, chopped	1 large onion, chopped
4 sticks celery, chopped	4 sticks celery, chopped	4 stalks celery, chopped
4 carrots, sliced	4 carrots, sliced	4 carrots, sliced
2 cloves garlic (optional)	2 cloves garlic (optional)	2 cloves garlic (optional)
½ pint chicken stock	3 dl. chicken stock	1¼ cups chicken stock
½ bottle white wine	½ bottle white wine	½ bottle white wine
1 teaspoon dried marjoram	1 teaspoon dried marjoram	1 teaspoon dried marjoram
1 15-oz. can Italian tomatoes	1 425-g. can Italian tomatoes	1 15-oz. can Italian tomatoes
2 small strips lemon rind	2 small strips lemon rind	2 small strips lemon rind
4 bay leaves	4 bay leaves	4 bay leaves
salt and pepper to taste	salt and pepper to taste	salt and pepper to taste
1 lb. ribbon noodles	450 g. ribbon noodles	1 lb. ribbon noodles

Cut the meat into 1-inch (2·5-cm.) cubes. Toss the cubes in the flour. Sauté in the hot oil (reserving 1 tablespoon) until brown, drain and place in a flameproof casserole. Add the prepared vegetables and crushed garlic to the pan and sauté lightly. Remove to casserole, add remaining flour to the oil in the frying pan, stir over gentle heat until beginning to brown. Gradually add the stock and wine, bring to the boil and cook until thickened stirring constantly. Add the herbs, tomatoes, lemon rind and bay leaves. Season well and pour over the meat and vegetables in the casserole. Cover and simmer for about 2 hours, or until the veal is tender. Cool. Meanwhile cook the noodles until just tender, drain well and toss with 1 tablespoon oil. Cool.

Flemish beef
(Belgium)

Makes 8 servings.
To freeze Pack sliced beef in shaped foil containers with a little of the sauce. Freeze remaining sauce in a small polythene container.
To serve Defrost, cover and reheat in a moderate oven (350°F, 180°C, Gas Mark 4) for 30 minutes. Reheat remaining sauce and spoon over the beef before serving.
Storage time 4–6 months.

IMPERIAL	METRIC	AMERICAN
8 oz. dried prunes	225 g. dried prunes	$\frac{1}{2}$ lb. dried prunes
2 tablespoons olive or corn oil	2 tablespoons olive or corn oil	3 tablespoons olive or corn oil
3-lb. piece topside of beef	$1\frac{1}{2}$-kg. piece topside of beef	3-lb. piece beef round
8 oz. button onions	225 g. button onions	$\frac{1}{2}$ lb. small onions
8 oz. carrots, sliced	225 g. carrots, sliced	$\frac{1}{2}$ lb. carrots, sliced
1 small bottle light ale	1 small bottle light ale	1 small bottle light beer
$\frac{1}{4}$ pint beef stock	$1\frac{1}{2}$ dl. beef stock	$\frac{2}{3}$ cup beef stock
salt and pepper to taste	salt and pepper to taste	salt and pepper to taste

Soak the prunes in cold water overnight. Heat the oil in a flameproof casserole and brown the meat on all sides. Remove and keep hot. Brown the onions and carrots in the fat, return the meat to the casserole and add the drained prunes, ale and sufficient stock to cover. Season, cover tightly and cook slowly on top of the cooker or in a moderate oven (325°F, 170°C, Gas Mark 3) for $2\frac{1}{2}$–3 hours, or until the meat is tender. Cool. Slice beef fairly thickly when cold.

Tournedos aux coeurs d'artichauts
(France)

(Illustrated on page 166)

Makes 4 servings.

IMPERIAL	METRIC	AMERICAN
4 thick fillet steaks	4 thick fillet steaks	4 thick tenderloin steaks
8 oz. frozen artichoke hearts	225 g. frozen artichoke hearts	$\frac{1}{2}$ lb. frozen artichoke hearts
2 tablespoons oil	2 tablespoons oil	3 tablespoons oil
4 slices white bread	4 slices white bread	4 slices white bread
2 oz. butter	50 g. butter	$\frac{1}{4}$ cup butter
2 oz. mushrooms	50 g. mushrooms	$\frac{1}{2}$ cup mushrooms
2 oz. liver pâté	50 g. liver pâté	2 oz. liver pâté
parsley sprigs and bunch of watercress to garnish	parsley sprigs and bunch of watercress to garnish	parsley sprigs and bunch of watercress to garnish

Defrost the steaks and the artichoke hearts completely. If not previously prepared, tie the steaks with string into a neat shape and remove the string just before serving. Brush the steaks on both sides with very little oil. Fry or grill for a few minutes on either side, according to how rare the steaks are preferred. Stamp out four rounds with a biscuit cutter from the slices of bread to fit the steaks. Fry golden brown on both sides in butter with a little oil added. Toss the sliced mushrooms in the remaining butter and oil, reserve the best slices for the garnish, chop the others and blend with the pâté. Drain the bread croûtes well, spread with the pâté mixture, place the steaks on top, arrange on a serving dish and garnish with the reserved mushroom slices. Reheat the artichoke hearts in their own moisture with 1 tablespoon of melted butter and arrange between the steaks. Place a parsley sprig on each steak and fill the centre of the dish with a bunch of watercress.

Note Croûtes and small croûtons can be fried and frozen in small quantities. Defrost and reheat spread out on a baking tray in a moderate oven for 20 minutes. Storage time 3 months.

Boeuf en croûte à la reine Marie

(France)

(Illustrated on page 166)

Makes 6 servings.
To freeze Open freeze on the baking tray until solid. Label.
To serve Thaw completely in the refrigerator for at least 24 hours, having removed the outer wrapping and opened the top seal of the foil slightly. To cook, unwrap and place on a dampened baking tray. Bake in a hot oven (450°F, 230°C, Gas Mark 8) for 12 minutes. Remove and brush with beaten egg. Reduce heat to 425°F, 220°C, Gas Mark 7 and continue baking for 25–30 minutes. If necessary protect with a sheet of foil to prevent the decorations from overbrowning. Serve garnished with a sprig of watercress.
Storage time 4–6 months.

Note Overwrap the foil parcel with a large polythene bag or place in a rigid-based polythene container as the pack is easily damaged while in the frozen state.

IMPERIAL	METRIC	AMERICAN
2-lb. piece fillet of beef	1-kg. piece fillet of beef	2-lb. piece beef tenderloin
2 oz. butter	50 g. butter	¼ cup butter
8 oz. button mushrooms, sliced	225 g. button mushrooms, sliced	½ lb. button mushrooms, sliced
salt and pepper to taste	salt and pepper to taste	salt and pepper to taste
12 oz. frozen puff pastry	350 g. frozen puff pastry	¾ lb. frozen puff paste

Brush the fillet with a little melted butter, place on a baking tray and seal in a moderately hot oven (400°F, 200°C, Gas Mark 6) for 15 minutes. Remove and cool. Meanwhile, heat the remaining butter and use to sauté the mushrooms until soft. Cool, season and chop very finely. Defrost the pastry until it is possible to roll it out to make a rectangle 14 inches (36 cm.) long and 12 inches (30 cm.) wide. It must be large enough to encase the steak completely. Spread half the mushroom mixture in the centre of the pastry, place the fillet on top, coat with the remaining mixture, bring up the pastry edges to form a parcel, damping them to seal together. Place a large piece of freezer foil on a clean baking tray and put the pastry parcel in the middle with the join underneath. Decorate with leaves made from the pastry trimmings and chill. Fold in the foil and seal to make a parcel.

Königsberger klopse

(Germany)

Makes 4 servings.
To freeze Pack meatballs and sauce together in a polythene container. Seal and label.
To serve Turn contents into a saucepan, reheat gently and simmer for 2 minutes. Stir frequently to prevent burning.
Storage time 4–6 months.

IMPERIAL	METRIC	AMERICAN
1 small can anchovy fillets	1 small can anchovy fillets	1 small can anchovy fillets
1 bread roll	1 bread roll	1 bread roll
1 lb. minced beef	450 g. minced beef	1 lb. ground beef
1 small onion, grated	1 small onion, grated	1 small onion, grated
2 tablespoons lemon juice	2 tablespoons lemon juice	3 tablespoons lemon juice
1 egg, beaten	1 egg, beaten	1 egg, beaten
salt and pepper to taste	salt and pepper to taste	salt and pepper to taste
pinch sugar	pinch sugar	pinch sugar
pinch nutmeg	pinch nutmeg	pinch nutmeg
1 oz. butter	25 g. butter	2 tablespoons butter
1 oz. flour	25 g. flour	¼ cup all-purpose flour
pinch ground cloves	pinch ground cloves	pinch ground cloves
1 tablespoon vinegar	1 tablespoon vinegar	1 tablespoon vinegar
1 tablespoon drained capers	1 tablespoon drained capers	1 tablespoon drained capers

Finely chop the anchovies, reserving the oil. Soak the bread roll in hot water for 10 minutes, squeeze dry and blend with the beef, onion, anchovies, oil, lemon juice and egg; season with salt, pepper, sugar and nutmeg. Form into walnut-sized balls and drop into lightly salted boiling water. Simmer until they rise to the surface, about 20 minutes. Remove with a slotted draining spoon and cool. Make a roux with the butter and flour, stir in ½ pint (3 dl.) of the stock, a pinch of ground cloves, the vinegar and capers. Cook, stirring, over gentle heat until thick and smooth, adding a little more stock if necessary to make a good pouring consistency. Cool.

Goulash with caraway dumplings

(Hungary)

Makes 6 servings.

To freeze Pack goulash into a polythene or shaped foil container. Pack dumplings separately in a polythene bag. Seal and label.

To serve Defrost goulash. Turn into an ovenproof casserole and place in a moderate oven (325°F, 170°C, Gas Mark 3) for 30 minutes. Defrost dumplings, add to goulash and raise oven heat to 375°F, 190°C, Gas Mark 5 for a further 30 minutes.

IMPERIAL	METRIC	AMERICAN
1½ lb. stewing steak or pork, diced	700 g. stewing steak or pork, diced	1½ lb. beef or pork stew meat, diced
3 oz. seasoned flour	75 g. seasoned flour	¾ cup seasoned flour
1 large red or green pepper	1 large red or green pepper	1 large red or green sweet pepper
1½ oz. pork or beef dripping	40 g. pork or beef dripping	3 tablespoons pork or beef drippings
3 large onions, chopped	3 large onions, chopped	3 large onions, chopped
6 tablespoons tomato purée	6 tablespoons tomato purée	7 tablespoons tomato paste
1½ tablespoons paprika pepper	1½ tablespoons paprika pepper	2 tablespoons paprika pepper
¾ pint beef stock	scant ½ litre beef stock	scant 2 cups beef stock
3 large tomatoes	3 large tomatoes	3 large tomatoes
1 teaspoon dried mixed herbs	1 teaspoon dried mixed herbs	1 teaspoon dried mixed herbs
dumplings	*dumplings*	*dumplings*
6 oz. self-raising flour	175 g. self-raising flour	1½ cups all-purpose flour sifted with 1½ teaspoons baking powder
1½ teaspoons salt	1½ teaspoons salt	1½ teaspoons salt
3 oz. shredded suet	75 g. shredded suet	scant ½ cup finely chopped suet
2 tablespoons caraway seeds	2 tablespoons caraway seeds	3 tablespoons caraway seeds
cold water to mix	cold water to mix	cold water to mix

Coat the meat in the seasoned flour. Deseed and chop the pepper. Melt the dripping and use to sauté the onion and pepper lightly until just beginning to brown. Add the meat and fry to seal on all sides. Stir in the tomato purée, paprika pepper and remaining seasoned flour. Add the stock, sliced tomatoes and herbs. Transfer mixture to a large ovenproof casserole, cover and cook in a moderate oven (325°F, 170°C, Gas Mark 3) for 2 hours. Cool. To make the dumplings, mix all the dry ingredients together and use just enough cold water to make a soft dough. With floured hands, shape into about 18 small dumplings.

Note It is not advisable to freeze cooked goulash and dumplings together. When dumplings are packed separately they can, of course, be used with other dishes.

Storage time 4–6 months.

Pork stroganoff

(Russia)

Makes 6 servings.
To freeze Pack into a polythene container. Seal and label.
To serve Defrost, turn contents into a saucepan. Add ¼ pint (1½ dl.) chicken stock and reheat gently to boiling point. Add ¼ pint (1½ dl.) soured cream, stir well and remove from heat.
Storage time 4–6 months.

IMPERIAL	METRIC	AMERICAN
2 lb. pork fillet	1 kg. pork fillet	2 lb. pork tenderloin
salt and pepper to taste	salt and pepper to taste	salt and pepper to taste
1 oz. flour	25 g. flour	¼ cup all-purpose flour
2 oz. butter	50 g. butter	¼ cup butter
2 tablespoons oil	2 tablespoons oil	3 tablespoons oil
2 large onions, chopped	2 large onions, chopped	2 large onions, chopped
8 oz. mushrooms, sliced	225 g. mushrooms, sliced	½ lb. mushrooms, sliced
2 tablespoons tomato purée	2 tablespoons tomato purée	3 tablespoons tomato paste
¾ pint chicken stock	scant ½ litre chicken stock	scant 2 cups chicken stock

Cut the meat into strips, sprinkle with salt and pepper and toss in flour. Heat half the butter and the oil in a heavy frying pan and use to fry the onions gently until transparent. Remove and keep warm. Add the meat, and fry until sealed on all sides, and golden brown. Remove with a slotted draining spoon and keep warm with the onion. Add rest of butter and the mushrooms to the frying pan and cook gently for a further 5 minutes. Add remaining flour and stir until smooth. Blend the tomato purée with the hot stock, add to the pan and stir thoroughly. Return the onions and meat to the pan, bring to the boil, cover and simmer gently for 2–3 minutes. Cool.

Pork fillet Cracova

(Poland)

Makes 4 servings.
To freeze Pack sliced fillet in a shaped foil container and pour the sauce over. Seal and label.
To serve Defrost and reheat in a moderately hot oven (375°F, 190°C, Gas Mark 5) for 30 minutes.
Storage time 4–6 months.

IMPERIAL	METRIC	AMERICAN
2 oz. butter	50 g. butter	¼ cup butter
1½ lb. pork fillet	700 g. pork fillet	1½ lb. pork tenderloin
1 large onion, chopped	1 large onion, chopped	1 large onion, chopped
4 oz. mushrooms, sliced	100 g. mushrooms, sliced	¼ lb. mushrooms, sliced
salt and pepper to taste	salt and pepper to taste	salt and pepper to taste
2 teaspoons cornflour	2 teaspoons cornflour	2 teaspoons cornstarch
¼ pint soured cream	1½ dl. soured cream	⅔ cup sour cream

Melt the butter and use to fry the fillet quickly on all sides to seal. Add the onion, mushrooms and seasoning, bring to the boil, cover and cook gently for 20 minutes. Remove fillet, cool and slice. Moisten the cornflour with a little cold water and add sufficient water to the sauce to make up to approximately ¼ pint (1½ dl.). Stir in the cornflour, bring to the boil, stirring constantly, until thickened. Add the soured cream and reheat but do not allow to boil. Cool.

Agneau aux cerises

(Switzerland)

Makes 8 servings.
To freeze Pack meat and sauce together in a large shaped foil container. Seal and label.
To serve Defrost and place in a moderately hot oven (375°F, 190°C, Gas Mark 5) for 30–40 minutes until heated through. Transfer to a heated dish and immediately before serving pour 3 tablespoons heated brandy over the dish and ignite.
Storage time 4–6 months.

IMPERIAL	METRIC	AMERICAN
16 lamb cutlets	16 lamb cutlets	16 lamb loin chops
salt and pepper to taste	salt and pepper to taste	salt and pepper to taste
2 oz. butter	50 g. butter	¼ cup butter
3 tablespoons oil	3 tablespoons oil	4 tablespoons oil
sauce	*sauce*	*sauce*
1 15-oz. can Morello cherries	1 425-g. can Morello cherries	1 15-oz. can bing cherries
finely grated zest and juice of ½ lemon	finely grated zest and juice of ½ lemon	finely grated zest and juice of ½ lemon

Season the cutlets on each side with salt and pepper. Heat the butter and oil together, add the cutlets and brown well on both sides, then cook gently for 8–10 minutes, until done. Remove meat from pan. Remove excess fat from pan. Drain the cherries and add the syrup to the pan, together with the lemon zest and juice. Boil for 5 minutes until the sauce thickens. Cool. Stone cherries and stir into cold sauce.

Noisettes bordelaise

(France)

(Illustrated on page 109)

Makes 4 servings.

IMPERIAL	METRIC	AMERICAN
3 oz. butter	75 g. butter	6 tablespoons butter
1 tablespoon flour	1 tablespoon flour	1 tablespoon all-purpose flour
5 tablespoons red wine	5 tablespoons red wine	6 tablespoons red wine
5 tablespoons beef stock	5 tablespoons beef stock	6 tablespoons beef stock
1 clove garlic, crushed	1 clove garlic, crushed	1 clove garlic, crushed
2 sprigs rosemary	2 sprigs rosemary	2 sprigs rosemary
1 bay leaf	1 bay leaf	1 bay leaf
salt and pepper to taste	salt and pepper to taste	salt and pepper to taste
8 frozen noisettes of lamb, defrosted	8 frozen noisettes of lamb, defrosted	8 frozen English lamb chops, defrosted
1 lb. hot duchesse potato (see page 179)	450 g. hot duchesse potato (see page 179)	1 lb. hot duchesse potato (see page 179)

Melt half the butter, add the flour and stir over gentle heat until pale golden brown. Stir in the wine, stock, crushed garlic, 1 sprig rosemary, bay leaf and seasoning. Bring to the boil, stirring constantly, cover and simmer gently for 12 minutes. Melt the remaining butter and use to brush the noisettes on both sides. Place under a hot grill and cook for about 5 minutes on each side. Pipe a potato border around a hot serving dish, place the noisettes, overlapping, in two rows down the centre, strain the sauce over them to glaze and garnish with the remaining rosemary sprig.

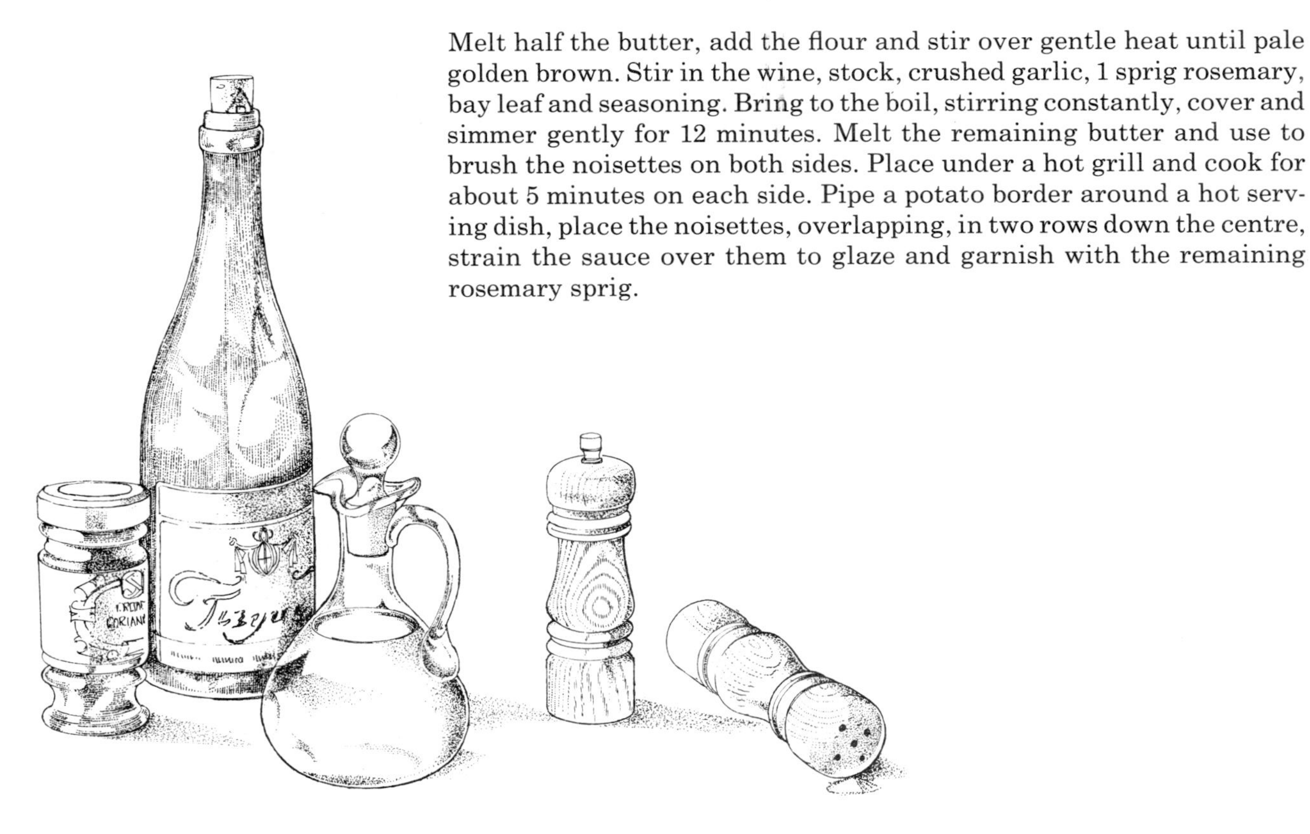

Brochettes d'agneau

(Greece)

Makes 8 servings.
To freeze Pack meat and marinade in polythene containers. Seal and label.
To serve Defrost and thread meat onto skewers alternating with small par-boiled button onions, squares of green pepper and halved bay leaves. Brush with the marinade and grill, turning several times, until meat is cooked.
Storage time 4–6 months.

IMPERIAL	METRIC	AMERICAN
4 tablespoons olive oil	4 tablespoons olive oil	5 tablespoons olive oil
4 tablespoons red wine	4 tablespoons red wine	5 tablespoons red wine
salt and pepper to taste	salt and pepper to taste	salt and pepper to taste
2 teaspoons ground coriander	2 teaspoons ground coriander	2 teaspoons ground coriander
1 tablespoon clear honey	1 tablespoon clear honey	1 tablespoon clear honey
2 lb. leg of lamb, cubed	1 kg. leg of lamb, cubed	2 lb. boneless lamb leg, cubed

Mix together the olive oil, wine, seasoning, coriander and honey. Place the meat in a shallow dish and pour the marinade over. Allow to stand for several hours in a cool place.

Hazelnut torte

(Austria)

Makes 8 portions.
To freeze Open freeze until solid and pack carefully into a polythene bag. Seal and label. Overwrap or place in a rigid-based container to prevent damage.
To serve Unwrap while frozen, place on serving dish and allow to defrost for about 6 hours.
Storage time 4–6 months.

IMPERIAL	METRIC	AMERICAN
4 eggs	4 eggs	4 eggs
6 oz. castor sugar	175 g. castor sugar	$\frac{3}{4}$ cup granulated sugar
1 oz. hazelnuts, ground	25 g. hazelnuts, ground	$\frac{1}{4}$ cup ground hazelnuts
1 oz. plain chocolate, grated	25 g. plain chocolate, grated	1 square semi-sweet chocolate, grated
2 oz. flour	50 g. flour	$\frac{1}{2}$ cup all-purpose flour
1 oz. butter, melted	25 g. butter, melted	2 tablespoons butter, melted
decoration	*decoration*	*decoration*
4 oz. plain chocolate	100 g. plain chocolate	4 squares semi-sweet chocolate pieces
$\frac{1}{2}$ pint double cream	3 dl. double cream	$1\frac{1}{4}$ cups whipping cream
2 tablespoons icing sugar	2 tablespoons icing sugar	3 tablespoons confectioners' sugar
2 oz. hazelnuts	50 g. hazelnuts	about $\frac{1}{4}$ cup hazelnuts

Whisk the eggs and sugar together until thick and creamy. Gently stir in the ground hazelnuts, chocolate, flour and lastly the melted butter. Divide the mixture between two greased and lined 7-inch (18-cm.) cake tins and bake in a moderate oven (350°F, 180°C, Gas Mark 4) for 20 minutes. Turn out onto a clean teatowel on a wire tray and when cool strip off the paper. To assemble and decorate the torte, slice each cake to make four layers. Melt the chocolate in a basin over hot water. Whip the cream and add the icing sugar and melted chocolate. Spread one layer of cake with the filling and place on a baking tray. Cover with another layer of cake and build up the layers of cake and filling ending with a layer of cake. Swirl or pipe the remaining cream over the top. Decorate with the whole hazelnuts and mark into 8 portions with a knife.

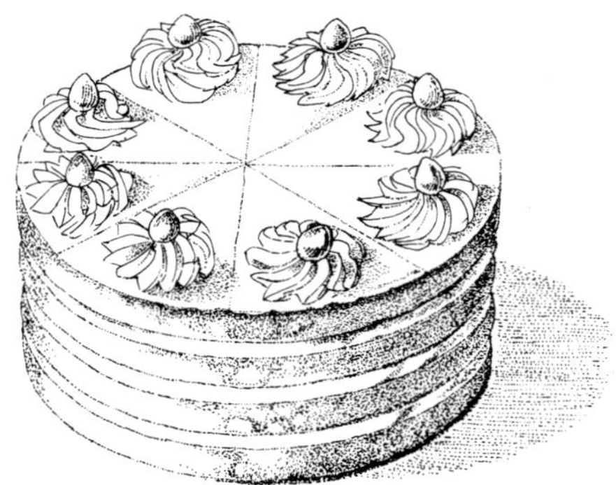

Redcurrant and raspberry mousse

(Denmark)

Makes 6 servings.
To freeze Pack into polythene containers. Seal and label.
To serve Remove lid and defrost in refrigerator for 2 hours. Turn into a glass bowl and sprinkle with flaked almonds.
Storage time 3–6 months.

Note Without the addition of egg whites this can be served frozen as an ice cream.

IMPERIAL	METRIC	AMERICAN
1 lb. raspberries	450 g. raspberries	1 lb. raspberries
8 oz. redcurrants	225 g. redcurrants	$\frac{1}{2}$ lb. red currants
4 tablespoons water	4 tablespoons water	$\frac{1}{3}$ cup water
sugar to taste	sugar to taste	sugar to taste
$\frac{1}{4}$ pint double cream	$1\frac{1}{2}$ dl. double cream	$\frac{2}{3}$ cup whipping cream
$\frac{1}{4}$ pint cold custard	$1\frac{1}{2}$ dl. cold custard	$\frac{2}{3}$ cup cold custard
6 egg whites	6 egg whites	6 egg whites

Place the raspberries, redcurrants, water and sugar in a saucepan and stew until the fruit is soft. Cool and sieve. Whip the cream lightly until it just holds its shape. Fold the cream and custard into the fruit purée and mix thoroughly. Beat the egg whites stiffly and fold evenly into the mixture.

Tortoni al cioccolato
(Italy)

Makes 12 servings.
To freeze Open freeze on a tray and when decoration is firm, cover with moulded freezer foil. Pack all together in one container, such as a Tupperware jumbo keeper to prevent damage to these rather delicate desserts.
To serve Remove covers while still frozen and allow to defrost in the refrigerator. These desserts are very rich so only give a small serving.
Storage time 4–6 months.

IMPERIAL	METRIC	AMERICAN
12 oz. plain bitter chocolate	350 g. plain bitter chocolate	12 squares bitter cooking chocolate
1 oz. butter	25 g. butter	2 tablespoons butter
8 eggs, separated	8 eggs, separated	8 eggs, separated
3 tablespoons rum, brandy or coffee liqueur	3 tablespoons rum, brandy or coffee liqueur	scant ¼ cup rum, brandy or coffee liqueur
2 tablespoons strong black coffee	2 tablespoons strong black coffee	3 tablespoons strong black coffee
¾ pint double cream	scant ½ litre double cream	scant 2 cups whipping cream
4 medium almond macaroons or 20 ratafia biscuits	4 medium almond macaroons or 20 ratafia biscuits	4 medium almond macaroons
decoration	*decoration*	*decoration*
chocolate curls or hazelnuts	chocolate curls or hazelnuts	chocolate curls or hazelnuts

Put the chocolate, broken into small pieces, and the butter in a bowl over a pan of hot water. Allow to melt, stirring occasionally. When smooth, remove from heat, beat in the egg yolks one at a time, 2 tablespoons rum or brandy, and the coffee. Whisk the egg whites stiffly and fold into the chocolate mixture. Whip the cream until fairly stiff and reserve one-third for decoration. Crumble the macaroons without reducing to crumbs and fold into the larger portion of cream. Put a spoonful of cream mixture into each of 12 miniature soufflé dishes or small freezer dishes. Cover with a spoonful of chocolate mixture and repeat the process, so that you have four layers, ending with chocolate. This should clear the top by at least a ½ inch (1 cm.) to allow space for the final decoration. Beat the remaining cream more stiffly, pipe or spoon a flat rosette on top of each sweet and top with a curl of chocolate or a hazelnut. Make sure it does not come above edge of container.

Greek honey pastries
(Greece)

Makes approximately 12 pastries.

IMPERIAL	METRIC	AMERICAN
1 lb. frozen puff pastry, defrosted	450 g. frozen puff pastry, defrosted	1 lb. frozen puff paste, defrosted
4 oz. hazelnuts	100 g. hazelnuts	¾ cup hazelnuts
½ pint double cream	3 dl. double cream	1¼ cups whipping cream
2 tablespoons clear honey	2 tablespoons clear honey	3 tablespoons clear honey
2 tablespoons brandy	2 tablespoons brandy	3 tablespoons brandy
clear honey to glaze	clear honey to glaze	clear honey to glaze

Roll out the pastry fairly thickly and cut into large fingers. Place on a dampened baking tray and bake in a hot oven (425°F, 220°C, Gas Mark 7) for 12 minutes, or until well risen and golden brown. Cool on a wire tray. Meanwhile, chop the hazelnuts finely. Whip the cream until fairly stiff and fold in evenly three-quarters of the chopped nuts, the honey and brandy. Split the pastry fingers in half and sandwich together with the filling. Brush with more honey and sprinkle with the remaining nuts.

Note The filling can be frozen but tends to separate slightly and will require to be whipped smooth again.

Basic choux pastry puffs
(France)

Makes Approximately 100 puffs.
To freeze Pack in rigid-based containers in layers with foil dividers.
To serve Reheat (from the frozen state) and crisp in a moderately hot oven (375°F, 190°C, Gas Mark 5) for 5–7 minutes.
Storage time 4–6 months.

IMPERIAL	METRIC	AMERICAN
5 oz. plain flour	150 g. plain flour	1¼ cups all-purpose flour
¼ teaspoon salt	¼ teaspoon salt	¼ teaspoon salt
½ pint water	3 dl. water	1¼ cups water
4 oz. butter	100 g. butter	½ cup butter
4 eggs	4 eggs	4 eggs

Sieve flour with salt. Put water and butter into a saucepan and bring slowly to the boil. Remove from heat and beat in flour to form a ball which leaves sides of pan clean. Cool mixture to blood heat, then beat in eggs, one at a time. Using a ½-inch (1-cm.) tube and large piping bag, force balls of paste, about the size of a small walnut onto wetted baking trays. Bake in a moderately hot oven (375°F, 190°C, Gas Mark 5) for about 20 minutes, until crisp and pale golden brown. Cool on a wire tray.

Choux aux marrons
(France)

Makes 6 servings.

IMPERIAL	METRIC	AMERICAN
4 tablespoons sweetened chestnut purée	4 tablespoons sweetened chestnut purée	5 tablespoons sweetened chestnut purée
1 tablespoon brandy	1 tablespoon brandy	1 tablespoon brandy
¼ pint double cream	1½ dl. double cream	⅔ cup whipping cream
4 tablespoons single cream	4 tablespoons single cream	⅓ cup coffee cream
50 choux puffs	50 choux puffs	50 choux puffs
icing sugar to sprinkle	icing sugar to sprinkle	confectioners' sugar to sprinkle

Blend the chestnut purée with the brandy, whip the two creams together and fold in the chestnut mixture. Make a hole in the base of each puff and pipe in the chestnut cream. Serve piled in a glass serving dish and sifted with icing sugar.

Choux diablotons
(France)

Makes 6 servings.

IMPERIAL	METRIC	AMERICAN
4 oz. soft cream cheese	100 g. soft cream cheese	½ cup soft cream cheese
1 oz. butter	25 g. butter	2 tablespoons butter
1 tablespoon mayonnaise	1 tablespoon mayonnaise	1 tablespoon mayonnaise
1 tablespoon finely chopped hazelnuts	1 tablespoon finely chopped hazelnuts	1 tablespoon finely chopped hazelnuts
salt and cayenne pepper to taste	salt and cayenne pepper to taste	salt and cayenne pepper to taste
50 choux puffs	50 choux puffs	50 choux puffs

Cream the cheese, butter and mayonnaise well, stir in the nuts and season highly. Make a hole in the base of each puff and pipe in the cheese mixture. Serve as a savoury.

Menus for all occasions

When you start to plan the menu for an important occasion you can be far more ambitious with the freezer waiting to help. Only one word of warning – you must run down your stocks of staple foods to accommodate vast quantities of food for a big party. For smaller occasions, there should always be a choice on your shelves or in your baskets, sufficient to serve three or four courses to guests. There is no feeling of soullessness about a meal composed largely of frozen food, providing you embellish it with fresh salads, fruits, crisp and curly garnishes, and other touches your friends know have not been frozen to await the happy day of their arrival. A green salad, consisting of crisp lettuce, chopped cucumber and watercress goes with most dishes.

Finger food buffet party
Piquant avocado dip*
Pâté de luxe*
Hazelnut cheese spread*
Hot garlic bread* and biscuits
Choux diablotons*
Toasted sandwiches*
Asparagus flan*
Mocha gâteau*
Blackcurrant surprise*
Wines – Muscadet. Nicolas Vieux Ceps

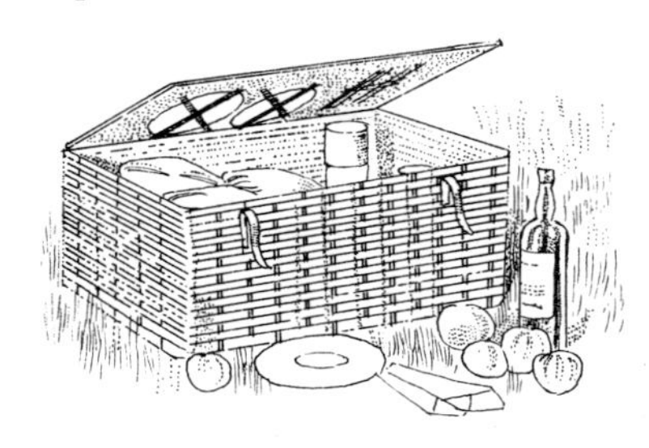

Picnic party
Devonshire pasties*
Pinwheel sandwiches*
Choc-o-nut nibbles*
Spice cakes*
Picnic pie*
Fried chicken portions*
Moulded raspberry soufflé*
Assorted fresh fruit
Orange concentrate and
Bottled lemonade

Tea party
Six-in-one biscuit batch*
Apricot nut bread*
Australian boiled fruit cake*
Orange coconut cake*
Herbed celery scones*
Sliced bumpy loaf* and butter
Raspberry jam*
Tea with milk or lemon

Italian fork supper
Vitello milanese*
Pizzas*
Italian stuffed pancakes*
Tortoni al cioccolato*
Assorted green salads
Cassata*
Compote of peaches and melon balls
Hot fancy rolls*
Gorgonzola, Bel Paese and grape platter
Wines – Orvieto Secco. Chianti

Impromptu party
Fish shells*
Beefy supper snacks*
Cheese and meat strudel* with petits pois
Spanish salad*
Orange butter pancakes*
Lemon and lime sorbet
Cob rolls and butter
Wine – Mateus Rosé

Coffee party
Orange and date twist*
Honey fruit scones*
Hazelnut torte*
Sliced cheese plait* with peanut butter
Greek honey pastries*
Black coffee or coffee with cream

Christmas party
LUNCH
Grilled grapefruit
Roast stuffed turkey or goose*
Roast potatoes*
Brussels sprouts
Green beans with almonds*
Iced Christmas pudding*
Mincemeat plait*
Wine – Veuve du Vernay

TEA
Cream cheese and gherkin sandwiches*
Christmas jewel cake*
Pineapple sherbert*

SUPPER
Giblet soup
Cold turkey with spicy cauliflower salad*
Mincemeat shortbread*
Wine – Chambertin

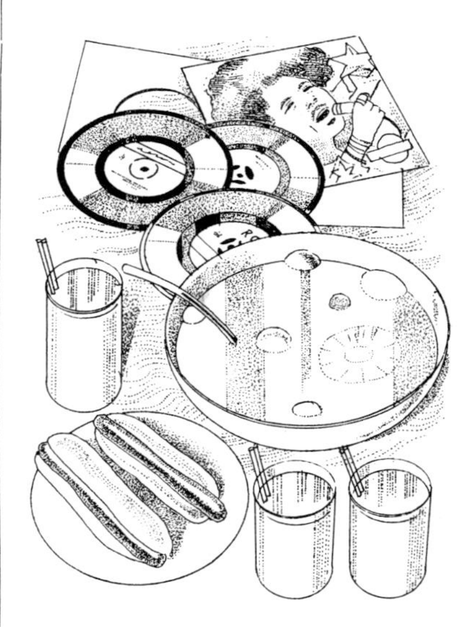

Young idea party
Corn and tomato chowder*
Hot dog rolls*
Bacon grill fritters*
Ham and apple pasties*
Sweet and sour meatballs*
Big baked potatoes
Blackcurrant cheesecake*
Caramel ice cream*
Gingered dry cider
Fruit punch

Informal dinner parties

A
Potage St. Germain*
Apricot-baked bacon*
Baked parsnips
Duchesse potatoes*
Chocolate walnut flan*

B
Lemon avocado soup*
Pork chops baked in foil*
garnished with fresh tomatoes
Cucumber salad
Orange and lemon cream pie*

C
Ratatouille basquaise*
Brochettes d'agneau* on rice
Petits pois
Peach flan* with cream

D
Danish liver pâté* and toast
Somerset chicken in cider*
Peppered saffron rice*
Buttered carrots
Frosted coffee gâteau*

E
Jellied consommé
Chicken Maryland*
French fried potatoes*
Mixed green salad
Pear condé*

F
Skipper pâté in lemons* and
brown bread and butter
Loin of lamb with fruity sauce*
Roast potatoes*
Broccoli spears
Orange, prune and apple ice cream*

Formal dinner parties

A
Potted shrimps*
Brown bread and butter
Faisan à la crème*
Duchesse potatoes*
Red cabbage with apple
Choux aux marrons*
English cheese platter

B
Prawn cocktail
Brown bread and butter
Crème Dubarry*
Pollo sauté alla cacciatora*
Buttered noodles
Hazelnut torte* or
Fruit salad with cream

C
Duck pâté*
French bread and butter
Stuffed fish turbans with oriental
sauce*
Noisettes bordelaise*
French fried potatoes*
Broad beans
Lemon flummery*

D
Pigeon pâté*
Hot toast and butter
Tournedos aux coeurs d'artichauts*
French fried potatoes*
Endive salad
Layered ice cream gâteau*
Brisling mousse*

E
Terrine of game*
Hot toast and butter
Truites aux amandes*
Boeuf en croûte à la reine Marie*
Button sprouts
Pineapple sherbert*

F
Lemon avocado soup*
Raisin-stuffed duck with vermouth*
Minted new potatoes
Mange tout peas
Frozen orange syllabub*
French cheese platter

Wedding reception for 50 people

Menu

Recipes marked * are given in the book.

Pierre's vichyssoise*
Grapefruit and cherry cocktails*
Two roast turkeys* with lemon and
Walnut stuffing* and mixed sausage stuffing*
Hot peppered saffron rice*
Mixed green salad
Coleslaw
Hot quiche belle bretonne*
Strawberry ice cream sundaes*
Pineapple cheesecake*
Champagne Wedding cake Coffee

Preparation day (1–6 weeks before the reception)

Make and freeze six quantities Pierre's vichyssoise.
Freeze 10 pints (6 litres) grapefruit segments in sugar syrup.
Make up 4 packets sage and onion stuffing with 2 oz. (50 g.) walnuts, finely chopped, finely grated zest and juice of 1 lemon and sufficient hot water to moisten. Freeze in two polythene bags.
Purchase two frozen 12-lb. (6-kg.) turkeys.
Purchase $1\frac{1}{2}$ lb. (700 g.) beef sausagemeat and $1\frac{1}{2}$ lb. (700 g.) pork sausagemeat, mix together well and freeze in two polythene bags.
Make and freeze 6 quiche belle bretonne, 4 pineapple cheesecakes and 4 quantities peppered saffron rice.

Work plan

Three days before

Remove turkeys from freezer; defrost, wrapped, in a cold room.

Two days before

Remove bags of lemon and walnut stuffing and mixed sausage stuffing from freezer and defrost in refrigerator.

One day before

Stuff turkeys and cook.
Make coleslaw and refrigerate in covered container. Remove grapefruit segments from freezer and defrost.

On the day

Remove cheesecakes from freezer, unwrap, place on serving dishes and allow to defrost at room temperature.
Mix packs of grapefruit with 2 pints (generous 1 litre) cherries in maraschino syrup and serve in glasses.
Slice and arrange turkey and stuffings on serving dishes.
Serve coleslaw and green salad in salad bowls.
Mix creams with Pierre's vichyssoise and keep chilled in two large tureens.
Chop a large bunch of chives and place in a small dish with spoon for sprinkling over soup.

One hour before reception

Reheat peppered saffron rice on baking trays in oven (see recipe). Keep hot.
Unwrap quiches and reheat as instructions in recipe, reversing positioning on shelves halfway through. Prepare ingredients for strawberry ice cream sundaes (see recipe). While guests enjoy the main course, start serving out the individual sundaes.

A to Z of hostess hints

Some of the useful extras which have found their way into my freezer do not fall into any special category, but are worth gathering together and describing in detail. You may discover many more, since half the fun of owning a freezer is expanding the range of its usefulness in just this way. For example, those accustomed to buying wine in generous litre bottles are less likely to drink the last precious drop, but will spare it for freezing and later use in cooking. Tupperware tumblers come in all sizes suitable to take just those few tablespoons which will make all the difference to your next beef stew (if the wine is red) or a fish stock (if white). Just add the frozen wine cube to the rest of the liquid.

Beurre manié This useful form of thickening can be used in the frozen state, and without any last-minute mess and mixing. Cream together equal quantities of butter and flour, or as much flour as the butter will easily take up, chill the mixture slightly and form into small lumps. These will freeze in a bag, and a few can be taken out and added to a stew or sauce that is not thick enough, and stirred in gently until melted. Gentle simmering, to cook the flour, is required for a few minutes longer.

Butter pats Making butter pats takes some time, and they need to be chilled afterwards. It is simpler to take them straight from the freezer when required. Flavoured butter pats can be prepared as follows and served as a garnish to grilled and fried fish or meat. They should be 1½ inches (3·5 cm.) in diameter and less than a ½ inch (1 cm.) thick. To make **maître d'hôtel butter,** cream the butter and beat in grated lemon zest, a little lemon juice and plenty of chopped parsley, to taste, season if liked. For short term storage, grated onion or crushed garlic may be added. Chill until the butter can be formed into a long sausage shape of a suitable diameter, rolling it inside a sheet of wet freezer film so that it does not stick. When firm enough, cut into thin slices with a sharp knife. Reform into a roll with film dividers between the slices, pack in small polythene bags or Tupperware tumblers, and take out as required. Butters flavoured with other herbs can be frozen in the same way and stored up to 3 months but those containing garlic for only 1 month.

Chocolate shapes When you feel in the mood for delicate work make up supplies of these decorations for freezing. As well as chocolate leaves (see page 210), chocolate diamonds, squares and curls are useful. Cooking chocolate (the special couverture type or chocolate chips) is the best to use. Scrape large curls off the flat side of the block with a potato peeler onto a sheet of freezer film. Or melt a block of chocolate, pour onto a sheet of waxed paper and allow to spread out thinly and set. Mark into diamonds or squares and cut through with a ruler and sharp knife. Move the edge of the paper out over the edge of the table and peel it away from under each shape, one at a time. Pack in layers, with foil dividers. These decorations store for 2–3 months and the chocolate does not discolour while frozen. Take out and use to decorate sweets and cakes while still in the semi-frozen state, as they are less inclined to melt when handled.

Chocolate cases Melt 1 lb. (450 g.) cooking chocolate, with a small nut of butter, in a basin over a pan of hot water. Avoid stirring. Cool slightly. Use small fluted paper baking cases, each made stronger by being used doubled. Spread chocolate inside each case with the back of a teaspoon in a layer thick enough to make a firm case. Arrange the prepared cases in bun trays, which helps to keep them a good shape, and open freeze until hard. Use any fruit-flavoured creamy filling (see page 98) to fill the cups up to edge of the chocolate. Open freeze, and pack in layers in a rigid-based container with dividers. To serve, transfer to refrigerator for about 30 minutes, then peel off paper cases. Decorate with piped rosettes of whipped cream or sprinkle with finely chopped pistachio nuts.

Fancy ice cubes Ice cubes made with slightly diluted fruit drink concentrates (orange, lemon, grapefruit, lime, blackcurrant) have the advantages of looking pretty, flavouring and cooling a drink without diluting it. For colour alone, water flavoured with a very little lemon juice can be tinted any shade with a few drops of food colouring. The transparent ice cube looks most attractive if a sliver of lemon or orange zest, or a maraschino cherry, has been frozen in the centre. Pop the decoration in and position in the centre with a skewer after you have filled the trays with water. Pack as for sauce cubes, but the storage time is longer, 3–4 months. The addition of soda water seems to make little difference to the cubes sticking together. The important thing is to pack the cubes and refreeze without any delay.

Fruit sugar When making citrus-flavoured mousses and creams, the recipe sometimes calls for the juice of several oranges or lemons, and grated zest of only one. Grate the zest from all the fruit before squeezing out the juice and pack any not required, mixed with castor sugar in proportions of one part zest to two parts sugar. This topping is useful for cakes, to strew over a custard flan or on rice puddings. Children love it.

Frozen flower decorations After many tests, the most suitable flower seems to be the garden's queen, the rose. Choose buds which are just beginning to open, and give them a long drink, up to their necks in water for several hours, or the stems will tend to collapse when defrosted. Lay them flat in a shallow polythene container (the Tupperware double diner holds eight comfortably) head to tail. Pour in 1 inch (2·5 cm.) of water, open freeze until partially frozen, fill up with water until the flower heads are completely covered. To use, thaw the covered container at room temperature or until you can remove the roses and arrange them in a vase. They will last for 1–2 days, but will come out more rapidly than if freshly picked. Those glorious fully blown heads of roses which can only be picked to float in a bowl can also be frozen. I use discarded foil steamed pudding basins, covering with only just enough water to cover the bloom, and a foil cap. When required, I dip the container quickly into hot water to release the ice shape, turn out, re-invert and slide into a glass finger bowl. I place one at each place setting when the centrepiece on the table is to be a fruit platter and/or candles. If the last course requires the finger bowls to be used in earnest, as it sometimes does, it is better to freeze a handful of well scented rose petals in each foil basin. By the time they are needed the bowls are filled with water delicately scented with the petals.

Herb garnishes Herbs, such as parsley, can be frozen in sprigs in small polythene bags, and rubbed briskly between the hands to crush them in the frozen state. But to use as a chopped parsley garnish, or chopped chive garnish, they look much better if neatly chopped before freezing. A nice amount for one serving can be frozen in a 2-oz. (50-g.) Tupperware tumbler, or in a foil tartlet case with a foil cap. If required merely to flavour stews, the chopped herbs (and this includes ones with less tender stalks than parsley) can be mixed with very little water, and frozen into ice cubes, each representing the addition of 1 teaspoon of fresh herbs to the recipe. Herbs such as rosemary with woody stems, can be frozen on the stem, and the leaves just rubbed off while still frozen for cooking. Cubes of frozen mint make excellent mint sauce if frozen with castor sugar in vinegar, and about three dissolved in tablespoons of boiling water. All herbs retain a brighter colour if blanched for 1 minute before freezing, but be prepared for them to become very limp.

Horseradish and beetroot cream Blend together 1 tablespoon grated mild onion, 4 tablespoons liquidised or sieved cooked beetroot, contents of a small pot of creamed horseradish and the same amount of soured cream. Freeze in small containers, foil or polythene, suitable to be put on the table and used up at one meal. Serve with hot or cold roast meat.

Horseradish sherbert Blend together 2 tablespoons lime juice cordial, contents of a small pot of creamed horseradish, and the same amount of half whipped double cream. Freeze as above, in small portions, and serve only partially defrosted, broken up with a fork. Very good with fish, roast or fried chicken.

Pastry leftovers When batch baking pastry, there are bound to be small pieces which cannot be gathered up to make a nice flan or pie crust. Use small fancy cutters, and cut out as many pretty shapes as possible. Brush with milk, sprinkle with grated Parmesan cheese, seasoned salt and pepper, or sesame seeds, or poppy seeds. Bake on a floured baking tray at 400°F, 200°C, Gas Mark 6 for 12 minutes or until golden brown. Cool, pack in layers in a rigid-based container, or open freeze them in bags. Serve with drinks, or if small enough use as garnishes for soups. Store 2–3 months.

Sauce cubes When making a large quantity of an elaborate sauce, pour some into ice cube trays, freeze, turn out and pack quickly in any suitable container, then refreeze at once. If the cubes have not had time to begin melting, they will not stick together. Use as required, to give dishes extra flavour, or to serve small amounts of sauce, reheat very carefully with a tablespoon of water to prevent it from sticking and burning. Store 2–3 months.

Serving advice When planning a meal, especially a dinner party, which is to come mainly from the freezer, work out the order of events to a very careful countdown. Dishes which need overnight defrosting may have to be transferred to the refrigerator the day before. If a number of dishes require defrosting and reheating in the oven, make sure the same temperature will suit all of them, and that the oven will not be overcrowded. Choose vegetables which can be cooked on the hob, or some item which can be cooked under the grill. With care, you can arrive at a freezer menu which requires your urgent attention for only 30 minutes before the meal. See suggested dinner menus, A, B, and F on page 237. One golden rule which the wise hostess always observes is that the meal should never appear to come totally from the freezer. Fresh fruits and salads dispel this impression. Fragile green salads never come amiss as part of an elaborate menu, and a bowl of fresh fruit, or merely a decoration of russet apples, golden pears or purple grapes on the cheese board reassure your guests that the goodies on the table are not all reawakened 'sleeping beauties'.

Four-season freezing

Have you ever studied the calendar to pick out those months when it would be most profitable to freeze down pork, pears or pheasant? For instance, the short season when game is plentiful can be extended by buying for your freezer as well as for the occasional dinner party. This applies to many other items which only make a brief appearance each year in the shops. If you garden, even on a small scale, it is worth devoting space to fruit and vegetable varieties which grow easily and crop abundantly. If picked in their prime and frozen without delay these provide delicacies all round the year which are quite impossible to buy. Tiny new carrots and broad beans straight from your plot, or finger-length courgettes are a revelation in the enjoyment of flavour.

If you have no vegetable or fruit garden, a weekend expedition into the country at harvest time gives you an opportunity to pick your own which farmers are often glad to permit in these days of labour shortage.

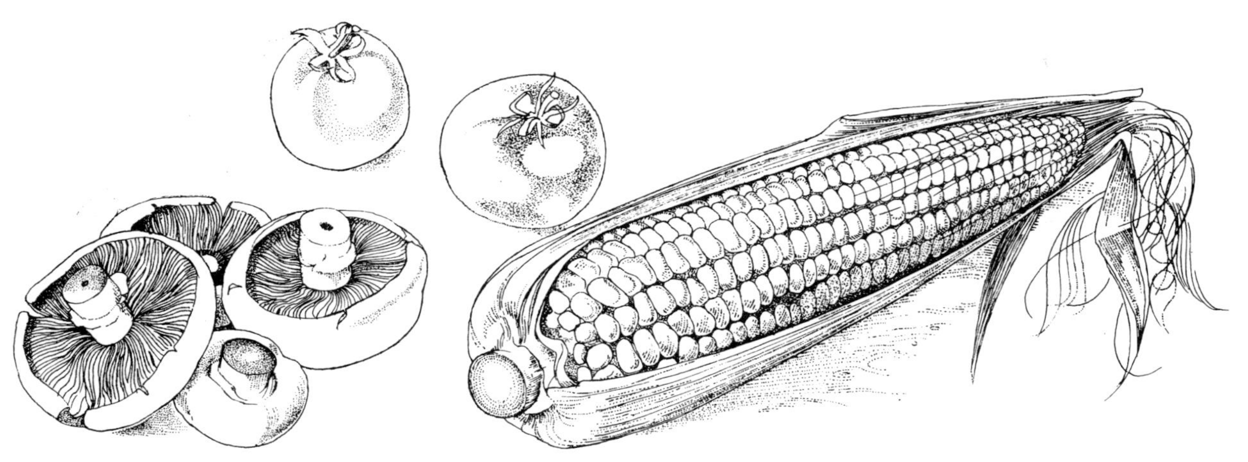

Growing for freezing

In the following chart you will find details of what to grow, when to sow and plant, and the best varieties to choose where this is applicable. Then there are special hints on harvesting at the right stage of growth.

Seasonal freeze plan

Planning is the key to all truly successful use of the freezer. Just as you should look ahead and plan to allow space for your own garden produce or a glut in the local market or farmer's fields, you should be ready for each season's special bargain as it comes along.

In the four-season freezing guide the year is divided into four periods of three months each. Most of the items mentioned under each month will be plentiful throughout the period although some are limited just to those few weeks. This also applies to the recipe suggestions. Seasonal availability in the northern areas may be a little later than in the south – this particularly applies to soft fruits.

Items such as chicken are produced to meet demand all round the year and are not likely to be cheap because of a seasonal over-supply. Others, such as apples, pears and grapes are deliberately imported to give a year-round supply and only locally grown produce is likely to be cheap in the autumn months.

Keeping a freezer log

A secondary advantage of keeping a complete record of all the food which goes in and out of your freezer is that it enables you to plan next year's seasonal buying better. It is a fascinating record and studied carefully it yields a lot of valuable information. For example, I was once tempted to buy 10 lb. (5 kg.) of red and green peppers when they were remarkably cheap. I prepared and bagged them in 8-oz. (225-g.) quantities. Fortunately I blanched them all, although more recent tests have shown unblanched peppers have a satisfactory storage life of six months. However, after a year I still had several bags in store which meant that the peppers were past their prime and that valuable storage space had been wasted. A glance at my freezer log reminds me exactly when peppers are likely to be at their cheapest and a purchase of 6 or 7 lb. (3 or 3½ kg.) would fill my requirement for a year.

Golden guide number

to growing vegetables and fruit for freezing

Vegetables to grow for freezing

Varieties	Sowing and harvesting times	Harvesting hints
Beetroot Detroit	sow – April to June harvest – July to October	Pull when the size of golf balls and pack whole. Larger beetroots can be sliced.
Broad beans Leviathan, Gillet's Imperial, Green Longpod, Masterpiece Longpod, Colossal Longpod, Dreadnought (early)	sow – March/April harvest – June to August	Pick before the beans are fully mature, that is before a black eye has formed at the end of the bean. Otherwise the skin becomes tough during freezing.
Broccoli Calabresse, Kangaroo and Barrier Reef for Autumn. Purple or white Sprouting Broccoli and Thanet Winter Cauliflower for cutting following spring	sow – March to May transplant – May to July harvest in spring	Cut compact heads with even-length stalks for easy packing.
Brussels sprouts Peer Gynt, Sanda, Leda, Jade Cross	sow – March/April transplant – May/June harvest – September to December	Pick button sprouts when solid. Older sprouts with loose leaves tend to create an unpleasant aroma in the freezer.
Carrots Early Horn, Early Market, Early Nantes, Perfect Gem, Chantenay (red cored), Amsterdam Forcing	sow – March to mid July harvest – June onwards	Pull when finger length for freezing whole. Larger carrots can be sliced.
Cauliflower Improved Snowball	sow – January to April transplant – April to June harvest – June to September	Cut small heads, freeze florets only and use remainder of stump for cauliflower soup.
Kangaroo, Barrier Reef	sow – late April transplant – June/July harvest – September to November	
Corn-on-the-cob First of all, North Star, John Innes Hybrid, Earliking, Kelvedon Glory	sow – April under glass and transplant – June in several short rows, or sow – May in situ harvest – August/September (6 weeks after silk appears on cobs)	Pick when the contents of the grains are like very thick cream. Freeze on day of harvesting.
Courgettes Courgette (F1 Hybrid) or zucchini (latter will grow to full-size marrow if allowed)	sow – April under glass and transplant – June or sow – May in situ harvest – July to October	Cut fruits when finger-length for freezing whole. Larger courgettes can be sliced.

Varieties	Sowing and harvesting times	Harvesting hints
Green beans – French Masterpiece, Processor, Blue Lake, Tendergreen, The Prince, Sprite	sow – April/May harvest – July to October	Pick while still young enough to freeze without cutting in half.
Green beans – Runner Early Blue Lake, White Emergo, Streamline, Prizewinner, White Achievement, Enorma, Achievement	sow – May/June harvest – July to October	Pick frequently when the plants are cropping and before beans form inside the pods. The beans should be picked while still flat and tender enough to be snapped in the hand.
Peas Show Perfection, Kelvedon Wonder, Progress, Miracle, Early Onward, Onward, Raynes Park, Victory Freezer	sow – March to June harvest – June to October	Pick when pods look just filled, peas themselves should be bright green with no film of white and easy to squash between finger and thumb.
Spinach Greenmarket, Longstanding Round	sow – March to May and August to October harvest – as ready	
Sugar peas Mange Tout (Carouby de Maussane)	sow – April to June harvest – July to September	Pick as soon as peas are visible through pods and before they begin to swell.

Fruits to grow for freezing

Varieties

Apples
Bramley Seedling, Arthur Turner, Lane's Prince Albert, George Neal, Wellington, Golden Noble, Encore, Woolbrook Russet

Blackberries
Oregon Thornless, Himalaya Giant

Blackcurrants
Laxton's Giant, Malvern's Cross and Blacksmith, Boskoop Giant, Baldwin, Wellington, Westwick Choice

Gooseberries
Careless, Golden Drop, Green Gem, Lancer, Leveller

Loganberries
L.Y. 59

Plums and gages
Victoria, Cambridge Gage, Jefferson, Count Althann's Gage, Greengage, Oullin's Gage, Yellow Egg, Warwickshire Drooper

Raspberries
Norfolk Giant, Lloyd George, Malling Enterprise, Malling Jewel, Malling Promise, Glen Clova

Varieties

Redcurrants
Red Lake, Laxton's No. 1

Rhubarb
Champagne, The Sutton, Victoria, Glaskin's Perpetual

Strawberries
Cambridge Vigour, Cambridge Favourite, Royal Sovereign, Cambridge Rival, Talisman, Gorella, Cambridge Late Pine

Tomatoes
Potentate, Moneymaker

Fruit – harvesting hints
Pick just before or when fully ripe, discard all damaged and over-ripe fruit. After picking, sort out firm, unblemished best specimens and reserve for freezing. Use remainder at once.

Herbs
Pick just before flowering, strip off and discard any woody stalks.

Freezer jams

Fresh berry fruits make delicious uncooked jams which can be stored in the freezer for six months at least. They are so easy to prepare, give you a higher yield of jam per 1 lb. ($\frac{1}{2}$ kg.) of fruit and lose nothing of the natural sweet juicy flavour. This method can be used for any of the following fruits: blackberries, blackcurrants, loganberries, raspberries, redcurrants, strawberries.

Put the prepared fruit in a bowl. To 2 lb. (1 kg.) of fruit add $3\frac{1}{2}$ lb. ($1\frac{3}{4}$ kg.) castor sugar and 4 tablespoons lemon juice. Stir well and leave to stand until the sugar has completely dissolved, which may take from 20 minutes to 2 hours. Add scant $\frac{1}{2}$ pint ($2\frac{1}{2}$ dl.) commercial pectin, stir until beginning to set. This may be delayed overnight but if necessary leave at warm room temperature in the bowl. Pot in glass jars with screw or self-sealing lids, or in Tupperware 6-oz. (175-g.) tumblers. This jam defrosts within 30 minutes but retains its easy spreading consistency.

Spring

March to May

Fruit Rhubarb, pineapple, apricots.

Vegetables Young carrots, cauliflower, spinach, seakale, chicory.

Fish Salmon, mackerel, plaice or dab, whitebait.

Meat Duckling, poussin, chicken, beef.

Extras Eggs (in cooked dishes or raw beaten with salt or sugar).

Rich chocolate mousse

Makes 12 servings.
To freeze Cover with moulded freezer foil and smooth down edges or snap on lid. Label.
To serve Uncover and defrost at room temperature for 30 minutes.
Storage time 4–6 months.

IMPERIAL	METRIC	AMERICAN
1 lb. plain chocolate	450 g. plain chocolate	1 lb. semi-sweet chocolate
8 eggs, separated	8 eggs, separated	8 eggs, separated
1 oz. butter	25 g. butter	2 tablespoons butter
2 tablespoons brandy or strong black coffee	2 tablespoons brandy or strong black coffee	3 tablespoons brandy or strong black coffee

Place the chocolate in a basin over a pan of hot water and allow to melt. Remove from the heat and stir in the egg yolks, butter and brandy or coffee and mix until smooth. Allow to cool. Whip the egg whites until stiff and fold into the chocolate mixture. Divide between 12 individual dishes or polythene containers.

Caramel custards

Makes 48 servings.
To freeze Place the cases on baking trays and open freeze until solid. Tape face together, two at a time, with freezer tape. Label.
To serve Remove tape, arrange custards in a bain-marie of hot water at least half way up the sides, and cook in a cool oven (300°F, 150°C, Gas Mark 2) for about 35 minutes. (If using several shelves in the oven change the containers round half way through baking time.) Serve in the cases, sprinkled with grated nutmeg, or turn out onto small glass plates.
Storage time 4–6 months.

Note This recipe can be scaled down in proportion and is based on 1 pint (generous ½ litre) milk, 4 eggs and 1½ oz. (40 g.) castor sugar for the custard and 4 oz. (100 g.) castor sugar and ¼ pint (1½ dl.) water for the caramel.

IMPERIAL	METRIC	AMERICAN
16 eggs	16 eggs	16 eggs
6 oz. castor sugar	175 g. castor sugar	¾ cup granulated sugar
4 pints warm milk	2¼ litres warm milk	10 cups warm milk
caramel	*caramel*	*caramel*
1 lb. castor sugar	450 g. castor sugar	2 cups granulated sugar
1 pint water	generous ½ litre water	2½ cups water

Whisk the eggs and sugar together lightly, pour onto the warm milk and continue whisking for 1 minute. Make the caramel by dissolving the sugar slowly in the water, then boiling it steadily until pale golden brown in colour. Remove saucepan from the heat at once, and pour a little of the caramel into 48 foil tartlet cases. As the caramel cools and thickens, put the saucepan for a few seconds into a larger pan of boiling water to prevent it from setting. Swirl the caramel around to coat the insides of the foil cases. Now strain sufficient custard into each to nearly fill them.

Summer

June to August

Fruit Cherries, peaches, raspberries, strawberries, blackcurrants, fresh figs, seedless grapes, loganberries, redcurrants, nectarines, greengages.

Vegetables Aubergines, courgettes, asparagus, new potatoes, peas, French beans, globe artichokes, broad beans, red and green peppers.

Fish Lobster, prawns, shrimps, trout, crab, salmon trout.

Meat Duck, guinea fowl, pigeon, English lamb, Welsh lamb.

Extras Herbs.

Pigeon pâté

Makes 8 servings.
To freeze Cover with moulded foil and smooth down edges. Seal and label.
To serve Uncover and defrost at room temperature for 4 hours, or turn out while frozen and defrost on serving dish.
Storage time 6 weeks.

IMPERIAL	METRIC	AMERICAN
3 pigeons	3 pigeons	3 pigeons
scant ¼ pint vinegar	1¼ dl. vinegar	scant ⅔ cup vinegar
generous ¼ pint red wine	2 dl. red wine	generous ⅔ cup red wine
½ teaspoon ground nutmeg	½ teaspoon ground nutmeg	½ teaspoon ground nutmeg
½ teaspoon dried thyme	½ teaspoon dried thyme	½ teaspoon dried thyme
1 bay leaf	1 bay leaf	1 bay leaf
salt and pepper to taste	salt and pepper to taste	salt and pepper to taste
2 onions, chopped	2 onions, chopped	2 onions, chopped
8 oz. sausagemeat	225 g. sausagemeat	1 cup sausagemeat
1 thick slice brown bread	1 thick slice brown bread	1 thick slice brown bread

Cut each pigeon into four joints and place in a shallow dish. To make the marinade, mix together the vinegar, wine, nutmeg, thyme, bay leaf, seasoning and onions. Pour this over the pigeons, cover and leave in the refrigerator for 2–3 days. Cut the meat from the bones and chop finely or mince. Mix in the sausagemeat and bread soaked in the marinade. Blend thoroughly and press into a greased pâté dish, loaf tin or shaped foil container. Cover with foil and place in a baking tin half filled with water. Cook in a moderate oven (350°F, 180°C, Gas Mark 4) for 1½ hours. Cool.

Autumn

September to November

Fruit Blackberries, plums, damsons, apples, pears, melons, bilberries, pumpkins, quinces, cranberries, coconut (grated).

Vegetables Tomatoes, avocados, old carrots, celery, baby onions, mushrooms, beetroot, old potatoes, runner beans, marrow, leeks.

Fish Oysters, herrings, halibut, haddock.

Meat Venison, partridge, pheasant, Scotch lamb.

Extras Chestnuts (purées and stuffings).

Ratatouille basquaise

Makes 8 servings.
To freeze Divide the mixture between four polythene containers. Seal and label.
To serve Defrost overnight still wrapped at room temperature. If required hot, turn contents into a heavy saucepan and reheat gently to boiling point, stirring frequently to prevent burning. Adjust seasoning.
Storage time 4–6 months (without garlic).

IMPERIAL	METRIC	AMERICAN
3 green peppers	3 green peppers	3 green sweet peppers
2 lb. tomatoes	1 kg. tomatoes	2 lb. tomatoes
6 tablespoons olive oil	6 tablespoons olive oil	½ cup olive oil
6 cloves garlic, crushed	6 cloves garlic, crushed	6 cloves garlic, crushed
2 onions, chopped	2 onions, chopped	2 onions, chopped
6 courgettes, sliced	6 courgettes, sliced	6 small zucchini, sliced
4 aubergines, sliced	4 aubergines, sliced	4 eggplants, sliced
½ teaspoon dried basil	½ teaspoon dried basil	½ teaspoon dried basil
½ teaspoon dried oregano	½ teaspoon dried oregano	½ teaspoon dried oregano
salt and pepper to taste	salt and pepper to taste	salt and pepper to taste

Deseed and thinly slice the peppers and skin and slice the tomatoes. Heat the oil and fry garlic and onions until softened but not browned. Add the courgettes and aubergines and stir well. Cook for 10 minutes, add the peppers, stir and cook for a further 10 minutes. Add the tomatoes, herbs and seasoning and bring to the boil. Cover and simmer gently until the vegetables are just tender. Cool.

Favourite tomato sauce

Makes 4 pints ($2\frac{1}{4}$ litres).
To freeze Pack in polythene containers, leaving headspace. Seal and label.
To serve Turn contents into a saucepan and reheat gently to boiling point.
Storage time 4–6 months.

IMPERIAL	METRIC	AMERICAN
$\frac{1}{4}$ pint corn oil	$1\frac{1}{2}$ dl. corn oil	$\frac{2}{3}$ cup corn oil
1 lb. onions, chopped	450 g. onions, chopped	1 lb. onions, chopped
2 lb. tomatoes	1 kg. tomatoes	2 lb. tomatoes
1 5-oz. can tomato purée	1 150-g. can tomato purée	1 5-oz. can tomato paste
3 pints strong chicken stock	$1\frac{1}{2}$ litres strong chicken stock	$7\frac{1}{2}$ cups strong chicken stock
1 tablespoon brown sugar	1 tablespoon brown sugar	1 tablespoon brown sugar
1 tablespoon dried basil	1 tablespoon dried basil	1 tablespoon dried basil
1 teaspoon ground bay leaves	1 teaspoon ground bay leaves	1 teaspoon ground bay leaves
salt and pepper to taste	salt and pepper to taste	salt and pepper to taste
2 tablespoons cornflour	2 tablespoons cornflour	3 tablespoons cornstarch

Heat the oil in a large pan and use to cook the onion gently until softened but not browned. Roughly chop the tomatoes and add to the pan with the tomato purée, stock, sugar, herbs and seasoning. Bring to the boil, cover and simmer gently for 20 minutes. Cool, sieve and return to the pan. Moisten the cornflour with a little cold water, add to the pan and bring to the boil, stirring constantly until smooth and thick. Cool.

Pumpkin pie

Makes 1 pie.
To freeze Cover with freezer foil. Seal and label.
To serve Unwrap, place on a serving dish and allow to defrost at room temperature for 2 hours.
Storage time 4–6 months.

IMPERIAL	METRIC	AMERICAN
1-lb. slice pumpkin	450-g. slice pumpkin	1-lb. slice pumpkin
4 oz. sugar	100 g. sugar	$\frac{1}{2}$ cup sugar
$\frac{1}{2}$ teaspoon ground nutmeg	$\frac{1}{2}$ teaspoon ground nutmeg	$\frac{1}{2}$ teaspoon ground nutmeg
$\frac{1}{2}$ teaspoon ground ginger	$\frac{1}{2}$ teaspoon ground ginger	$\frac{1}{2}$ teaspoon ground ginger
$\frac{1}{2}$ teaspoon salt	$\frac{1}{2}$ teaspoon salt	$\frac{1}{2}$ teaspoon salt
$\frac{1}{4}$ pint double cream, whipped	$1\frac{1}{2}$ dl. double cream, whipped	$\frac{2}{3}$ cup whipping cream, whipped
1 8-inch pie case, baked	1 20-cm. pie case, baked	1 8-inch pie shell, baked
2 oz. pecan or other nuts, chopped	50 g. pecan or other nuts, chopped	$\frac{1}{2}$ cup chopped pecan or other nuts

Note Bake the pastry in a foil pie case and cool but do not remove. This gives added protection in the frozen state.

Peel and remove seeds from the pumpkin. Steam until tender then mash until smooth. Beat in the sugar, spices and salt. Cool and fold in the cream. Pour into the pie case, sprinkle with the nuts and chill.

Winter

December to February

Fruit Oranges, grapefruit, dried figs, dates, mandarins, tangerines, satsumas.

Vegetables Jerusalem artichokes, swedes, parsnips, Brussels sprouts, big onions, broccoli, turnips.

Fish Scallops, cod, sole, sprats, turbot, whiting.

Meat Hare, goose, turkey, pork, New Zealand lamb, Australian lamb.

Extras Seville oranges (for marmalade making).

Jerusalem artichoke soup

Makes 3 pints (1½ litres).
To freeze Pour into tumbler-shaped polythene containers leaving headspace. Seal and label.
To serve Turn contents into a saucepan, add a little milk or stock and reheat gently to boiling point. Adjust seasoning and serve sprinkled with chopped parsley.
Storage time 4–6 months.

IMPERIAL	METRIC	AMERICAN
2 lb. Jerusalem artichokes	1 kg. Jerusalem artichokes	2 lb. Jerusalem artichokes
juice of 1 lemon	juice of 1 lemon	juice of 1 lemon
1 pint chicken stock	generous ½ litre chicken stock	2½ cups chicken stock
1 pint milk	generous ½ litre milk	2½ cups milk
salt and pepper to taste	salt and pepper to taste	salt and pepper to taste
2 teaspoons cornflour	2 teaspoons cornflour	2 teaspoons cornstarch

Peel the artichokes and simmer until tender in the lemon juice and stock. Cool, liquidise in an electric blender or sieve and return to the pan. Add the milk, reheat and season to taste. Thicken if necessary with the cornflour moistened with a little cold water. Cool.

Roast and jugged hare

Makes 4 servings – roast and 8 servings – jugged.
To freeze Divide the meat and sauce between two shaped foil containers. Seal and label.
To serve Defrost and place in a moderately hot oven (375°F, 190°C, Gas Mark 5) for 30 minutes.
Storage time 4–6 months.

IMPERIAL	METRIC	AMERICAN
2 hares, skinned and jointed	2 hares, skinned and jointed	2 hares, skinned and jointed
2 oz. fat streaky bacon	50 g. fat streaky bacon	3 bacon slices
2 tablespoons seasoned flour	2 tablespoons seasoned flour	3 tablespoons seasoned flour
4 oz. meat dripping or bacon fat	100 g. meat dripping or bacon fat	½ cup meat or bacon drippings
4 tablespoons port	4 tablespoons port	⅓ cup port
¼ pint soured cream	1½ dl. soured cream	⅔ cup sour cream
salt and pepper to taste	salt and pepper to taste	salt and pepper to taste
2 large onions, halved	2 large onions, halved	2 large onions, halved
2 carrots, halved	2 carrots, halved	2 carrots, halved
½ lemon	½ lemon	½ lemon
1 teaspoon dried thyme	1 teaspoon dried thyme	1 teaspoon dried thyme
4 cloves	4 cloves	4 cloves
1 oz. butter	25 g. butter	2 tablespoons butter
1 oz. flour	25 g. flour	¼ cup all-purpose flour
2 tablespoons redcurrant jelly	2 tablespoons redcurrant jelly	3 tablespoons red currant jelly

To roast Use the four saddle portions and serve immediately. Lard the meat with strips of bacon and turn in the seasoned flour. Melt half the dripping in a roasting tin and use to brown the meat on all sides over high heat. Roast for 1½ hours in a moderately hot oven (375°F, 190°C, Gas Mark 5) or until tender, turning and basting the joints several times with the juices in the tin. Remove meat onto a serving dish and keep hot. Add half the port to the juices in the pan, scrape the sediment and boil gently for 2 minutes. Add the cream, season to taste, reheat gently to boiling point and strain over the roast hare. Serve with redcurrant jelly.

To jug Melt the rest of the dripping and use to fry the remaining hare joints and liver until brown all over. Place in an ovenproof casserole with the onions, carrots, lemon half, herbs, cloves and seasoning. Pour over sufficient water to cover. Put on lid and place in a moderate oven (325°F, 170°C, Gas Mark 3) for 2½–3 hours or until hare is tender. Cool completely in the liquid then remove joints. Melt the butter in a saucepan, stir in the flour and cook over gentle heat for 2 minutes or until light brown. Strain in the stock (add some of the blood if liked) and bring to the boil, stirring continuously. Add the remaining port and the redcurrant jelly and simmer until the jelly has melted. Cool.